F. Gibreel - iSewhit
mends

Hormones in Health and Disease

Series Editor
V. K. Moudgil

BOOKS IN THE SERIES

Introduction to Cellular Signal Transduction

Ari Sitaramayya

EDITOR

Birkhäuser

Boston • Basel • Berlin

Ari Sitaramayya
Eye Research Institute
Oakland University
422 Dodge Hall
Rochester, MI 48309-4480
USA

Library of Congress Cataloging-in-Publication Data

Sitaramayya, Ari.
    Introduction to cellular signal transduction / Ari Sitaramayya.
        p.   cm. — (Hormones in health and disease)
    Includes bibliographical references and index.
    ISBN 0-8176-3982-9 (hardcover : alk. paper)
    1. Cellular signal transduction. 2. Second messengers
(Biochemistry) 3. G proteins. I. Title. II. Series.
QP517.C45S58   1999   98-25749
572'.69—dc21   CIP

Printed on acid-free paper.
© 1999 Birkhäuser Boston   *Birkhäuser* ℬ ®

ISBN 0-8176-3982-9
ISBN 3-7643-3982-9

Typeset by Northeastern Graphic Services, Inc., Hackensack, NJ.
Printed and bound by Edwards Brothers, Inc., Ann Arbor, MI.
Printed in the United States of America.

9 8 7 6 5 4 3 2 1

For Usha, Vani, and Aruna

Contents

Foreword

Our understanding of biological communication has grown significantly during the past decade. The advances in knowledge about the chemical nature of signals and their corresponding reception by specialized cells have led to identification, characterization, purification, cloning, and expression of specific receptor molecules. While the earlier literature emphasized compartmentalized treatment of informational molecules and their interaction with receptors, the progress in the recent past has allowed cross-fertilization in the examination of the of actions and mechanisms of steroid and protein hormones and other messengers. Investigators now have an increased appreciation of the multiple effects of specific hormones and of the diverse responses by receptor proteins to closely related ligands. The task of compiling this enormous literature into a focused treatise was undertaken with the launching of the series Hormones in Health and Disease. This latest volume, *An Introduction to Cellular Signal Transduction*, complements the previous monographs in the series and brings to the fore recent developments in the field of biochemical communication. This volume combines discussions on the basic tenets of the signal transduction process and its relevance to health and disease. While various chapters provide exhaustive dissection of specific topics for researchers in the field, the book is also an excellent vehicle for introducing students and new investigators to the subject. The contributors of the chapters are active and accomplished scientists brought together on a common platform by the editor, Dr. Ari Sitaramayya, whose knowledge of the field and contributions to the subject made this project a success and a pleasurable experience.

V. K. Moudgil
SERIES EDITOR

Contributors

Ari Sitaramayya, Eye Research Institute, Oakland University, Rochester, MI 48309

Nigel W. Bunnett, Departments of Surgery and Physiology, University of California, San Francisco, CA 94143

Jeffrey L. Garvin, Division of Hypertension and Vascular Research, Henry Ford Hospital, Detroit, MI 48202

Narasimhan Gautam, Department of Anesthesiology, Washington University School of Medicine, St. Louis, MO 63110

Geoffrey H. Gold, Monell Chemical Senses Center, Philadelphia, PA 19104

Peter J. Kennelly, Virginia Polytechnic Institute and State University, Blacksburg, VA 24061-0308

Takashi Kurahashi, National Institute for Physiological Sciences, Myodaiji, Okazaki 444, Japan

Ponnal Nambi, Department of Renal Pharmacology, SmithKline Beecham Pharmaceuticals, King of Prussia, PA 19406

Uma Prabhakar, Department of Cellular Biochemistry, SmithKline Beecham Pharmaceuticals, King of Prussia, PA 19406

Maarten E. A. Reith, Department of Basic Sciences, College of Medicine, University of Illinois, Peoria, IL 61656

Mary F. Roberts, Merkert Chemistry Center, Boston College, Chestnut Hill, MA 02167

Dolores J. Takemoto, Department of Biochemistry, Kansas State University, Manhattan, KS 66508

Theodore G. Wensel, Department of Biochemistry, Baylor College of Medicine, Houston, TX 77030

Akio Yamazaki, Kresge Eye Institute, Departments of Opthalmology and Pharmacology, Wayne State University School of Medicine, Detroit, MI 48201

Part I

OVERVIEW

ARI SITARAMAYYA

Cells in the human body constantly interact with each other by sending signals, or messages. Specialized sensory cells monitor environmental signals. The type of signals used by the cells, the methods employed in detecting them, and the cellular mechanisms utilized in deciphering and responding to them constitute the subject matter of signal transduction.

The molecules used in communication between neurons, and between a nerve cell and a muscle fiber or another target cell, are called neurotransmitters. If both the communicating and receiving cells are non-neuronal, such messages between adjacent cells are usually called paracrine signals. The synthesis of neurotransmitters, their release and recognition, and cellular responses to them are among the topics generally discussed in neurochemistry. Messengers that travel between cells of distant tissues via the bloodstream are called hormones; adrenaline, which is released from chromaffin cells in the adrenal gland and elicits a response from the liver cells, falls into this category. The synthesis, release, and transportation of hormones between tissues, and their recognition and the physiological responses to them in the target cells, have been the subjects constituting endocrinology. Over the last few decades, it has become evident that there are many similarities in hormonal and neurotransmitter signaling. This realization, and the advances in pharmacology, molecular biology, and electrophysiology have uncovered the diversity in signal transduction molecules at all levels. The results of the preceding developments have propelled the emergence of signal transduction as an independent discipline.

It is virtually impossible to cover in any single book all aspects of signal transduction. However, attempts have been made in this volume to present a discussion of the major classes of proteins that participate in this process.

Introduction to Cellular Signal Transduction
A. Sitaramayya, Editor
©1999 Birkhäuser Boston

It is hoped that senior undergraduate and graduate students will gain from this book an understanding of the function and significance of the proteins that convert receptor-mediated signals into intracellular responses.

The subject matter of this book is divided into five sections (Parts II–VI). Part II describes the role of receptors and G-proteins. Signal transduction begins when a signal (first messenger) is recognized by a receptor molecule at a cell surface. The role of the receptor is to serve as a conduit for transfer of the information carried by the signal from the outside to the inside of the cell. In doing so, the receptor often amplifies the signal: in some cases the activated receptor becomes an ion channel, while in others it acts as an enzyme and catalyzes the synthesis of second messenger molecules or activates signal transducers such as GTP-binding proteins (G-proteins). Heterotrimeric G-proteins, those that contain three subunits referred to as alpha, beta, and gamma, mediate signaling by a large number of receptors. The inactive G-protein contains a GDP molecule bound to the alpha subunit, and the receptor-catalyzed exchange of GDP for a GTP activates the G-protein. The activated G-protein subunits in turn regulate the activities of effector proteins, which include ion channels, adenylate cyclases, and phospholipases. G-proteins of another class, the small molecular weight G-proteins, contain only one subunit. Functionally, these proteins resemble the alpha subunits of heterotrimeric G-proteins in that they become activated when the bound GDP is exchanged for GTP. Small molecular weight G-proteins are involved in transducing a wide range of signals, including those regulating cellular growth, differentiation, and survival. In recent years, signaling mechanisms that include participation of both heterotrimeric and small molecular weight G-proteins have been discovered. Part III focuses on second messengers and describes some of the enzymes involved in their synthesis and degradation. Adenylyl (adenylate) and guanylyl (guanylate) cyclases catalyze the formation of cyclic AMP and cyclic GMP, respectively. Adenylate cyclases are regulated by G-proteins as well as by changes in the intracellular calcium concentration. Guanylate cyclases, on the other hand, can be regulated without the mediation of G-proteins. There are two types of guanylate cyclases: one is integral to the plasma membrane and regulated directly by hormones or by accessory intracellular proteins responding to changes in calcium concentration; the other type is soluble proteins that are regulated by nitric oxide. Cyclic nucleotides are often referred to as "second messengers," a term indicating that extracellular signals (first messengers) do not enter the cells, but bring about cellular changes via production of intracellular messengers. For example, as Sutherland and colleagues discovered, detection of epinephrine by a liver cell causes production of cyclic AMP in the cell. In the absence of epinephrine, artificially elevating intracellular cyclic AMP concentration mimics the effects of epinephrine, suggesting that the former is a second messenger of the latter.

Phospholipases constitute a major group of second-messenger-generating enzymes that act on membrane lipids. Phospholipase D generates phosphatidic acid, which functions as a second messenger in a variety of cellular responses. Phospholipase C is activated by G-proteins, and its products, diacylglycerol and inositol triphosphate, serve as second messengers. Diacylglycerol activates protein kinase C leading to phosphorylation of specific proteins, and inositol triphosphate transiently raises intracellular calcium concentration resulting in activation of calcium-mediated reactions. Phospholipase A2 is regulated by intracellular calcium concentration, and its product, arachidonic acid, can diffuse into neighboring cells and serve as an activator of soluble guanylate cyclase. The preceding discussion centered on how cells generate second messengers. What happens to the second messengers after they bring about the desired effects? Calcium is pumped into internal stores or pumped out of the cell. Diacylglycerol is phosphorylated to phosphatidic acid. And the cyclic nucleotides, cyclic AMP and cyclic GMP, are inactivated by hydrolysis to AMP and GMP, respectively. The enzymes that catalyze the hydrolysis of the cyclic nucleotides, the phosphodiesterases, will be discussed at length. These enzymes are found in a wide variety, exhibiting both specific and nonspecific substrate preference. Phosphodiesterases are themselves regulated by other second messengers, such as calcium, or by G-proteins. Regulation of phosphodiesterases is critical to signal transduction because accelerating or slowing down the degradation of cyclic nucleotides can result in shortening or prolonging the cellular responses to the signals.

In recent years, nitric oxide (NO) has emerged as a novel messenger molecule. NO is not a hormone, for it is not carried to target tissues via the blood stream, nor is it a classical neurotransmitter, since it is not released into the synaptic cleft. Furthermore, NO does not represent a traditional second messenger, since it diffuses out of the cell of origin. It is, therefore, treated in Part IV as a novel messenger. NO is a membrane-permeant, dissolved gaseous molecule. It was first discovered as a product of endothelial cells with relaxing effects on muscle cells of the blood vessels. Further studies during the last decade have revealed that nitric oxide has several other important roles in the cardiovascular and central nervous systems. Also, NO produced by macrophages plays a pivotal role in fighting bacterial infections. In most cases, NO appears to bring about cellular responses by activating soluble guanylate cyclases and increasing the cellular concentration of cyclic GMP. Other targets of nitric oxide include ADP-ribosylating enzymes.

Part V describes the role of regulatory mechanisms and ion channels in signal transduction. The second messenger molecules activate specific protein kinases, which in turn phosphorylate and regulate the biological activities of proteins. Phosphorylation is most widely used for functional and structural modification of proteins because the process is so cost-effective and readily reversible: dramatic changes in a protein's activity can be brought about by phosphorylation, and the phosphorylated protein can be rapidly converted back to its former state of activity by simple removal of phosphate catalyzed by a protein phosphatase. Protein kinases and phosphatases are the workhorses of signal transduction.

For many years it was thought that cyclic nucleotide second messengers functioned solely by activating protein kinases. However, about a decade ago, Fesenko and colleagues discovered a novel role for them when they found that cyclic GMP influenced the cation channel activity in the plasma membrane of retinal photoreceptor cells without the mediation of protein kinases. It is now firmly established that light-induced changes in cyclic-GMP-gated ion channel activity represent a pivotal step in the vision process. Cyclic-nucleotide-gated ion channels have now been reported in many other cells. Recently, cyclic-AMP-gated ion channels were conclusively demonstrated to be the mediators of olfactory signal transduction.

The book concludes with two special chapters that highlight applications of signal transduction (Part VI). Since signal transduction pathways play such important roles in regulating the cellular activities in every tissue and organ, defects in signaling proteins could lead to disease. A chapter in this section is dedicated to this topic, with examples of diseases caused by somatic and heritable mutations in the genes coding signaling proteins. The discovery of causal relationships between a mutation and a disease has been described with references to a few cases. Interestingly, the investigations of some defective signaling proteins have provided valuable insights into the functioning of their normal counterparts and have led to the development of effective therapies. The other chapter covers the roles signal transduction proteins play in drug abuse. This is a subject in which our knowledge is still very limited, but there is sufficient evidence to suggest that dopaminergic neurons are involved in the rewarding action of the drugs abused. The chapter discusses the roles of cyclic AMP, NO, cyclic AMP response element binding protein, transcription factors, and protein kinases in drug tolerance and plasticity occurring in dopaminergic neurons in response to repeated exposure to drugs.

Thus, the volume in its entirety provides an introduction to the various components of signal transduction and illuminates their role in physiological processes and under pathological conditions.

Part II

RECEPTORS AND G-PROTEINS

.

1

Cell Surface Receptors: Mechanisms of Signaling and Inactivation

ARI SITARAMAYYA AND NIGEL W. BUNNETT

Cells in the human body recognize numerous signals coming from other cells and from the environment, and respond to them appropriately. Traditionally, signals coming from distant tissues via the bloodstream are called hormones, and the signaling molecules used by neurons to communicate with other neurons or non-neuronal cells are called neurotransmitters. Signals derived from non-neuronal cells that influence neighboring cells are known as paracrine substances. Although there are important exceptions, hormones, neurotransmitters, and paracrine substances do not directly enter their target cells; they are recognized by specific molecules, called receptors, present on the cell surface. A receptor is activated when it binds a signal molecule, and the activated receptor becomes a catalyst or attracts other proteins to it that serve as catalysts in bringing about the desired intracellular changes in the biochemical activity of the cell. In other words, the receptor is a conduit for transfer of information from the outside to the inside of the cell.

All cells in the body do not contain receptors for all possible signals that they are likely to encounter. Depending upon the function of the tissue in which it is present, a cell is dedicated to recognizing and responding to specific signals. Photoreceptor cells in the retina detect light, olfactory epithelial cells in the nose detect smells, gustatory epithelial cells in the tongue detect tastes, and so on.

MULTIPLE RECEPTORS FOR A GIVEN SIGNAL MOLECULE

Given the wide variety of signals—light, odorants, taste substances, amino acids and derivatives, and small and large peptides—one would expect each

Introduction to Cellular Signal Transduction
A. Sitaramayya, Editor
©1999 Birkhäuser Boston

Table 1-1. Two distinct types of receptors for acetylcholine

Type	Activator	Blocker	Cell	Effect
nicotinic	nicotine acetylcholine	curare	skeletal muscle	contraction
muscarinic	muscarine acetylcholine	atropine	cardiac muscle	less contraction

Modification of Table 9-1 from Irwin B. Levitan and Leonard K. Kaczmarek, 1991. The Neuron. Cell and Molecular Biology. Oxford University Press, New York.

of them to be recognized by a specific receptor whose structure is optimal for its detection. However, functional, pharmacological, and more recently molecular biological studies have revealed that there are multiple receptor types for virtually every signaling molecule. For example, acetylcholine is recognized by two different types of receptors called nicotinic and muscarinic (Table 1-1, Levitan and Kaczmarek 1991). The two receptors are pharmacologically distinct: nicotinic receptors are activated by nicotine as well as by acetylcholine, and their activation is blocked by curare; muscarinic receptors are activated by muscarine and acetylcholine but not by nicotine, and are blocked by atropine but not by curare. Activation of these receptors for the same signal molecule can have different effects, depending on the target cell. Activation of nicotinic receptors on a skeletal muscle cell signals contraction, while activation of muscarinic receptors on a cardiac muscle cell signals a decrease in the strength of contraction. There are several subtypes in the muscarinic receptors discovered by molecular biological techniques, though their functional specificities are not all clear. Similar diversity is found in receptors for other signaling molecules such as epinephrine, dopamine, serotonin, and adenosine. It is likely that cells in different tissues express a slightly different type of receptor to suit their specific needs.

MODES OF RECEPTOR ACTIVITY

On the basis of the mechanisms employed by the activated receptors in bringing about intracellular changes, receptors can be classified into a few broad categories (Figure 1).

1. Ionotropic receptors

Receptors for some neurotransmitters become ion channels upon activation. These receptors, also called ligand-gated receptors/ion channels, are the fastest acting of all receptors, with response times in submilliseconds. The nicotinic acetylcholine receptor present on skeletal muscle is a prototypical

a) Ion-channel-linked receptors (e.g.,receptors for excitatory and inhibitory neurotransmitters)

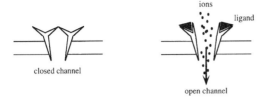

b) G-protein-coupled receptors (e.g.,receptors for neuropeptides, hormonal peptides)

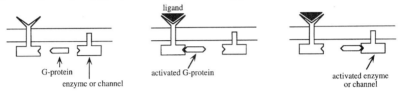

c) Enzyme-linked receptors (e.g.,receptors for growth factors and cytokines)

Figure 1-1. Modes of receptor activity. Modified from Figure 1-5 in the chapter "Gastrointestinal Hormones and Neurotransmitters," by Bunnett, N. W. and Walsh, J. H., from the sixth edition of Sleisenger and Fordtran's *Gastrointestinal and Liver Diseases*, edited by Drs. Mark Feldman, Bruce Scharschmidt, and Marvin Sleisenger, pages 3–18, Saunders, 1997.

ionotropic receptor. It becomes a cation channel upon activation, permitting the influx of sodium ions, which depolarizes the cell and sets off an action potential, leading to the contraction of the muscle fiber. Receptors for the inhibitory neurotransmitters γ-aminobutyric acid (GABA) and glycine also become ion channels upon ligand binding. In these cases the channels become permeable to chloride ions, thereby hyperpolarizing the cell membrane and inhibiting the cell's activity. Nicotinic, glycine, and GABA receptors are each made up of multiple subunits. In the nicotinic receptor there are two α subunits with acetylcholine binding sites and one each of β, γ, and δ subunits (Weill et al 1974, Lindstrom et al 1979). Each of these subunits has four membrane spanning domains, and the N- and C-terminal portions of the subunits are exposed to the outside of the cell (Noda et al 1983).

Glutamate, a neurotransmitter in the brain, is recognized by multiple ionotropic receptors classified into three groups based on agonist specificity:

α-amino-3-hydroxy-5-methyl-isoxazole-4-propionic acid (AMPA) receptors, kainic acid (KA) receptors, and N-methyl-D-aspartic acid (NMDA) receptors (Monaghan et al 1989). Each group has several subtypes, distinguished pharmacologically and by primary sequence. Activated AMPA and KA receptors become cation channels, permitting the entry of sodium and depolarization of the cell. NMDA receptors are more complex. Activation by NMDA requires the simultaneous presence of glycine, which also binds to the receptor (Kleckner and Dingledine, 1988). In addition, the unactivated NMDA receptor channel is blocked by magnesium. Prior depolarization of the cell, permitting expulsion of magnesium that is blocking the channel, is required for activation of NMDA receptors (Nowak et al 1984, Mayer et al 1984). Once activated, the receptors permit influx of both sodium and calcium. Calcium entry through NMDA receptors and the subsequent calcium-dependent reactions are thought to play a role in mechanisms pertaining to memory (Malenka 1991). Uncontrolled influx of calcium through NMDA receptors under pathological conditions causes neuronal death.

2. G-protein-coupled receptors

In addition to signaling via ionotropic receptors, acetylcholine and glutamate also bind to slower acting receptors, which bring about intracellular changes through the mediation of GTP- binding proteins, or G-proteins. This latter class of receptors is referred to as G-protein-coupled receptors (GPCRs). A partial list of signaling molecules that bind to GPCRs is shown in Table 1-2.

As is evident, a wide variety of molecules signal through GPCRs. They include neurotransmitters and large peptide hormones. In every case, however, GPCRs contain a single peptide chain. There are seven hydrophobic, helical sequences that traverse the membrane, forming three extracellular and three intracellular loops. The N-terminus is exposed to the extracellular space, and the C-terminus to the cytosol. At least in the case of receptors for small molecules such as epinephrine, the binding pocket for the ligand appears to be not on the surface of the receptor in the hydrophilic region, but in a hydrophobic pocket to which almost all of the transmembrane alpha helical domains contribute (reviewed by Savarase and Fraser 1992). Many neurotransmitters that signal through GPCRs have a positive charge center, which suggests the possibility that acidic amino acids may be involved in ligand binding. In the case of the beta-adrenergic receptor, an aspartic acid residue in transmembrane helix 3 and one in helix 2 appear to play that role. Other amino acids in helices 5 and 6 also appear to be important for binding epinephrine. Bigger ligands, in particular the peptides, may interact with the extracellular portions of their respective receptors.

Rhodopsin, the receptor for visible light in the human retina, is a prototypical GPCR. A great deal of what is known about GPCRs was learned

Table 1-2. A partial list of molecules signaling through GPCRs

small neurotransmitters	epinephrine
	dopamine
	serotonin
	histamine
	acetylcholine
	γ-aminobutyric acid
	glutamate
	adenosine
peptides	opioids
	tachykinins
	bradykinins
	vasoactive intestinal peptide
	neuropeptide Y
	thyrotropic hormone
	luteinizing hormone
	follicle-stimulating hormone
	adrenocorticotropic hormone
	cholecystokinin
	gastrin
	glucagon
	somatostatin
	endothelin
	vasopressin
	oxytocin
sensory signals	light
	odorants

From Bertil Hille, 1992. G-protein coupled mechanisms and nervous signaling. Neuron 9:187–195.

from research on this molecule. Rhodopsin differs from all other surface receptors in that it is actually present on intracellular membranes called disks in the photoreceptor cells of the retina (Figure 1-2). Its C-terminal domain is exposed to the cytoplasm, and the N-terminus is in the lumen of the disk vesicles. This arrangement does not interfere with the function of rhodopsin as a receptor, since light penetrates the plasma membrane and is absorbed by rhodopsin. It also differs in another way: while other receptors are activated by their ligands, rhodopsin has its ligand, 11-*cis*-retinal, bound covalently to a lysine in the 7th transmembrane helix (Wang et al 1980 Pellicone et al 1980). Absorption of light by the 11-*cis*- retinal isomerizes it to all-*trans*-retinal and activates rhodopsin, an event comparable to ligand binding in other receptors.

Activated GPCRs in turn activate G-proteins. As discussed at length in Chapter 3, there is a large family of G-proteins, and they are classified into different groups as G_s, G_i, and G_q, on the basis of the proteins/enzymes they influence in turn. G-proteins are heterotrimeric (consisting of one each of α,

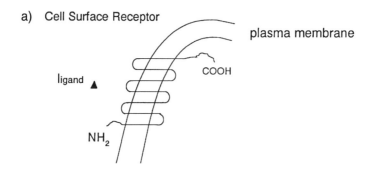

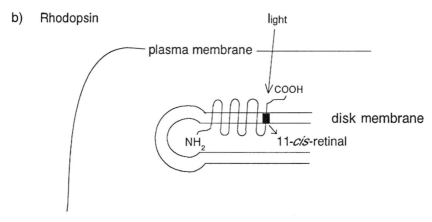

Figure 1-2. Receptors in surface and internal membranes. While most receptors are integral proteins of the plasma membrane of the cell, and activated by extracellular ligands, the light receptor, rhodopsin, is found in the intracellular membranes of the retinal photoreceptor cells. Light isomerizes 11-*cis*-retinal, a molecule covalently attached to rhodopsin, and activates rhodopsin.

β, and γ subunits). In the inactive state, the three subunits are held together, with the α subunit having a bound GDP. The inactive G-protein has high affinity for the activated GPCR. In the receptor-bound state, the α subunit has lower affinity for GDP, permitting the release of the nucleotide and making the site available for binding GTP. Once GTP replaces GDP in the receptor-G-protein complex, α-GTP dissociates from the βγ subunits and both dissociate from the receptor. The receptor becomes free to activate another molecule of G-protein, and the activated subunits of G-protein associate with their target proteins/enzymes and regulate their activities. Among the targets of G-proteins are adenylate cyclase, phospholipase C, and ion channels. Chapters 3 and 4 describe the role of G-proteins in signal transduction in greater detail.

A given signaling molecule can bind to a variety of GPCRs, each of which brings about intracellular changes through the mediation of a specific G-protein. For example, among the GPCRs for epinephrine are the $\alpha 1$, $\alpha 2$, and β receptors (reviewed by Lefkowitz and Caron 1988). The $\alpha 1$ receptors activate a G_q type of G-protein, which stimulates the activity of phospholipase C, leading to increase in the cytoplasmic calcium concentration and activation of protein kinase C; the $\alpha 2$ receptors activate a G_i type of G-protein, which in turn *inhibits* (the $_i$ in G_i) the activity of adenylate cyclase, leading to a decrease in the concentration of cyclic AMP; and the β receptors activate a G_s type of G-protein, which *stimulates* (the $_s$ in G_s) adenylate cyclase and brings about an increase in cellular cyclic AMP concentration. The enzymes (or other proteins) that are regulated by G-proteins are usually referred to as effector systems. The adrenergic receptors are therefore said to transduce signals through multiple effector systems. Other signaling molecules, such as adenosine and dopamine, also bind to multiple GPCRs that signal through different types of G-proteins and effector systems. This diversity permits different cells to respond differently to the same signal.

In recent years it has become evident that a given GPCR can transduce signals through more than one effector system. For example, $\alpha 2$ adrenergic receptors have been shown to stimulate phospholipase C as well as to inhibit adenylate cyclase, when expressed in Chinese hamster lung fibroblasts (Cotecchia et al 1990). Whether the two effectors are regulated by a single G-protein or by two different proteins is unclear, although it is quite likely that different G-proteins are involved. Similarly, tachykinin receptors and parathyroid hormone receptor were each shown to stimulate both phospholipase C and adenylate cyclase (Nakajima et al 1992, Abou-Samra et al 1992).

It now appears that signal molecules that differ only slightly in their structure can interact with a single GPCR and activate/regulate different effector systems (Evans et al 1995). For example, octopamine and tyramine were shown to cause release of calcium from intracellular stores and inhibit adenylate cyclase through binding to a single receptor. Tyramine was more effective in Ca release than octopamine, while the latter was more effective in inhibiting adenylate cyclase. This observation suggests that the same receptor may be transducing octopamine signal in one cell and tyramine in another or that the same cell could be manipulated to either release calcium or inhibit adenylate cyclase by means of selective signaling through octopamine or tyramine (Evans et al 1995).

The various ways in which signaling molecules interact with receptors to influence the effector systems are shown in Figure 1-3.

3. Receptor tyrosine kinases

Growth factors such as epidermal growth factor, insulin, and platelet-derived growth factor are slower acting signals than neurotransmitters. They bring

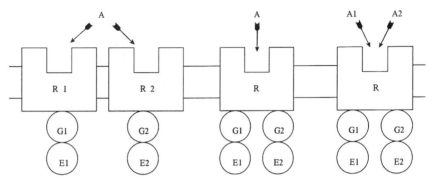

Figure 1-3. The receptor-effector relationship. A ligand can regulate two different effectors (E1 and E2) by activating two different receptors (left), each coupled to a different type of G-protein. A ligand may activate only one type of receptor, but the receptor may be able to interact with two different types of G-proteins, and thereby regulate two different effectors (middle). Finally, a single receptor may be able to detect two different types of ligands (A1, A2), and depending upon the activating ligand, it would preferentially interact with either G1 or G2 (right). Modified from Figure 1 of Evans et al, *Progress in Brain Research*, 106:259–268, 1995.

about their effects through regulation of gene transcription and therefore of the synthesis of proteins. Growth factors are identified by surface receptors, which are usually single polypeptide chains (except in the case of insulin receptor, which is made up of four peptides) that have just one transmembrane domain. Upon activation, two receptor molecules come together (dimerize), and one molecule in the dimer phosphorylates the other on tyrosine residues in the intracellular domain (Heldin 1995). The activated receptors function as tyrosine kinases and phosphorylate other proteins in the cell (Frantl et al 1993). Phosphotyrosine-bearing segments of the receptor also attract proteins containing the SH2, SH3, PTB, and PH domains (Zhou et al 1993, Kavanaugh et al 1995, Pawson 1995). These proteins either have enzymatic activities by themselves or in turn attract/recruit other proteins with similar domains. For example, phospholipase C-γ is an SH2-domain-bearing protein. It binds to activated receptor tyrosine kinases (RTKs) and becomes activated upon tyrosine phosphorylation (Valius and Kazlauskas 1993). Grb2 is an SH3-bearing protein that attaches to RTK and in turn attracts a guanine nucleotide exchange factor called *Sos*. *Sos* activates *Ras*, a small molecular weight G-protein, by facilitating exchange of bound GDP for GTP. (Aronheim et al 1994). Activated *Ras* in turn triggers the *Ras-Raf*-MAP kinase kinase-MAP kinase cascade, which influences gene transcription.

Phosphotidylinositol 3′-kinase is another enzyme that is activated by RTKs. This enzyme appears to play a role in RTK-mediated mitogenesis and transformation (Valius and Kazlauskas 1993).

Cytokines, such as interleukin, also bind to receptors like the RTKs and bring about changes in transcription of specific genes. However, these receptors do not have tyrosine kinase activity by themselves. Activated receptors attract cytosolic tyrosine kinases, which then phosphorylate each other and the receptor (Ziemiecki et al 1994). *Jak*, *Tyk*, *Yes*, and *Hck* kinases are thought to be involved. The phosphorylated receptor triggers the biological response through factors that influence gene transcription.

The receptors on lymphocytes which detect antigen molecules also seem to employ tyrosine kinases in the signaling pathways (June et al 1990). Unlike the RTKs, these receptors are complex and contain several subunits. The cytoplasmic portions of these receptors contain domains in which the tyrosine residues are phosphorylated and play an important role in the antigen signaling (Weiss 1993).

4. Receptor serine/threonine kinases

Signaling molecules belonging to the family of transforming growth factor-β (TGF-β) bind to receptors that have protein kinase activity like the receptor tyrosine kinases, but they phosphorylate serine and threonine residues. TGF-β is a potent inhibitor of growth and differentiation. Several types of tumor cells that are unresponsive to TGF-β do not have receptors for this molecule (Inagaki et al 1993, Kadin et al 1994, Sun et al 1994).

TGF-β receptors are of two types (type I and type II) (Derynck 1994, Massague et al 1994). Two copies of type II receptor and two copies of type I receptor bind to TGF-β. The type II receptor is constitutively active as a kinase and phosphorylates the type I in the ligand-receptors complex, transforming the latter into an active serine/threonine kinase (Wrana et al 1994). Type I receptor then phosphorylates intracellular proteins, but how these proteins are related to the biological action of TGF-β is not well understood.

5. Receptor guanylate cyclases

Unlike the adenylate cyclase system where a receptor activates G-proteins which in turn activate or inhibit adenylate cyclase, some membrane guanylate cyclases are directly activated by peptide hormones. Hormone binding and cyclase activity reside in the same protein (Paul et al 1987). The most extensively studied are the cyclases that serve as receptors for natriuretic peptides (NP) secreted by the heart. These peptides control excretion of sodium and regulation of urine volume and blood pressure. All the effects of natriuretic peptides are thought to result from hormone binding to membrane cyclases and stimulation of the synthesis of cyclic GMP in the target cells. NP receptor guanylate cyclases (NPRGCs) consist of a single polypeptide chain that contains a single transmembrane domain. The hormone binds to the extracellular portion of the cyclase and triggers an increase in the cyclase activity. Receptor guanylate cyclases appear to be similar to the RTKs in that they too appear to dimerize upon ligand binding. In fact, the intracellular portion of the spermatozoan membrane guanylate cyclase contains a domain with a high degree of sequence similarity with RTKs (Singh et al 1988). It is now known that all membrane guanylate cyclases have kinase-like domains.

TERMINATION OF RECEPTOR ACTIVITY

The binding of a signaling molecule to its receptor is a reversible reaction. The fraction of ligand-bound (activated) receptors on a cell surface depends upon the affinity of the ligand for the receptor and their concentrations. In neuronal communication the release of the transmitter by the presynaptic cell into the narrow confines of the synaptic cleft allows the buildup of a high concentration of the neurotransmitter, favoring rapid activation of receptors on the postsynaptic membrane. If the concentration of the signaling molecule remains high, the receptors will continue to be in the activated state, and the cell will be insensitive to further release of the same signal from the presynaptic cell. The activated receptors should therefore be inactivated after a desired length of time. This can be accomplished by reducing the concentration of the signaling molecule in the vicinity of the receptors to levels well below those that favor binding. In some instances, this is done by the destruction of the signaling molecules, and in other cases by their removal and reuse by the presynaptic cell (Figure 1-4). When both are not possible, as in the case of light, the activated receptors themselves are modified to terminate their activity.

1. Ligand uptake

Neurotransmitters such as norepinephrine, dopamine, GABA, and serotonin are removed from the synaptic cleft by specific transporters that carry them back into the presynaptic cell for reuse (Uhl and Hartig 1992). Uptake of the neurotransmitter via the transporters is supported by the transmembrane sodium gradient. The prototypical GABA transporter is a large protein of about 70 kDa with 12 transmembrane domains. Glutamate transporters are somewhat different, consisting of 6–10 transmembrane domains and dependent upon both sodium and potassium for their activity. Neurotransmitter transporters could be manipulated to regulate the concentration of transmitter in the synaptic cleft and thereby the receptor activity (Bohm 1997). GABA transport inhibitors are useful as anticonvulsant drugs; norepinephrine transport inhibitors and serotonin uptake inhibitors, such as Prozac, are used as antidepressants; the behavioral effects of cocaine and amphetamines are thought to occur via inhibition of dopamine and norepinephrine transporters.

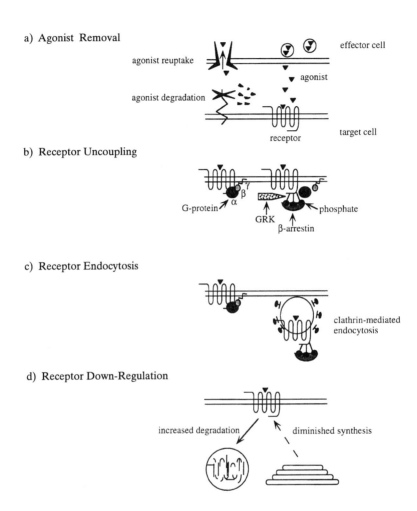

Figure 1-4. Mechanisms affecting receptor activity. Agonist removal by reuptake or degradation (a) terminates receptor activity. Phosphorylation and arrestin coupling terminates the activity of rhodopsin, the light receptor, and desensitizes the activity of other GPCRs (b). Agonist-induced endocytosis removes receptors from the cell surface (c). Desensitized receptors are often recycled by this mechanism. A reduction in the receptor response to ligand concentration can be brought about by increased degradation or diminished synthesis of the receptors (d). From Bohm et al, *Biochemical Journal*, 322:1–18, 1997.

2. Ligand degradation

Acetylcholine released into the synaptic cleft or the neuromuscular junction is not directly removed for reuse, but rapidly hydrolyzed by acetylcholinesterase. The enzyme can hydrolyze tens of thousands of molecules of acetylcholine per second, permitting a rapid lowering of its concentration. The reduced concentration favors dissociation of ligand from activated receptors and therefore the termination of receptor activity. Inhibition of acetylcholinesterase activity could lead to prolonged activity of acetylcholine receptors and result in death due to respiratory failure. Nerve gases such as Tabun, Sarin, and Soman, which are lethal in submilligram doses, are potent inhibitors of acetylcholinesterase. Less active inhibitors such as Parathion and Malathion are used as insecticides in agriculture (Taylor 1985).

Peptide hormones and neurotransmitters are either diluted in the extracellular medium to levels that are too low to activate receptors or degraded by peptidases. Neutral endopeptidase (NEP) is a prototypical neuropeptide-degrading enzyme present in a variety of tissues responding to neuropeptides. The enzyme is present in the plasma membranes and contains one transmembrane domain and a short intracellular domain. Most of the protein and the catalytic domain are in the extracellular medium (Devault et al 1987, Malfroy et al 1987). NEP degrades a variety of neuropeptides, including substance P, enkephalins, bradykinin, and neurotensin (Bohm et al 1997).

The adverse effects of limiting NEP activity on substance P and bradykinin induced proinflammatory responses were studied in mice made genetically deficient in NEP. These animals showed a high level of proinflammatory condition, reflecting a higher than normal concentration of substance P and bradykinin. The condition could be reversed by administering the antagonists of receptors for these neuropeptides (Lu et al 1997). As further evidence of the role of NEP in degrading inflammatory neuropeptides, its activity was found to be lower in inflamed tissues (Hwang et al 1993).

3. Receptor phosphorylation and desensitization

In the case of rhodopsin, the receptor for visible light, neither degradation nor removal of light is possible. Once bleached, the receptor continues to catalyze activation of G-proteins for tens of seconds until it naturally assumes an inactive conformation. It was, however, found that an ATP-mediated process drastically reduces the lifetime of the receptor activity (Liebman and Pugh 1980). This process was identified as phosphorylation by a specific kinase, the rhodopsin kinase, which phosphorylates the activated receptors on serines and threonines. Complete inactivation of the receptor by rhodopsin kinase is slow because it requires 7–9 phosphates/activated receptor, but can be accelerated by a protein called arrestin, which binds to the receptor with 1–2 phosphates and prevents further activation of G-proteins (Wilden et al 1986, Bennett and Sitaramayya 1988). Subsequent research has shown that receptor-specific kinases and arrestins exist for other GPCRs also (Haga et al 1994).

Similar mechanisms uncouple other G-protein-linked receptors from G-proteins. Luteinizing hormone (LH) stimulates adenylate cyclase activity in follicle membranes. However, when one observes the time course of the LH-stimulated cyclase activity, it appears to decrease gradually, while that in the unstimulated membranes remains steady (Bockaert et al 1976). A careful study of the attenuation of the LH-stimulated rate showed that it was hormone- and ATP-dependent. One or more elements in the adenylate cyclase system appeared to be undergoing hormone-dependent phosphorylation, which was causing a desensitization of the membranes to the hormone. Research over the last two decades has shown that desensitization is a general phenomenon: tissues briefly exposed to hormones usually react much less vigorously to a subsequent exposure to the same ligand. Desensitization is now known to be due to phosphorylation of the activated receptor by a kinase specific to the receptor or by one that also phosphorylates other proteins. β-adrenergic receptors are phosphorylated by the specific kinase β-adrenergic receptor kinase (BARK), by cyclic AMP-dependent kinase, and by protein kinase C (Pitcher et al 1992). Phosphorylation by any one of the kinases was shown to cause desensitization, though the extent of desensitization varied. Desensitization by BARK was strengthened by the presence of β-arrestin, a protein similar to the retinal arrestin, but the desensitization by cyclic AMP-dependent kinase was not. The former case is analogous to the inactivation of bleached rhodopsin by rhodopsin kinase and its acceleration by arrestin. While most of the work on receptor desensitization was done on β-adrenergic receptors, muscarinic cholinergic receptors were also found to undergo desensitization, and the same probably holds true for most of the GPCRs (Kwatra et al 1987).

In addition to GPCRs, the ionotropic receptors are also known to be desensitized following prolonged exposure to activating ligands. Here again the mechanism of desensitization is by phosphorylation of the receptors. Nicotinic acetylcholine receptors are phosphorylated on the γ and δ subunits by cyclic AMP-dependent kinase, on the δ and α subunits by protein kinase C, and on the β, γ, and δ subunits by tyrosine kinase (Huganir and Miles 1989). The control and phosphorylated receptors are not different in their initial rates of ion transport, but on prolonged exposure to acetylcholine, the phosphorylated receptors are desensitized 7–8 times. Similarly, GABA receptors are phosphorylated by cyclic AMP-dependent protein kinase, resulting in a decrease in the amplitude of the response of the receptors to GABA, and in a slowdown of the desensitization rate (Moss et al 1992). Regulation of the desensitization of acetylcholine and GABA receptors by protein kinases activated by second messengers such as cyclic AMP suggests that ligands that regulate intracellular second messenger concentrations would have profound effects on the excitability of cells expressing receptor ion channels (Moss et al 1992).

4. Receptor internalization

Another mechanism for lowering the responsiveness of the cell to a ligand is the removal of receptors from the plasma membrane. Stimulation of a cell responding to a growth factor results in endocytosis (internalization) of the activated receptors. The cell would therefore become less responsive to subsequent stimulation by the same ligand. A fraction of the internalized receptors is degraded in the lysosomes, and the rest are eventually returned to the plasma membrane. Internalization of growth factor (EGF) receptors was extensively studied (Sorkin 1996). While unoccupied EGF receptors also undergo slow internalization, EGF binding causes rapid accumulation of activated receptors in the coated pits, followed by endocytosis (Hanover 1984). What signals the endocytosis of activated receptors is not well understood. Growth factor receptors are known to have tyrosine kinase activity and to be phosphorylated on tyrosine residues. Both of these properties appear to favor internalization (Sorkin et al 1992). Inactivation of receptor kinase activity reduced the rate of EGF receptor internalization (Sorkin et al 1991). Mutations in the tyrosine phosphorylation sites on the receptor itself inhibited internalization of FGF receptor (Sorkin et al 1994). In addition, other parts of the receptor molecule devoid of kinase activity and phosphorylated tyrosines are also shown to be required for internalization. These may contain sequences that serve as codes for endocytosis (Chang et al 1993, Opresko et al 1995).

Since the activated receptors are endocytosed along with the activating ligand, the internalization mechanism serves not only to down-regulate the receptor activity but also to remove activating ligand from the vicinity of the receptors. In fact, since the receptors are for the most part recycled and the ligands are not, endocytosis could serve as a mechanism for reducing the concentration of activating ligand and thereby terminating or reducing receptor activity (Sorkin 1996).

Agonist-mediated endocytosis has been observed for many G-protein-coupled receptors. Epinephrine stimulates endocytosis of the β2-adrenergic receptor, and the peptide substance P induces endocytosis of the neurokinin-1 receptor (von Zastrow and Kobilka 1992, Grady et al 1995). Agonist-occupied receptors cluster at the plasma membrane and internalize through clathrin-coated pits into early endosomes. Proteins known as β-arrestins, which also contribute to desensitization, may serve as a bridge between activated receptors and clathrin, and thereby facilitate endocytosis (Ferguson et al 1997). Once internalized into early endosomes, receptors and their ligands are sorted into different pathways. Endosomes are acidified, which promotes dissociation of the receptor and its ligand. Receptors for hormones and neurotransmitters recycle to the plasma membrane, where they can be reused, whereas ligands such as substance P are degraded in lysosomes.

What is the importance of agonist-induced endocytosis of G-protein-coupled receptors? Endocytosis depletes the cell surface of receptors that are available to interact with hydrophilic ligands in the extracellular fluid, and could thus contribute to desensitization. However, β2-adrenergic receptors and neurokinin-1 receptors desensitize even if endocytosis is blocked, which indicates that endocytosis is not the principal mechanism of desensitization (Yu et al 1993, Garland et al 1996). Most receptors desensitize by agonist-induced phosphorylation by G-protein receptor kinases and interaction with arrestins, which disrupt activation of G-proteins and thereby terminate the signal. Rather than mediating desensitization, endocytosis is required for recovery of responsiveness, or resensitization (Yu et al 1993, Garland et al 1996). Resensitization of β2-adrenergic receptors and neurokinin-1 receptors requires receptor endocytosis, processing (which may involve dissociation of ligand, removal of phosphates, and dissociation of arrestins), and recycling back to the plasma membrane (Figure 1-5).

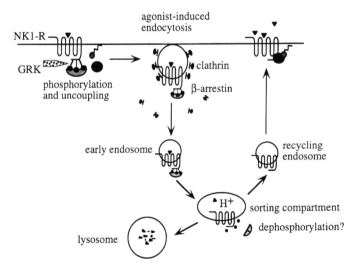

Figure 1-5. Receptor trafficking. Activated GPCRs such as neurokinin-1 receptor (NK1-R) are desensitized by kinases (GRK) and removed from the cell surface by endocytosis in clathrin-coated pits. After the dissociation of clathrin, the vesicles are acidified to remove the bound ligand, and the ligand-free receptors are either degraded in the lysosomes, or dephosphorylated and recycled back to the plasma membrane. Taken from Figure 6 in Bohm et al, *Biochemical Journal*, 322:1–18, 1997.

Movement of receptors throughout cells requires interaction of specific receptor sequences, or motifs, with proteins that direct receptor trafficking (Trowbridge et al 1993). Clathrin-mediated endocytosis requires interaction of endocytic motifs with components of the endocytic machinery. Tyrosine-containing endocytic motifs, in which tyrosine is of critical importance, have been identified for many single transmembrane domain proteins. They usually contain six residues forming an exposed β-turn. Positions 1, 3, and 6 are frequently occupied by aromatic or large hydrophobic residues, and residues in positions 2, 4, and 5 tend to be polar and often found in turns. The importance of these motifs for endocytosis has been established by demonstration that they interact with clathrin-associated proteins. For example, tyrosine- containing motifs in the C-tails of receptors for epidermal growth factor, asialoglycoproteins, and polymeric immunoglobulins interact directly with the clathrin-associated adaptor protein complex AP-2 (Bohm et al 1997). Equivalent tyrosine-containing endocytic motifs may exist for some G-protein-coupled receptors. Mutation of conserved tyrosine residues within potential endocytic domains of the angiotensin II receptor and the neurokinin-1 receptor impairs agonist-induced endocytosis without substantially affecting agonist binding or signaling. In contrast, tyrosines in the C-tails of the β2-adrenergic receptor and m2 muscarinic receptor are not critical for internalization. Clearly, there is much still to be learned about the nature of endocytic motifs of GPCRs (Bohm et al 1997).

5. Receptor down-regulation

A reduction in the number of receptors on the plasma membrane is called down-regulation. Down-regulation could result either from a decreased synthesis of receptors or from accelerated degradation. A prolonged exposure of the cell to a ligand would cause repeated internalization and return of the receptors to the plasma membrane, a process in which a fraction of the receptors is lost in each cycle to lysosomal degradation. This could eventually deplete the receptor number severely and cause down-regulation. Mutations in adrenergic and muscarinic receptors that impaired down-regulation did not abolish desensitization, suggesting that internalization and down-regulation are independent processes, probably relying on independent signals (Bohm et al 1997). Prolonged activation could result in modification of the receptor to assume a conformation that targets it to lysosomal degradation rather than recycling. Alternately, prolonged receptor activity might generate a downstream signal in sufficient concentration to signal degradation of the receptor. Either way, it is unlikely under physiological conditions that cells will be exposed to activating ligands for such long periods of time that down-regulation of receptors is required. Under pathological conditions, however, down-regulation may serve as a protective mechanism in the face of continuous secretion of hormones or neurotransmitters from tumors (Bohm et al 1997).

SUMMARY

Surface receptors recognize signaling molecules in the cell's environment and convey the information to the inside of the cell, triggering a response to the signal. The information transfer is accomplished by different mechanisms: (1) receptors becoming ion channels, (2) receptors generating second messengers, (3) receptors phosphorylating other proteins or activating protein kinases, which in turn phosphorylate other proteins, and (4) receptors activating G-proteins, which regulate the second-messenger-generating enzymes or ion channels. A given signaling molecule could be detected on different cell types by different types of receptors, each one transferring the information by one of the above mechanisms. In every case, the cell's response to a signaling molecule is restricted to a desired length of time because the signaling molecules are either rapidly destroyed or removed by uptake mechanisms, or the activated receptors are inactivated. The cells have also developed mechanisms to deal with the prolonged presence of signaling molecules in their environment. These include desensitization and down-regulation of receptors.

ACKNOWLEDGMENTS

Supported by grants EY 07158, EY 05230, DK 39957, and DK 43207.

REFERENCES

Abou-Samra AB, Juppner H, Force T, Freeman MW, Kong XF, Schipam E, Ureva P, Richards J, Bonaventre JV, Potts JT, Kronenberg HM, and Signe GV. 1992. Expression cloning of a common receptor for parathyroid hormone and parathyroid hormone-related peptide from rat osteoblast-like cells - a single receptor stimulates intracellular accumulation of both cAMP and inositol trisphosphate. Proc Natl Acad Sci USA 89:2732–2736.

Aronheim A, Engleberg D, Li N, al Alawi N, Schlessinger J, and Karin M. 1994. Membrane targeting of the nucleotide exchange factor Sos is sufficient for activating the Ras signaling pathway. Cell 78:949–961.

Barak LS, Menard L, Fergusson SS, Colapietro AM, and Caron MG. 1995. The conserved seven-transmembrane sequence NP(X)2,3Y of the G-protein-coupled receptor superfamily regulates multiple properties of the β_2-adrenergic receptor. Biochemistry 34:15407–15414.

Bennett N and Sitaramayya A. 1988. Inactivation of photoexcited rhodopsin in retinal rods: the role of rhodopsin kinase and 48-kDa protein (arrestin). Biochemistry 27:1710–1715.

Bockaert J, Hunzicker-Dunn M, and Birnbaumer L. 1976. Hormone-stimulated desensitization of hormone-dependent adenylyl cyclase. Dual action of luteinizing hormone on pig graaffian follicle membranes. J Biol Chem 251:2653–2663.

Bohm SK, Grady EF, and Bunnett NW. 1997. Regulatory mechanisms that modulate signalling by G-protein-coupled receptors. Biochem J 322:1–18.

Bohm SK, Khitin LM, Smeekens SP, Grady EF, Payan DG, and Bunnett NW. 1997a. Identification of potential tyrosine-containing endocytic motifs in the carboxyl-tail and seventh transmembrane domain of the neurokinin 1 receptor. J Biol Chem 272:2363–2372.

Chang CP, Lasar CS, Walsh BJ, Komuro M, Collawn JF, Kuhn LA, Tainer JA, Trowbridge IS, Farquhar MG, Rosenfeld MG, Wiley HS, and Gill GN. 1993. Ligand-induced internalization of epidermal growth factor receptor is mediated by multiple endocytic codes analogous to the tyrosine motif found in constitutively internalized receptors. J Biol Chem 268:19312–19320.

Cotecchia S, Kobilka BK, Caniel KW, Nolan RD, Lapetina EY, Caron MG, Lefkowitz RJ, and Regan JW. 1990. Multiple second messenger pathways of alpha adrenergic receptor subtypes expressed in eukaryotic cells. J Biol Chem 265:63–69.

Derynck R. 1994. TGF-β-receptor-mediated signaling. Trends Biochem Sci 19:548–553.

Devault A, Lazure C, Nault C, LeMoual H, Seidah NG, Chretien M, Kahn P, Powell J, Mallet J, Beaumont A, Roques BP, Crine P, and Boileau G. 1987. Amino acid sequence of rabbit kidney neutral endopeptidase 24.11 (enkephalinase) deduced from a complementary DNA. EMBO J. 6:1317–1322.

Evans PD, Robb S, Cheek TR, Reale V, Hannan FL, Swales LS, Hall LM, and Midgley JM. 1995. Agonist-specific coupling of G-protein-coupled receptors to second messenger systems. Prog Brain Res 106:259–268.

Ferguson S, Zhang J, Barak L, and Caron M. 1997. Pleiotropic role for GRKs and β-arrestins in receptor regulation. NIPS 12:145–151.

Figini M, Emanueli C, Geppetti P, Lovett M, Gamp PD, Kirkwood K, Payan DG, Gerard C, and Bunnett NW. 1996. Spontaneous plasma extravasation in the gastrointestinal-tract of neutral endopeptidase (NEP) knockout mice is mediated by bradykinin (BK) and substance-P (SP). Gastroenterol 110:A1071.

Frantl WJ, Johnson DE, and Williams LT. 1993. Signaling by receptor tyrosine kinases. Annu Rev Biochem 62:453–481.

Garland AM, Grady EF, Lovett M, Vigna SR, Frucht MM, Krause JE, and Bunnett NW. 1996. Mechanisms of desensitization and resensitization of G-protein-coupled neurokinin 1 and neurokinin 2 receptors. Mol Pharmacol 49:438–446.

Grady EF, Garland AG, Gamp PD, Lovett M, Payan DG, and Bunnett NW. 1995. Delineation of the endocytic pathway of substance P and the seven transmembrane domain NK1 receptor. Mol Biol Cell 6:509–524.

Haga, T, Haga, K, and Kameyama, K. 1994. G-Protein-coupled receptor kinases. J. Neurochem 63:400–412.

Hanover JA, Willingham MC, and Pastan I. 1984. Kinetics of transit of transferrin and epidermal growth factor through clathrin-coated membranes. Cell 39:283–293.

Heldin C-H, 1995. Dimerization of cell surface receptors in signal transduction. Cell 80:213–223.

Huganir RL and Miles K. 1989. Protein phosphorylation of nicotinic acetylcholine receptors. CRC Critical Rev Biochem Mol Biol 24:183–215.

Hwang L, Leichter R, Okamoto A, Payan D, Collins SM, and Bunnett NW. 1993. Downregulation of neutral endopeptidase (EC 3.4.24.11) in the inflamed rat intestine. Am J Physiol 264:G735–G743.

Inagaki M, Moustakas A, Lin HY et al. 1993. Growth inhibition by transforming growth factor β (TGF-β type I is restored in TGF-β-resistant hepatoma cells after expression of TGF-β receptor type II cDNA. Proc Natl Acad Sci USA 90:5359–5363.

June CH, Fletcher MC, Ledbetter JA, Schieven GL, Siegel JN, Phillips AF, and Samelson LE. 1990. Inhibition of tyrosine phosphorylation prevents T-cell receptor-mediated signal transduction. Proc Natl Acad Sci USA 87:7722–7726.

Kadin ME, Cavaille-Coll MW, Gertz R, Massague J, Cheifetz S, and George D. 1994. Loss of receptors for transforming growth factor β in human T-cell malignancies. Proc Natl Acad Sci USA 91:6002–6006.

Kavanaugh WM, Turck CW, and Williams LT. 1995. Binding of PTB domains to signaling proteins through a motif containing phosphotyrosine. Science 268:1177–1179.

Kleckner NW and Dingledine R. 1988. Requirement for glycine in activation of NMDA-receptors expressed in Xenopus oocytes. Science 241:835–837.

Kwatra MM, Leung E, Maan AC, McMahon KK, Ptasienski J, Green RD, and Hosey MM. 1987. Correlation of agonist-induced phosphorylation of chick heart muscarinic receptors with receptor desensitization. J Biol Chem 262:16314–16321.

Lefkowitz RJ and Caron MG. 1988. Adrenergic receptors. Models for the study of receptors coupled to guanine nucleotide regulatory proteins. J Biol Chem 263:4993–4996.

Levitan IB and Kaczmarek LK. 1991. The neuron. Cell and molecular biology. Oxford University Press: New York.

Liebman PA and Pugh Jr EN. 1980. ATP mediates rapid reversal of cyclic GMP phosphodiesterase activation in visual receptor membranes. Nature 287:734–736.

Lindstrom J, Merlie J, and Yogeeswaran G. 1979. Biochemical properties of acetylcholine receptor subunits from Torpedo Californica. Biochemistry 18:4465–4470.

Lohse MJ, Benovic JL, Caron MG, and Lefkowitz RJ. 1990. Multiple pathways of rapid β_2-adrenergic receptor desensitization. Delineation with specific inhibitors. J Biol Chem 265:3202–3211.

Lu B, Figini M, Gepetti P, Grady EF, Gerard NP, Ancel JC, Payan DG, Gerard C, and Bunnett NW. 1997. The control of microvascular permeability and blood pressure by neutral endopeptidase. Nature Medicine 3:904–907.

Malenka RC. 1991. Postsynaptic factors control the duration of synaptic enhancement in area CA1 of the hippocampus. Neuron 6:53–60.

Malfroy B, Schofield PR, Kuang WJ, Seeburg PH, Mason AJ, and Henzel WJ. 1987. Molecular cloning and amino acid sequence of rat enkephalinase. Biochem Biophys Res Commun 144:59–66.

Massague J, Attisano L, and Wrana JL. 1994. The TGF-β family and its composite receptors. Trends Cell Biol 4:172–178.

Mayer ML, Westbrook GL, and Guthrie PB. 1984. Voltage-dependent block by Mg2+ of NMDA responses in spinal cord neurons. Nature 309:261–263.

Monaghan DT, Bridges RJ, and Cotman CW. 1989. The excitatory amino acid receptors: their classes, pharmacology and distinct properties in the functions of the central nervous system. Annu Rev Pharmacol Toxicol 29:365–402.

Moss SJ, Smart TG, Blackstone CD, and Huganir RL. 1992. Functional modulation of GABA receptors by cAMP-dependent protein phosphorylation. Science 257:661–665.

Nakajima Y, Tsuchida K, Negishi M, Ito S, and Nakanishi S. 1992. Direct linkage of three tachykinin receptors to stimulation of both phosphatidylinositol hydrolysis and cyclic AMP cascade in transfected Chinese hamster ovary cells. J Biol Chem 267:2437–2442.

Noda M, Takahashi H, Tanabe T, Toyosato M, Kikyotani S, Furutani Y, Hirose T, Takashima H, Inayama S, Miyata T, and Numa S. 1983. Structural homology of Torpedo Californica acetylcholine receptor subunits. Nature 302:528–532.

Nowak L, Bregestovski P, Ascher P, Herbert A, and Prochiantz A. 1984. Magnesium gates glutamate-activated channels in mouse central neurons. Nature 307:462–465.

Opresko LK, Chang CP, Will BH, Burke PM, Gill GN, and Wiley HS. 1995. Endocytosis and lysozomal targeting of epidermal growth factor receptors are mediated by distinct sequences independent of the tyrosine kinase domain. J Biol Chem 270:4325–4333.

Paul AK, Marala RB, Jaiswal RK, and Sharma RK. 1987. Coexistence of guanylate cyclase and atrial natriuretic factor receptor in a 180-kD protein. Science 235:1224–1226.

Pawson T. 1995. Protein modules and signaling networks. Nature 373:573–580.

Pellicone C, Bouillon P, and Virmaux N. 1980. Purification et sequence partielle d'un polypeptide hydrophobe de la rhodopsine bovine fragmentée par le BNPS-Scatole. C R Seances Acad Sci Paris D 290:567–569.

Pitcher J, Lohse MJ, Codina J, Caron MG, and Lefkowitz RJ. 1992. Desensitization of the isolated β2-adrenergic receptor by β-adrenergic receptor kinase, cAMP-dependent protein kinase, and protein kinase C occurs via distinct molecular mechanisms. Biochemistry 31:3193–3197.

Savarese TM and Fraser CM. 1992. In vitro mutagenesis and the search for structure-function relationships among G-protein coupled receptors. Biochem J. 283:1–19.

Singh S, Lowe DG, Thorpe DS, Rodriguez H, Kuang WJ, Dangott LJ, Chinkers M,

Goeddel DV, and Garbers DL. 1988. Membrane guanylate cyclase is a cell surface receptor with homology to protein kinases. Nature 334:708–712.

Sorkin A, Helin K, Waters CM, Carpenter G, and Beguinot L. 1992. Multiple auto-phosphorylation sites of the epidermal growth factor receptor are essential for receptor kinase activity and internalization. J Biol Chem 267:8672–8678.

Sorkin A, Mohammadi M, Huang J, and Schlessinger J. 1994. Internalization of fibroblast growth factor receptor is inhibited by point mutation at tyrosine 766. J Biol Chem 269:17056–17061.

Sorkin A, Westermark B, Heldin CH, and Clesson-Welsh L. 1991. Effect of receptor kinase inactivation on the rate of internalization and degradation of PDGF and PDGF β-receptor. J Cell Biol 112:469–478.

Sorkin A. 1996. Receptor-mediated endocytosis of growth factor. In: Heldin CH, Purton M, eds. Signal Transduction. London: Chapman & Hall. 109–123.

Sun L, Wu G, Willson JKV, Zborowska E, Yang J, Rajkarunanayake I, Wang J, Gentry LE, Wang XF, and Brattain MG. 1994. Expression of transforming growth factor β type II receptor leads to reduced malignancy in human breast cancer MCF-7 cells. J Biol Chem 269:26449–26455.

Taylor P. 1985. Acetylcholinesterase agents. In: Gilmam AG, Goodman LS, Rall TW, Murad F, eds. The pharmacological basis of therapeutics. 7th edition. New York: Macmillan Publishing Company. 110–129.

Trowbridge IS, Collawn JF, and Hopkins CR. 1993. Signal-dependent membrane protein trafficking in the endocytic pathway. Annu Rev Cell Biol 9:129–161.

Uhl GR and Hartig PR. 1992. Transporter explosion: update on uptake. Trends Pharmacol Sci 13:421–425.

Valius M and Kazlauskas A. 1993. Phospholipase C-gamma 1 and phosphatidylinositol 3 kinase are the downstream mediators of the PDGF receptor's mitogenic signal. Cell 73:321–334.

von Zastrow M and Kobilka BK. 1992. Ligand-regulated internalization and recycling of human β2-adrenergic receptors between the plasma membrane and endosomes containing transferrin receptors. J Biol Chem 267:3530–3538.

Wang JK, McDowell JH, and Hargrave PA. 1980. Site of attachment of 11-cis-retinal in bovine rhodopsin. Biochemistry 19:5111–5117.

Weill CL, McNamee MG, and Karlin A. 1974. Affinity labeling of purified acetylcholine receptor from Torpedo Californica. Biochem Biophys Res Commun 61:997–1003.

Weiss A. 1993. T-cell antigen receptor signal transduction: a tale of tails and cytoplasmic protein-tyrosine kinase. Cell 73:209–212.

Wilden U, Hall SW, and Kuhn H. 1986. Phosphodiesterase activation by photoexcited rhodopsin is quenched when rhodopsin is phosphorylated and binds the intrinsic 48-kDa protein of rod outer segments. Proc Natl Acad Sci USA 83:1174–1178.

Wrana JL, Attisano L, and Wieser R. 1994. Mechanism of activation of the TGF-β receptor. Nature 370:341–347.

Yu SS, Lefkowitz RJ, and Hausdorff WP. 1993. β-adrenergic receptor sequestration. A potential mechanism of receptor resensitization. J Biol Chem 268:337–341.

Zhou S, Shoelson SE, Chaudhuri M, Gish G, Pawson T, Haser WG, King F, Roberts
    T, Ratnofsky S, Lechleider RJ, Neel BG, Birge RB, Fajardo JE, Chou MM,
    Hanafusa H, Schaffhausen B, and Cautley LC. 1993. SH2 domains recognize
    specific polypeptide sequences. Cell 72:767–778.
Ziemiecki A, Harpur AG, and Wilks AF. 1994. MAP kinase protein tyrosine kinases:
    their role in cytokine signaling. Trends Cell Biol 4:207–212.

RECOMMENDED READING

1. Zach W. Hall. 1992. An Introduction to Molecular Neurobiology. Sinauer Associates Inc., Sunderland, MA.

2. Irwin B. Levitan and Leonard K. Kaczmarek. 1991. The Neuron. Cell and Molecular Biology. Oxford University Press, New York.

3. Carl-Henrik Heldin and Mary Purton, editors. 1996. Signal Transduction. Chapman & Hall, London.

4. A collection of reviews on signal transduction in the journal *Cell*, Volume 80, Number 2, 1995.

2

Heterotrimeric G-proteins: Structure, Regulation, and Signaling Mechanisms

THEODORE WENSEL

G-proteins are transducers of a wide array of extracellular signals, conveying information from transmembrane receptors to the regulatory and metabolic machinery of the cell. While there is tremendous variety among the pathways using G-proteins as transducers, there are certain common features of G-protein-based transduction mechanisms that are conserved in eukaryotic organisms ranging from yeast to human beings, and these are the focus of this chapter. There are so many different G-protein-coupled receptors, so many commonly used drugs that act mainly on G-protein-dependent pathways, and so many fundamental biological processes that depend on them, that G-protein-mediated pathways could be considered the most important class of signaling pathways in the human body.

FUNDAMENTAL CONCEPTS OF G-PROTEIN SIGNALING: RECEPTORS, EFFECTORS, AND GAPS

G-protein-mediated signaling is usually described in terms of the sequential interactions, or "information flow", (Fung et al 1981) among three classes of proteins: receptor → G-protein → effector (Figure 2-1). The receptor is one of a large number of transmembrane proteins, normally but not always found on the cell surface, and able to bind a ligand delivered usually, but again not always, from outside the cell. Upon ligand binding it undergoes a conformational change that dramatically enhances its interactions with the G-protein. The G-protein is generally depicted as consisting of three kinds of subunits: α (the GTP- or GDP-binding polypeptide), β, and γ, with β and

Introduction to Cellular Signal Transduction
A. Sitaramayya, Editor
©1999 Birkhäuser Boston

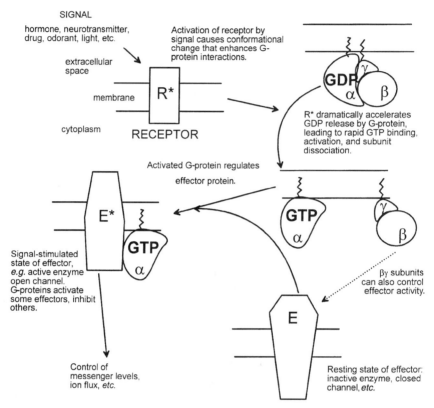

Figure 2-1. Information flow in G-protein signaling. The mechanism by which G-proteins convey information about extracellular signals to the intracellular biochemical apparatus is shown schematically. When G-protein α subunits bind GTP, they are themselves converted to an activate state, capable of regulating effector enzymes and ion channels, and they also release βγ subunits, which can also regulate effector proteins. In the absence of activated receptors, the G-proteins remain predominantly in the "inactive" GDP state. The main reason is that GDP is very slow relative to the rates at which hydrolysis converts the GTP state back to the GDP state. The triggering role of receptors is based on their conversion, upon binding agonist, into a conformation that catalyzes GDP release by the G-protein.

γ acting together as an effectively inseparable unit. Either the α or βγ unit may interact with the effector. "Effector" is not a particularly descriptive term, but it is in universal use. It is generally meant to refer to a catalytic protein (enzyme or ion channel) capable of changing levels of "second messengers," usually fairly small molecules that can regulate the activity of macromolecules. An interesting variation on this theme is the role of potassium channels as effectors that possess real G-protein-regulated catalytic activity (ion flux), but whose major functional role consists of manipulating membrane potential rather than significantly altering "second messenger" concentrations in a direct way.

Fundamental to the information carried by a G-protein is that it is transient. GTP hydrolysis provides a mechanism for returning to the preactivation state. As discussed below, we now know that just as activation is catalyzed by receptors, specific proteins (GTPase Accelerating Proteins, or GAPs) catalyze this inactivation step as well.

AMPLIFICATION

One of the most important features of G-protein signaling is that its use of multiple sequential catalytic steps results in tremendous amplification, enabling signals communicated by G-proteins to elicit cellular responses with speed and great sensitivity. Indeed, cases have been documented of a measurable cellular response elicited by a single excited receptor molecule. The way in which amplification is achieved is depicted schematically in Figure 2-2. For any step in the transduction pathway to achieve amplification "gain," it must be catalytic. That is, each molecule carrying information into an amplified step must generate multiple information-carrying molecules as output. Thus, for example, binding of a ligand such as a hormone to the receptor does not represent an amplified step. At least one hormone molecule must reach the cell surface for every receptor activated. However, once the receptor is activated by a hormone-induced conformational change, it becomes a potent catalyst whose activation is greatly amplified through its ability to catalyze activation of many G-proteins. The amplification at this step has been most accurately measured for the visual system, in which each activated receptor (R*) has been found to activate on the order of hundreds of G-proteins in the early stages of the light response (Kahlert and Hofmann 1991; reviewed by Pugh and Lamb 1993). In cells with lower G-protein concentrations, the amplification may be somewhat less because of diffusional delay, while R* repeatedly encounters inactive G-proteins. A twofold increase in amplification may be achieved at this step if both βγ and α-GTP are activators of the effector, as in the case of phospholipase C β, for example.

The next step, effector activation by G-protein is not generally an amplified one. At least one G-protein unit (α-GTP or βγ) is thought to be required to activate one effector molecule, and presumably it must stay bound to maintain the effector's activated state. However, the effectors are often potent catalysts whose ability to generate large numbers of signaling molecules in a short time outshines even that of the receptor. Ion channels can also generate amplification gain, by virtue of their catalysis of transmembrane flux of many ions per channel protein.

Very often the catalytic activity of the effector immediately regulated by a G-protein is not the last amplified stage in transduction. The signaling

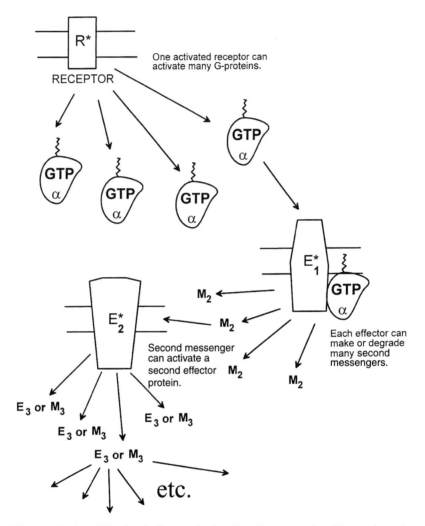

Figure 2-2. Amplification in G-protein signaling. Every step in a G-protein pathway that involves catalytic production of signaling molecules gives rise to amplification, as depicted schematically here. E_n refers to effector molecules (enzymes or ion channels), and E_n * indicates the activity state induced by binding the activated G-protein. M_n refers to diffusible messenger molecules whose levels are regulated by the E_n. Examples of E_1 include adenylyl cyclase, phospholipase C, and cGMP phosphodiesterase. Examples of E_2 include cAMP-dependent protein kinase, protein kinase C, InsP$_3$-gated Ca^{2+} release channels and cGMP-gated cation channels. Glycogen phosphorylase kinase represents an example of E_3. Examples of M_2 (second messengers, with hormone or other extracelluar ligand considered the "first" messenger) include cAMP, cGMP, diacylglycerol, InsP$_3$, etc. Ca^{2+} could be considered an example of M_3.

molecules generated (or degraded) by an effector generally serve to regulate the activity of yet another catalytic moiety. The classic example is activation of cAMP-dependent protein kinase (PKA) by cAMP generated by adenylyl cyclase in response to G-protein stimulation. While activation of the kinase by cAMP binding is not itself an amplified step, the ability of PKA to catalyze phosphorylation of many substrate polypeptides makes this phosphorylation step an amplified one, and the substrates themselves are most often catalytic proteins, and so introduce additional gain into the signaling cascade. Similar behavior is observed with InsP$_3$ generated by the effector phospholipase C; it activates the InsP$_3$ receptor calcium channel, and in turn released Ca^{2+} regulates the activity of many catalytic proteins in the cytoplasmic compartment. Thus, multiple stages of amplification are the rule.

G-PROTEIN STRUCTURE

The structures of G-proteins are well characterized in terms of subunit composition, posttranslational modifications, amino acid sequence, and three-dimensional structure. A complete G-protein is usually thought of as consisting of a heterotrimeric complex of α, β, and γ subunits.

α Subunits

The α subunit binds GTP or GDP and displays significant sequence similarity in its nucleotide binding domain to the Ras superfamily of small GTP-binding proteins, as well as to prokaryotic translation factors such as EF-Tu and to other GTP-regulated proteins. The α subunits range in mass from 40 to 46 kDa, and are frequently modified posttranslationally. In all known cases the N-terminal methionyl residue is proteolytically removed and replaced with an N-acyl group. A number of Gα subunits contain glycine as their N-terminal residue following this first modification, and in those cases (e.g., G$_t$, G$_i$, G$_{o\alpha}$) the acylating group is a fatty acid, normally (with interesting exceptions, Kokame et al 1992; Neubert et al 1992) a 14-carbon myristoyl group (Buss et al 1987). A distinct but overlapping set of G$_\alpha$ subunits contains a cysteine residue near the N-terminus that serves as a site for acylation by another fatty acid, most commonly palmitate, linked via a thioester (Degtyarev et al 1993). Both fatty acyl modifications appear to enhance membrane binding of G$_\alpha$ (for a review of protein lipidation in signaling, see Resh 1996), but they differ in the mechanisms of attachment and in their reversibility. Myristoylation occurs cotranslationally (Yang & Wensel 1992; reviewed by Gordon et al 1991) and is essentially irreversible, while palmitoylation is more dynamic, palmitate being readily added and removed in response to changes in association with other proteins (receptors and βγ subunits) and membranes. Palmitoylation can occur without intervention of a specific enzyme (Duncan and

Gilman 1996; O'Brien et al 1987), and this nonenzymatic acylation may be important in modulating G_α function.

γ Subunits: Sequences and Posttranslational Modifications

The γ subunits are much smaller than the α and β (molecular masses ~6 kDa–8 kDa) and show a fair degree of sequence diversity. Eleven different γ subunit types have been identified in mammals, with sequence identity ranging from 30 to 80% (Ray et al 1995). They γ subunits are subject to an interesting sequence of posttranslational processing events that is rare for proteins in general, but much more common for those involved in membrane signaling events. It involves addition of either a farnesyl or geranylgeranyl group to a cysteine residue located four residues away from the carboxyl terminus, proteolytic removal of the carboxyl terminal three residues, and methylation of the new carboxyl terminus to form a methyl ester (reviewed by Higgins and Casey 1996). In the three-dimensional structure of heterotrimers of the G_i subfamily, the carboxyl terminus of γ is located near the lipidated amino terminus of the α subunit, presumably allowing both lipid moieties to cooperate in membrane attachment.

β Subunits: Sequence, WD-40 Repeats, Lack of Lipidation

The beta subunits are characterized by an amino acid sequence motif of seven "WD repeats" (Neer et al 1994; Neer 1995; Neer and Smith 1996) found in a wide range of proteins known to be involved in regulatory interactions with other proteins. The rules defining this motiff have been defined as: $\{X_{6-94}\text{-}[GH\text{-}\text{-}X_{23-41}\text{-}\text{-}WD]\}_{4-8}$, where for Gβ subunits the number of repeats, 4–8 in the general case, is equal to seven. This general motif is not uncommon, and many of the proteins featuring it are likely to have little in common with β subunits in terms of function, other than perhaps an involvement of this structural motif in protein-protein interactions. Although the β subunits appear to be even more tightly membrane-associated than the lipidated α subunits, they themselves have not been found to contain covalently attached lipids. Rather, their membrane association is thought to be largely conferred by very tight protein-protein binding to the γ subunits, which are universally lipidated. Interaction of β and γ is partly mediated by coiled-coil interactions involving an N-terminal helical domain in G_β.

X-RAY CRYSTALLOGRAPHY OF G-PROTEINS

α Subunits

The three-dimensional structures of $G_{t\alpha}$ and $G_{i\alpha}$ subunits bound to various nucleotides and analogues (Noel et al 1993; Lambright et al 1994; Sondek et al 1994; Coleman et al 1994; Mixon et al 1995) reveal a common structural organization in which the structural differences among different α subunits and among different nucleotide-regulated activity states are of a fairly subtle (but extremely important) nature. In all cases there are two distinct domains, termed the "GTPase domain" or "*ras* domain" and the "helical domain." The α carbon backbone of the GTPase domains can be aligned very closely with that of the small GTP-binding protein *ras*. This domain consists of a six-stranded β sheet (β1–β6) surrounded by five α helices (α1–α5). The helical domain consists of a long central helix (αA) with five shorter helices (αB–αF) packed around it, and is formed by sequence elements inserted within interruptions in the conserved GTPase sequences, and connected by two connecting stretches of polypeptide, linker 1 and linker 2. The GTP binding site is found in a deep cleft between these domains, and is packed so tightly within the protein that bound GTPγS is almost completely inaccessible. A network of many hydrogen bonds and other contacts clamps the nucleotide tightly in place. A total of nine residues is involved in binding the triphosphate moiety, another nine help hold the guanine base in place, and two make hydrogen bonds to the 3′ and 5′ hydroxyls of the ribose. Mutual binding of the protein and the phosphate to Mg^{2+} also helps hold the nucleotide in place. These multiple interactions and nearly complete steric hindrance help to explain the very tight binding of GTP and GTPγS, and especially, their extremely slow rates of dissociation. In the GDP complex, most of these contacts are preserved, but those to the γ phosphoryl are lost, and the nucleotide is more accessible to solvent. These differences may account for the fact that GDP, while very slow to dissociate, is much less tightly bound than GTP or GTPγS.

Switch Domains

Fundamental to the function of G-proteins are subtle structural alterations that occur upon switching from the GDP- to GTP-bound state, inferred from structures solved for the GDP- and GTPγS-bound states. These changes occur primarily in subdomains referred to as "switch" regions: switch I, corresponding to the second linker domain connecting the helical domain (N-terminal side) to the GTPase domain (C-terminal side) and part of the β2 strand of the GTPase domain in $G_{t\alpha}$; switch II, the C-terminal end of β3, the α2 helix (actually a 3_{10} helix in $G_{t\alpha}$) and the loop connecting α2 and β4; and switch III, the loop connecting β4 and α3. Switch I and switch II resemble switch domains in Ras and other GTP-binding proteins, but switch III has only been observed in G_α. In $G_{i\alpha}$ a switch IV domain was also identified, which occurs within the helical domain between αB and αC, a region notable for being quite far from the phosphate binding domain that presumably triggers all the conformational "switches." It seems likely that these changes account for the dramatic differences in affinities for other molecules (receptor, βγ subunits, effectors) that occur upon switching between the GDP and GTP forms. Indeed, residues in the switch I and switch II regions are involved in, or close to, the interface between G_α and $G_{\beta\gamma}$ in the heterotrimer structures (see below).

Models for Receptor and Effector Interactions

Structural information has been combined with information about conserved residues and results of mutagenesis experiments and peptide competition experiments to formulate models for assigning interactions with the receptor and effector to specific regions within the G_α structure. For example, the "evolutionary trace" technique was used to map conserved residues onto the surface of $G_{t\alpha\beta\gamma}$ and to propose that certain parts of the structure participate in the receptor or effector interfaces (Lichtarge et al 1996). Site-directed mutagenesis has identified residues in $G_{s\alpha}$, $G_{i\alpha}$, and $G_{q\alpha}$ important for effector interactions (Berlot and Bourne 1992; Medina et al 1996) and has led to proposals for actual sites of interaction. While there is considerable evidence in support of these models, they remain hypotheses to be tested by solution of structures for the relevant complexes.

Heterotrimer and βγ Structures

Structures have also been solved for βγ subunits, either alone (Sondek et al 1996) or complexed to $G_{i\alpha}$ (Wall et al 1995) or to a $G_{t\alpha}/G_{i\alpha}$ chimera (Lambright et al 1996). The most striking feature of these is the organization of the β subunit into a seven-blade beta propeller structure, with the WD-40 repeats making up the blades with four-stranded antiparallel beta sheets. The polypeptide chain of the γ subunit is completely extended without any tertiary contacts other than those with β, with which it has extensive hydrophobic interactions, including a coiled coil formed by the N-terminal helix of γ and the N-terminal helix of β. The G_α subunit interacts with β primarily through residues at or near the switch I and switch II regions, contacting loops and turns on one face of the β propeller. There are additional contacts between the side of the β propeller and the N-terminal helix of α. The isoprenylated C-terminus of the γ subunit (missing the last three residues and the isoprenoid in the crystal structure) is appropriately placed with respect to the myristoylated N-terminus of α (lacking the myristoyl moiety in the crystal structure) to allow simultaneous insertion in the membrane of both lipids, although it remains to be determined whether or not this insertion actually occurs.

INTERACTIONS WITH GUANINE NUCLEOTIDES

The functions of G-proteins depend critically on their interactions with guanine nucleotides. The kinetics and thermodynamics (energetics) of these interactions are both important in signal transduction. Confusion about the nature of these interactions has resulted from two features of the GTP/GDP cycle of G-proteins that complicate interpretation of experimental results. One is that the multistep nature of the process makes it impossible to apply classical Michaelis-Menten analysis to GTP hydrolysis, except under very specialized conditions. The other is that the affinity of $G\alpha$ subunits for GTP or GTPγS is so high that accurate measurements of dissociation constants (K_d) can only be carried out at very low protein concentrations, or by calculating them indirectly from kinetic measurements.

For G-proteins of the G_i subfamily, $G_{i\alpha}$, $G_{o\alpha}$, and $G_{t\alpha}$, the interactions with GTPγS have been analyzed using techniques that account for these features. Nucleotide-free proteins prepared in 1 M NH_4SO_4 and stabilized by glycerol were used to measure second-order rate constants, k_{on}, for GTPγS binding to $G_{o\alpha}$ and $G_{i\alpha}$ of $> 10^7$ $M^{-1}\cdot s^{-1}$ and 10^6 $M^{-1}\cdot s^{-1}$, respectively (Ferguson et al 1986). Dissociation data (Higashijima et al 1987c) revealed that in the presence of $[Mg^{2+}]$ above 100 nM, the first-order dissociation rate constants, k_{off}, for GTPγS were $< 2 \times 10^{-5}$ s^{-1} for both $G_{o\alpha}$ and $G_{i\alpha}$, implying equilibrium dissociation constants of $K_d = k_{off}/k_{on} < 2 \times 10^{-12}$ M and $< 2 \times 10^{-11}$ M, respectively. Similar high-affinity binding, with $K_d = 3 \times 10^{-11}$ M, was demonstrated for $G_{t\alpha}$, using equilibrium binding assays at very low protein concentrations (Malinski et al 1996).

Results with G_s are similar. For a complex of G_s and β-adrenergic receptor (presumably GDP-free), k_{on} for GTPγS is 5.5×10^5 $M^{-1}\cdot s^{-1}$ (May and Ross 1988). Just as observed for the G_i family, $k_{off} < 2 \times 10^{-5}$ s^{-1} (Lee et al 1992), implying a K_d value $< 3.6 \times 10^{-11}$ M. The numerous publications describing apparent K_d values that are orders of magnitude above these high affinity values can be easily explained as inevitable consequences of the use of protein concentrations in excess of the true K_d values. Under such conditions, apparent K_d values on the same order as the total protein concentrations can always be expected (Goody et al 1991).

Interactions with GDP are quite different, and of lower affinity, even though they occur at the same binding site, and differ only in the presence or absence of the γ phosphate group. GDP dissociates at measurable, although slow, rates even in the absence of Mg^{2+}. For $G_{o\alpha}$ and $G_{i\alpha}$, k_{off} values are ~5×10^{-3} s^{-1} and ~1×10^{-3} s^{-1}, respectively (Ferguson et al 1986), while for $G_{s\alpha}$ the value is ~3×10^{-3} s^{-1} (Lee et al 1992), and for $G_{t\alpha}$, it is in the range of 10^{-4}–10^{-5} s^{-1} (Ramdas et al 1991; Fawzi and Northup 1990). For $G_{s\alpha}$, competition binding experiments indicate that GDP binds with about 50-fold lower affinity than does GTPγS (Northup et al 1982), while for $G_{t\alpha}$, the ratio is at least 300 (Zera et al 1996). Thus, high affinity (subnanomolar K_d values) binding of GTP and GTPγS, and much lower affinity for GDP, with extremely slow dissociation of all three nucleotides seem to be the general rule for G_α.

GTP Hydrolysis

GTP hydrolysis is thought to be essential both for inactivating G-proteins following termination of the external signal, and for resetting them to the resting state, in which they are capable of receiving and transducing a new signal. Isolated G-proteins do this rather slowly; k_{cat} values range from 0.001 to 0.07 s^{-1} (Angleson and Wensel 1994 and references therein, Brandt and Ross 1986, Casey et al 1990). For a number of the physiological processes mediated by G-proteins, these rates of inactivation are far too slow to account for the observed temporal resolution of signal responses. This dilemma has been solved in recent years by the discovery of GTPase-activating proteins, which dramatically accelerate the rate at which G_α can hydrolyze bound GTP. In one well-documented case, the GAP is an effector molecule, phospholipase C β (Berstein et al 1992), but there are also GAPs,

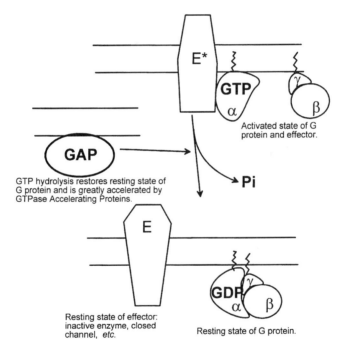

Figure 2-3. Returning G-proteins to the resting state by GTP hydrolysis. G_α subunits possess an intrinsic ability to catalyze hydrolysis of bound GTP that allows them to return to the resting state upon signal termination. This mechanism for inactivation not only stops the signaling process, but also prepares the G-protein to respond to the next signal. In many cases the intrinsic hydrolytic rate is too slow for physiological function and must be accelerated by the action of GTPase accelerating proteins, or GAPs.

called RGS proteins (Koelle and Horvitz 1996, Berman et al 1996a), which do not appear to be effectors.

Solution of the structure of GDP-bound $G_{t\alpha}$ formed in the presence of Al^{3+} and F^- (Sondek et al 1994) revealed AlF_4^- binding with a geometry suggestive of a pentavalent intermediate in GTP hydrolysis, and led to a proposal for the hydrolytic mechanism. Interestingly, at least some of the RGS-type GAPs appear to bind to the AlF_4^- GDP complex of some G_α better than to the GDP or GTPγS forms (Berman et al 1996b).

DIVERSITY OF G-PROTEINS

α Subunits

Structural and functional diversity within the G-protein family in a given eukaryotic organism is conferred by expression of multiple α, β, and γ subunits that form the functional heterotrimers. In mammals, 17 G_α genes have been identified, and they give rise to 23 distinct G_α proteins, including some generated through alternative splicing (Gudermann et al 1997). The G_α proteins, whose sequence identities range from ~50 to 80%, may be grouped into subfamilies based on sequence similarity: The G_i family is the most abundantly expressed, and includes three isoforms of $G_{i\alpha}$ ($G_{i\alpha 1}$, $G_{i\alpha 2}$, $G_{i\alpha 3}$) encoded by three different genes, two isoforms of $G_{o\alpha}$ resulting from alternative slicing, $G_{z\alpha}$, and the transducins ($G_{gus t\alpha}$, $G_{t\alpha 1}$, $G_{t\alpha 2}$) encoded by different genes and found almost exclusively in rod ($G_{t\alpha 1}$) or cone ($G_{t\alpha 2}$) photoreceptors or in taste cells ($G_{gus t\alpha}$, $G_{t\alpha 1}$, $G_{t\alpha 2}$; McLaughlin et al 1992, 1994). The $G_{s\alpha}$ subfamily includes $G_{s\alpha}$ and $G_{olf\alpha}$, both activators of adenylyl cyclases. The G_q subfamily, including $G_{q\alpha}$, $G_{\alpha 11}$, $G_{\alpha 14}$, and $G_{\alpha 16}$, act as activators of phospholipase C β. The G_{12} subfamily includes $G_{\alpha 12}$ and $G_{\alpha 13}$, important in vascular development (see below).

β and γ Subunits

The β subunits are less diverse. Five different genes are found in mammals, and the amino acid sequences and functional properties of the proteins they encode do not differ greatly (~80% identity). The much smaller γ subunits come in at least 11 different mammalian forms (Ray et al 1995) and show more divergence in their sequences (33–77% identity).

If there were no restrictions on mixing and matching of subunits, well over 1000 different heterotrimers could be assembled, bringing their diversity into line, at least in order of magnitude, with that of the receptors to which they couple. Experimental evidence suggests, however, that a much smaller subset of such combinations is actually used, largely, perhaps, because of cell-type-specific expression patterns, or possibly because of poorly understood mechanisms for subcellular compartmentalization of specific subunits.

DIVERSITY OF RECEPTORS

In contrast to G-protein subunits whose diversity is relatively modest (less than two dozen varieties of each type), the number of distinct receptors encoded in a mammalian genome is staggering (Baldwin 1994, Strader et al 1994). The number currently entered in the database is in the hundreds, and it is generally assumed that there will be well over a thousand identified ultimately. The ligands recognized span a wide range of structural entities as well, ranging from small hydrophobic ligands, such as retinal, eicosanoids, catecholamine hormones, and neurotransmitters, etc., to amino acids and nucleotides such as glutamate and adenosine, to small peptides, such as vasopressin, opiate peptides, or chemokines, to sizable proteins. The widest range of diversity in ligands recognized occurs in the huge family of olfactory receptors (Buck and Axel 1991). While the great majority of receptors are triggered into an activated state by binding of diffusible and dissociable ligands, the transducins respond to visible light through a constitutively bound covalent ligand (11-cis-retinal), and it has been suggested that thrombin receptors may actually respond to proteolytic cleavage of the receptor itself.

DIVERSITY OF EFFECTORS

The list of effector proteins known to be regulated by direct interaction with G-protein subunits has not grown in recent years as rapidly as might have been expected a decade ago, when in a relatively short period of time three new effectors were added to the list. cGMP-specific phosphodiesterases, phospholipase C β, and potassium channels have been shown convincingly to be regulated directly by binding G-protein subunits (either G_α or $G_{\beta\gamma}$, or both) in addition to adenylyl cyclase, the traditional effector whose study led to the discovery of G-proteins. Despite several suggestive reports for additional candidate effectors, such as additional phospholipases, Ca^{2+} channels, and ion-motive pumps, the list of confirmed effectors remains small. How a greater variety of G-proteins and a much greater array of receptors could communicate their messages effectively through such a tiny bottleneck of effector catalysts is far from obvious. One possibility is that many more effectors remain to be discovered but have not been observed so far, simply because their catalytic activities fall outside the range of those commonly assayed.

NEGATIVE REGULATION OF G-PROTEIN PATHWAYS

Nature has paid no less attention to turning off G-protein-coupled pathways than to turning them on. Complex and effective mechanisms exist for inactivation or desensitization at the level of receptors, G-proteins, and effectors. Several broad categories have been identified. These include (1) receptor phosphorylation by kinases specific for activated receptors, as well as by more generic kinases, (2) binding and blocking of G-protein activating sites on phosphorylated receptors by the arrestin family of receptor capping proteins, (3) GTPase acceleration by the action of GTPase accelerating proteins (GAPs), and (4) feedback or crosstalk from calcium-mediated signaling pathways onto effectors (e.g., adenylyl cyclase), enzymes that antagonize their actions (e.g., photoreceptor guanylyl cyclase) or kinases that in turn may inactivate the cascade at various levels.

Recently the role of GTPase accelerating proteins, or GAPs, has received increasing attention. It has long been known that GTP-binding proteins of the Ras superfamily hydrolyze GTP at an appreciable rate only when bound to specific GAPs. However, in the last several years it has become apparent that a distinct set of GAPs plays an important role in terminating G-protein-mediated signals.

There has been particular interest in the RGS (regulators of G-protein signaling) family of GAPs (Koelle and Horvitz 1996, Berman et al 1996a), but the effector phospholipase C β has also been shown to act as a GAP for $G_{q\alpha}$ (Berstein et al 1992), and an inhibitory subunit of the $G_{t\alpha}$ effector cGMP phosphodiesterase has been shown to enhance the activity of a GAP in rod photoreceptors (Angleson and Wensel 1994, Arshavsky et al 1994). Additional GAPs have been identified biochemically, if not yet in terms of gene or protein sequence (Angleson and Wensel 1993, 1994; Wang et al 1997). These and other mechanisms for negative regulation of G-protein pathways are likely to receive increasing attention over the next several years.

G-PROTEINS AND HUMAN HEALTH AND DISEASE

The pivotal role of G-proteins in human health is underscored by the observation that drugs targeting G-protein-mediated pathways make up the largest class of drugs when all are grouped by biochemical mechanisms. Most of these do not target G-proteins directly, but rather are usually antagonists or agonists for the receptors, e.g., the extremely popular H_2-antagonists, cimetidine (Tagamet®) and famotidine (Pepcid®). Others target the biosynthesis of such agonists, e.g., aspirin and other nonsteroidal anti-inflammatory drugs. G-proteins themselves, specifically the α subunits, are the targets of two important bacterial toxins: pertussis toxin and cholera toxin. These have also served as important experimental tools because of their high degree of specificity. Pertussis toxin modifies a number of different G-proteins by catalyzing transfer of an ADP-ribosyl moiety from NAD^+ to a cysteine residue four amino acids from their carboxyl termini (West et al 1985). The functional consequence of this ADP-ribosylation is uncoupling from the receptor. G-proteins lose their ability to be stimulated to release GDP by activated receptors, without losing their abilities to bind guanine nucleotides, to hydrolyze GTP, or to activate effectors when in the GTP form

(Ramdas et al 1991). Cholera toxin, for which the major target is G_s, also catalyzes ADP-ribosylation, but the modification occurs on an arginine residue, and the functional consequence is a dramatic reduction in the rate of hydrolysis of bound GTP (Abood et al 1982).

A number of genetic defects in genes encoding G-protein-coupled receptors have been described (reviewed by Spiegel 1996). These appear to be much more common than those affecting G-protein genes, as might be expected simply from their relative numbers. However, genetic defects in $G_{s\alpha}$ have been described (reviewed by Spiegel 1996). These include germline mutations giving rise to a form of pseudohypoparathyroidism, termed Albright hereditary dystrophy (AHD), and somatic mutations implicated in acromegaly in patients with somatotroph tumors, and in hyperfunctional thyroid nodules. Somatic $G_{s\alpha}$ mutations occurring early in embryogenesis may account for McCune-Albright syndrome (MAS).

G-PROTEIN KNOCKOUTS IN MICE

New insights into the true biological role of G-proteins are being gained through the use of gene inactivation strategies based on embryonic stem cell technology and mouse genetics. For example, disruption of the gene encoding $G_{\alpha 13}$ (Offermanns et al 1997) led to embryonic lethality, attributed to severe defects in development of the vascular system, as well as cellular defects in ligand-induced cell movement. Inactivation of the gene for gustducin ($G_{gust\alpha}$) revealed its role in both bitter and sweet taste transduction (Wong et al 1996). Many more G-protein genes have been successfully inactivated recently, and a much deeper understanding of G-protein function can be expected as the phenotypes are fully characterized.

ADDITIONAL SOURCES OF INFORMATION

The literature of G-proteins is enormous, and I have made no attempt here to supply a complete list of references. The interested reader is referred to an extensive compendium (Pennington 1995) that lists over 3000 references, and to recent reviews (Gudermann et al 1997, Coleman and Sprang 1996, Dessauer et al 1996, Spiegel 1996, Birnbaumer and Birnbaumer 1995, Bourne 1995, Neer 1995) for an overview of recent developments.

REFERENCES

Abood ME, Hurley JB, Pappone MC, Bourne HR, and Stryer L. 1982. J Biol Chem 257:10540–10543.
Angleson JK and Wensel TG. 1993. A GTPase accelerating factor for transducin, distinct from its effector phosphodiesterase, in rod outer segment membranes. Neuron 11:939–949.

Angleson JK and Wensel TG. 1994. Enhancement of rod outer segment GTPase accelerating protein activity by the inhibitory subunit of cGMP phosphodiesterase. J Biol Chem 269:16290–16296.

Arshavsky V Yu, Dumke CL, Zhu Y, Artemyev N, Skiba NP, Hamm HE, and Bownds MD. 1994. Regulation of transducin GTPase activity in bovine rod outer segments. J Biol Chem 269:19882–19887.

Baldwin JM. 1994. Structure and function of receptors coupled to G-proteins. Curr Opin Cell Biol 6:180–190.

Berlot CH and Bourne HR. 1992. Identification of effector-activating residues of Gs alpha Cell 68:911–922.

Berman DM, Wilkie TM, and Gilman AG. 1996a. GAIP and RGS4 are GTPase-activating proteins for the G_i subfamily of G-protein α subunits. Cell 86:445–452.

Berman DM, Kozasa T, and Gilman AG. 1996b. The GTPase-activating protein RGS4 stabilizes the transition state for nucleotide hydrolysis. J Biol Chem 271:27209–27212.

Berstein G, Blank JL, Jhon DY, Exton JH, Rhee SG, and Ross EM. 1992. Phospholipase C-β1 is a GTPase-activating protein for $G_{q/11}$, its physiological regulator. Cell 70:411–418.

Birnbaumer L and Birnbaumer M. 1995. Signal transduction by G-proteins: 1994 edition. J Rec Signal Transduct Res 15:213–252.

Bourne HR. 1995. GTPase: a family of molecular switches and clocks. Philos Trans R Soc Lond B Biol Sci 349:283–289.

Brandt DR and Ross EM. 1986. Catecholamine-stimulated GTPase cycle. Multiple sites of regulation by β-adrenergic receptor and Mg^{2+} studied in reconstituted receptor-G_s vesicles. J Biol Chem 260:266–272.

Buck L and Axel R. 1991. A novel multigene family may encode odorant receptors: a molecular basis for odorant recognition. Cell 65:175–187.

Buss JE, Mumby SM, Casey PJ, Gilman AG, and Sefton BM. 1987. Myristoylated α subunits of guanine nucleotide regulatory proteins. Proc Natl Acad Sci USA 84:7493–7497.

Casey PJ, Fong HK, Simon MI, and Gilmen AG. 1990. Gz, a guanine nucleotide binding protein with unique biochemical properties. J Biol Chem 265:2383–2390.

Coleman DE, Berghuis AM, Lee E, Linder ME, Gilman AG, and Sprang SR. 1994. Structures of active conformations of $G_{i\alpha 1}$ and the mechanism of GTP hydrolysis. Science 265:1405–1412.

Coleman DE and Sprang SR. 1996. How G-proteins work: a continuing story. Trends Biochem Sci 21:41–44.

Degtyarev MY, Spiegel AM, and Jones TLZ. 1993. The G-protein $α_s$ subunit incorporates [^{3}H]palmitic acid and mutation of cysteine-3 prevents this modification. Biochemistry 32:8057–8061.

Dessauer CW, Posner BA, and Gilman AG. 1996. Visualizing signal transduction: receptors, G-proteins, and adenylate cyclases. Clinical Science 91:525–537.

Duncan JA and Gilman AG 1996. Autoacylation of G-protein alpha subunits. J Biol Chem 271:23594–23600.

Fawzi AB and Northup JK. 1990. Guanine nucleotide binding characteristics of transducin: essential role of rhodopsin for rapid exchange of guanine nucleotides. Biochemistry 29:3804–3812.

Ferguson KM, Higashijima T, Smigel MD, Gilman AG, 1986. The influence of bound GDP on the Kinetics of guanine nucleotide binding to G-proteins. J Biol Chem, 261:7393–7399.

Fung BK-K, Hurley JB, and Stryer L. 1981. Flow of information in the light triggered cyclic nucleotide cascade of vision. Proc Natl Acad Sci USA 78:152–156.

Goody RS, Frech M, and Wittinghofer A. 1991. Affinity of guanine nucleotide binding proteins for their ligands: facts and artefacts. Trends Biochem Sci 16:327–328.

Gordon JL, Duronio RJ, Rudnick DA, Adams SP, and Gokel GW. 1991. Protein N-myristoylation. J Biol Chem 266:8647–8650.

Gudermann T, Schoneberg T, and Schultz G. 1997. Functional and structural complexity of signal transduction via G-protein-coupled receptors. Ann Rev Neurosci 20:399–427.

Higashijima T, Ferguson KM, Sternweis PC, Smigel MD, and Gilman AG. 1987. Effects of Mg^{2+} and the beta gamma subunit complex on the interactions of guanine nucteotides with G-proteins. J Biol Chem 262:762–766.

Higgins JB and Casey PJ. 1996. The role of prenylation in G-protein assembly and function. Cell Signal 8:433–437.

Kahlert M and Hofmann KP 1991. Reaction rate and collisional efficiency of the rhodopsin-transducin system in intact retinal rods. Biophys J 59:375–386.

Koelle MR and Horvitz HR. 1996. EGL-10 regulates G-protein signaling in the C elegans nervous system and shares a conserved domain with many mammalian proteins. Cell 84:115–125.

Kokame K, Fukada Y, Yoshizawa T, Takao T, and Shimonishi Y. 1992. Lipid modification at the N terminus of photoreceptor G-protein α subunit. Nature 359:749–752.

Lambright DG, Noel JP, Hamm HE, and Sigler PB. 1994. Structural determinants of the a subunit of a heterotrimeric G-protein. Nature 369:621–628.

Lambright DG, Sondek J, Bohm A, Skiba NP, Hamm HE, and Sigler PB. The 2.0 A crystal Structure of a heterotrimeric G-protein. Nature, 379:311–319, 1996.

Lichtarge O, Bourne HR, and Cohen FE. 1996. Evolutionary conserved $G_{\alpha\beta\gamma}$ binding surfaces support a model of the G-protein-receptor complex. Proc Natl Acad Sci USA 93:7507–7511.

Malinski JA, Zera EZ, Angleson JK, and Wensel TG. 1996 High affinity interactions of GTPγS with the heterotrimeric G-protein, transducin: Evidence at low and high protein concentrations. J Biol Chem 271:12919–12924.

May DC and Ross EM. 1988. Rapid binding of guanosine 5'-O-(3-thiotriphosphate) to an apparent complex of β-adrenergic receptor and the GTP-binding regulatory protein G_s. Biochemistry 27:4888–4893.

McLaughlin SK, McKinnon PJ, and Margolskee RG. 1992. Gustducin is a taste-cell-specific G-protein closely related to the transducins. Nature 357:563–569.

McLaughlin SK, McKinnon PJ, Spickofsky N, Danho W, and Margolskee RG. 1994. Molecular cloning of G-proteins and phosphodiesterases from rat taste cells. Physiology and Behavior 565:1157–1164.

Medina R, Grishina G, Meloni EG, Muth TR, and Berlot CH. 1996. Localization of the effector-specifying regions of Gi2alpha and Gqalpha. J Biol Chem 271:24720–24727.

Mixon MB, Lee E, Coleman DE, Berghuis AM, Gilman AG, and Sprang SR. 1995. Tertiary and quaternary structural changes in Gial induced by GTP hydrolysis. Science 270:954–960.

Neer EJ. 1995. Heterotrimeric G-proteins: organizers of transmembrane signals. Cell 80:249–257.

Neer EJ, and Smith TF. 1996. G-protein heterodimers: new Structures propel new questions. Cell, 84:175–178.

Neer EJ, Schmidt CJ, Nambudripad R, and Smith TF. 1994. The ancient regulatory-protein family of WD-repeat proteins. Nature 371:297–300.

Neubert TA, Johnson RS, Hurley JB, and Walsh KA. 1992. The rod transducin α subunit amino terminus is heterogeneously fatty acylated. J Biol Chem 267:18274–18277.

Noel JP, Hamm HE, and Sigler PB. 1993. The 2.2 Å crystal structure of transducin-α complexed with GTPγS. Nature 366:654–663.

Northup JK, Smigel MD, Sternweis PC, and Gilman AG. 1983. The subunits of the stimulatory regulatory of adenylate cyclase. Resolution of the activated 45,000-dalton a subunit. J Biol Chem 258:11369–11376.

Northup JK, Smigel MD, and Gilman AG. 1982. The guanine nucleotide activating site of the regulatory component of adenylate cyclase. Identification by ligand binding. J Biol Chem 257:11416–11423.

O'Brien PJ, St. Jules RS, Reddy TS, Bazan NG, and Zatz M. 1987. Acylation of disc membrane rhodopsin may be nonenzymatic. J Biol Chem 262:5210–5215.

Offermanns S, Mancino V, Revel JP, and Simon MI. 1997. Vascular system defects and impaired cell chemokinesis as a result of $G_{\alpha 13}$ deficiency. Science 275:533–536.

Pennington SR. 1995. GTP binding proteins. Protein Profile 2:167–315.

Pugh EN and Lamb TD. 1993. Amplification and kinetics of the activation steps in phototransduction. Biochim Biophys Acta 1141:111–149.

Ramdas L, Disher RM, and Wensel TG. 1991. Nucleotide exchange and cGMP phosphodiesterase activation by pertussis toxin inactivated transducin. Biochem 30:11637–11645.

Ray K, Kunsch C, Bonner LM, and Robishaw JD. 1995. Isolation of cDNA clones encoding eight different human G-protein γ subunits, including three novel forms designated the γ4, γ10, and γ11 subunits. J Biol Chem 270:21765–21771.

Resh MD. 1996. Regulation of cellular signalling by fatty acylation and prenylation of signal transduction proteins. Cell Signal. 8:403–412.

Sondek J, Lambright DG, Noel JP, Hamm HE, and Sigler PB. 1994. GTPase mechanism of G-proteins from the 1.7 A crystal structure of transducin a GDPA1F4. Nature 372:276–279.

Spiegel AM. 1996. Defects in G-protein-coupled signal transduction in human disease. Ann. Rev Physiol 58:143–170.

Strader CD, Fong TM, Tota MR, and Underwood D. 1994. Structure and function of G-protein-coupled receptors. Ann Rev Biochem 63:101–132.

Wall MA, Coleman DE, Lee E, Iniquez-Lluhi JA, Posner BA, Gilman AG, and Sprang SR. The Structure of the G-protein heterotrimer G_i alpha 1 beta 1 gamma 2. Cell 83:1047–1058, 1995.

Wang J, Tu Y, Woodson J, Song X, and Ross EM. 1997. A GTPase-activating protein for the G-protein G_{az}. J Biol Chem 272:5732–5740.

West RE, Jr., Moss J, Vaughan M, Liu T, and Liu T-Y. 1985. Pertussis toxin-catalyzed ADP-ribosylation of transducin. Cysteine 347 is the ADP-ribose acceptor site. J Biol Chem 260:14428–14430.

Wong GT Gannon KS, and Margolskee RF. 1996. Transduction of bitter and sweet taste by gustducin. Nature 381:796–800.

Yang Z and Wensel TG. 1992. N-myristoylation of rod outer segment G-protein, transducin, in cultured retinas. J Biol Chem 267:23197–23201.

Zera EZ, Molloy DP, Angleson JK, Lamture JB, Wensel TG, and Malinski JA. 1996. Low affinity interactions of GDPT and ribose- or phsophoryl-substituted GTP analogues with heterotrimetric G-protein, transducin. J Biol Chem 271:12925–12931.

3

Ras- and Rho-Related Small Molecular Weight G-proteins: Structure and Signaling Mechanisms

UMA PRABHAKAR AND PONNAL NAMBI

INTRODUCTION

The transmission of extracellular signals to the appropriate intracellular targets is mediated by a complex network of many interacting proteins. The binding (or interaction) of the first messenger (hormones, neurotransmitters, cytokines, and growth factors) to the appropriate cell surface receptor triggers a cascade of biological reactions as follows. The specific cell surface receptors activate a number of intracellular effector systems, resulting in the release of a host of second messengers such as inositol trisphosphate, calcium, diacylglycerol, cyclic AMP, cyclic GMP, arachidonic acid, prostaglandins, etc. These second messengers, in turn, activate various kinases and phosphatases, which phosphorylate or dephosphorylate appropriate key regulatory proteins, resulting in the final physiological responses.

It is now well established that a number of receptors are coupled to their effector systems through transducers called GTP-binding proteins. GTP-binding proteins, also called G-proteins, bind guanine nucleotides, especially guanosine triphosphate (GTP) and guanosine diphosphate (GDP) (Rodbell 1992, 1995a, 1995b, 1996), and in addition have GTPase activity. A wide variety of G-proteins has been identified in many cell types, and these have been cloned and expressed. In general, two major groups of G-proteins are recognized:

1. The first group of G-proteins comprises the superfamily of G_s, G_i, G_o, G_q, and the transducin families (Bourne 1986, Stryer 1986). These proteins

Introduction to Cellular Signal Transduction
A. Sitaramayya, Editor
©1999 Birkhäuser Boston

are heterotrimeric and are important for signal transduction from membrane receptors with seven transmembrane domains to various effectors such as adenylate cyclase and cyclic GMP-phosphodiesterase. Typically, the heterotrimeric G-proteins consist of α, β, and γ subunits. This class of G-proteins is reviewed elsewhere in this book.

2. The second group is the monomeric low or small molecular weight G-proteins (LMWG) with molecular masses ranging from 18 to 32 kDa. In this review we will focus on LMWG-proteins, a rapidly growing family of proteins represented by the *ras* oncogene as the prototype.

The review is divided into three sections. In the first section we will provide a brief overview of what is known about the general structure, function, and regulatory aspects of low molecular weight G-proteins. The second section will integrate a more detailed account of the Ras superfamily and the Rho subfamily. Finally, we will conclude with the recently identified role of LMWG-proteins in integrin-mediated signaling. Throughout the text, the terms "small" and "low" molecular weight G-proteins may be used interchangeably.

OVERVIEW OF SMALL MOLECULAR WEIGHT G-PROTEINS

General Structure

LMWG-proteins of molecular weight 18–32 kDa include *ras*-related proteins and many oncogene products (Barbacid 1987, Burgoyne 1989, Santos and Nebreda 1989). These proteins regulate various aspects of cell growth and differentiation, gene expression, cytoskeletal assembly and cell motility, protein and lipid trafficking, nuclear transport, and host defense (Hall 1992, Nuoffer and Balch 1994, Bokoch 1995, Lowy and Willumsen 1993, and Marshall 1995). They lack a site for ADP ribosylation with either pertussis or cholera toxins, although some of them can be ADP-ribosylated with botulinum toxins (Quilliam et al 1989, Nobes and Hall 1995).

LMWG proteins comprise more than 50 members, and can be categorized into several subfamilies such as Ras, Ral, Rab, Rho, Rac, etc. (Capon et al 1983, Schmidt et al 1986, Yamamoto et al 1988). A more simplified and general classification of these proteins, however, is based on sequence homology to Ras, and includes four groups, namely (1) Ras/Rap/Ral,(2) Rhol/Rac,(3) Ypt/Rab, and (4) Arf. These proteins share 20–50% overall homology to *ras*, and in addition have several common features, including four conserved domains, as shown in Figure 3-1.

In this simplified model of small molecular weight G-proteins, the open boxes represent conserved sequence motifs that are necessary for guanine

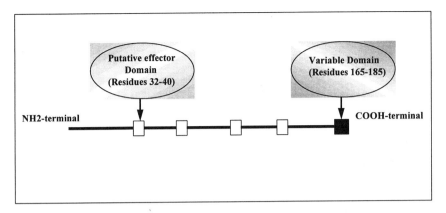

Figure 3-1. Common features of small molecular weight G-proteins.

nucleotide binding and hydrolysis. Residues 32–40 specifically represent the effector domain, whereas the black box represents the variable carboxyl domain.

Unlike the heterotrimeric G-proteins, not much is known about the effectors of the LMWG-proteins. In yeast and NIH/3T3 cells, mutational analysis has shown that residues 32–40 are essential for biological activity of Ras. Mutations in this region completely destroyed the transforming ability of oncogenic *ras* (Barbacid 1987, Sigal et al 1986, Willumsen et al 1986). Furthermore, on the basis of crystallographic studies, it has been demonstrated that these residues are exposed on the surface and that exchange of GDP for GTP is accompanied by conformational changes in this area (Milburn et al 1990, Pai et al 1989, Pai et al 1990, and Schlichting et al 1990). Given these findings, it is quite likely that residues 32–40 may represent a putative effector domain.

A feature common to most LMWG-proteins includes a highly variable region close to the COOH-terminal. Not only is the amino acid sequence highly variable in this region, but there is also significant variability in the types of posttranslational modifications that can occur in this region (Andreas et al 1994). Some of these modifications include prenylation of cysteine residues (Schafer and Rhine 1992, Marshall 1993, Khosravi-Far et al, 1992, Bokoch 1993), farnesylation of methionine or serine residues, or even geranylgeranylation of leucine residues (Bokoch 1993). Since LMWG-proteins are specifically targeted towards a great number of cellular locations, posttranslational modification may play a role in correct/specific targeting of these proteins to the appropriate cellular locations. For example, the function of prenylation could be simply to nonspecifically increase lipophilicity (Chavrier et al 1991).

Model of LMWG-Protein Function

GTPases, in general, alternate between two distinct conformations in response to the phosphorylation state of the bound nucleotide. This state varies as the proteins go through a cycle of guanine nucleotide exchange and GTP hydrolysis. The binding of GTP renders the protein "active," in which state it can interact with the downstream effector and elicit a functional response. Return to the "inactive" GDP-bound form is a consequence of the eventual hydrolysis of GTP.

The guanine nucleotide exchange cycles for heterotrimeric G-proteins and LMWG-proteins are diagramatically represented in Figures 3-2 and 3-3, respectively. As shown in Figure 3-2, the α subunits of the heterotrimeric G-proteins bind guanine nucleotides, and the βγ subunits function as a single unit. When an agonist binds to the receptor, it enhances the interaction between the receptor and G-proteins. This interaction promotes the dissociation of bound guanosine diphosphate (GDP) from the α subunit of the G-proteins, and stimulates the association of guanosine triphosphate (GTP) to the α subunit.

As a result of this interaction, the α subunit dissociates from the βγ subunit, and the dissociated α subunit modulates the activity of effector molecules [E]. This cycle is stopped by the hydrolysis of GTP to GDP by the α subunit, which also displays intrinsic GTPase activity. Once hydrolyzed, the α subunit is inactivated and recycles back to associate with the βγ subunit.

In contrast to the heterotrimeric G-proteins, most LMWG-proteins have low intrinsic hydrolytic activities. GTP hydrolysis for these proteins is regu-

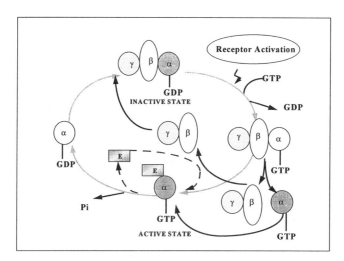

Figure 3-2. Model of heterotrimeric G-protein activation.

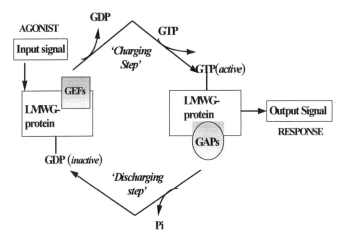

Figure 3-3. Model of LMWG-protein activation.

lated by a class of proteins called GTPase activating proteins (GAPs). Similarly, binding of GTP to LMWG-proteins is controlled by another class of regulatory molecules named guanine-nucleotide exchange factors (GEFs), which regulate dissociation of GDP from the GTPase either as GDP-dissociation stimulator (GDS) or as GDP-dissociation inhibitor (GDI).

As shown in Figure 3-3, LMWG-proteins are "inactive" when bound to GDP. The release of GDP, mediated by GEFs, frees up a binding site on these LMWG-proteins that is then occupied by GTP. This conformation renders the LMWG-proteins "active" and triggers downstream effector targets to be activated in turn. The GTPase activating proteins subsequently hydrolyze GTP, and the exchange cycle continues.

Several regulatory peptides acting as either GAPs or GEFs have been identified and characterized in recent years (Andreas et al 1994), and these will be discussed in more detail later. However, not much is known regarding the nature of the effectors of LMWG-proteins.

THE *RAS* SUPERFAMILY

Ras is a 189 amino acid GTP-binding protein that is ubiquitously expressed and is involved in the regulation of cell growth and differentiation (Bollag and McCormick 1991). Ras proteins play a key role in the signal transduction pathway connecting extracellular signals to intracellular processes. Three *ras* genes (H-*ras*, N-*ras*, and K-*ras*) are present in mammalian cells. K-*ras* has two splice variants, A and B, which differ in their C-terminal sequences. At the protein level, Ras is present in all cells, although proliferating cells display very high levels of Ras. The highest levels of mRNA for

H-*ras* are found in skin and muscle, K-*ras* in gut and thymus, and N-*ras* in testis and thymus (Wagner and Williams 1994).

The transforming ability of Ras has been shown to be antagonized by a family of proteins called *Rap* proteins, which are found in the golgi apparatus and endoplasmic reticulum. Rap 1A is also called Krev-1 because it reverses (suppresses) the action of K-*ras* oncogene. Other Ras-related proteins are Ral A and Ral B, which regulate the activity of exocytic and endocytic vesicles. Thus, the Ras/Rap/Ral group of proteins has been shown to be important for growth and development, anabolic processes, exocytosis, and regulation of oxidative burst.

All members of the Ras superfamily have several common features, as described in the earlier section. More specifically, they possess a highly conserved catalytic domain consisting of amino acids 1–164, GTP binding site (dispersed throughout the sequence), and the two "switch" regions between amino acids 30–38 and 60–76. Since the switch regions are close to the γ phosphate of the guanine nucleotide, different conformations exist, depending on whether GTP or GDP is bound (Lowy and Willumsen 1993, Polakis and McCormick 1993). The Ras proteins undergo posttranslational modifications that include the addition of farnesyl group at the C-terminal end, and the resulting proteins are bound to the inner surface of the plasma membrane (Wagner and Williams 1994, Lowy and Willumsen 1993). In addition, some Ras proteins are also palmitoylated.

As discussed earlier, the activation states of Ras proteins are under the regulatory control of GAP and GEF proteins. Two GAPs, p120GAP and neurofibromin, the product of the type I neurofibromatosis (*NF-1*) gene, that specifically act on Ras proteins have been reported in mammalian cells (Stang et al 1996). More recently, however, several novel Ras GAPs have been added to this list. These include a 98-kDa R-Ras GAP purified from bovine brain cytosol (Yamamoto et al 1995), GAP 1m and GAPIII/GAP 1IP4BP (Hattori and Baba 1996). Although the majority of GAPs have been shown to act as negative regulators of their respective LMWG-proteins, there is some evidence to suggest that p120GAP may have a dual role as negative regulator and downstream effector (Bokoch and Der 1993).

Among the GEFs, a number of putative GTP/GDP dissociation stimulators (GDSs) for Ras proteins have recently been isolated (Downward 1992). CDC25 gene product is the exchange factor for yeast Ras. SDC25, a second yeast product, has also been shown to be active for both yeast and mammalian Ras proteins.

While Rab and Arf subfamilies have been suggested to regulate different aspects of vesicular transport (Pryer et al 1992), Ras- and Rho-related proteins are mainly involved in the regulation of signal transduction pathways mediated by tyrosine kinase receptors such as epidermal growth factor and nerve growth factor receptors (Mulcahy et al 1985, Szeberenyi et al 1990). Binding of growth factors to the receptors leads to the activation of

Ras in its GTP-bound form. Activated Ras functions as a regulated, membrane-bound anchor for Raf. Whereas Raf is normally present in the cytosol, in cells expressing activated Ras, Raf is associated with the plasma membrane (Leavers et al 1994, Hunter and Karin 1992). In addition to growth factor receptors, seven-transmembrane G-protein-coupled receptors such as thrombin, lysophosphatidic acid (LPA), α2-adrenergic, and muscarinic M2 receptors have also been shown to activate Ras (LaMorte et al 1993, Van Corren et al 1993, Winitz et al 1993, Alblas et al 1993). Some of these pathways have been shown to be sensitive to pertussis toxin pretreatment, suggesting the involvement of the G_i G_o class of heterotrimeric G-proteins. In addition, G_q-coupled α1-adrenergic receptor activation has also been shown to increase the proportion of GTP-bound Ras (Thornburn and Thornburn 1994). Collectively, these data indicate that Ras can be activated by a number of heterotrimeric G-protein-coupled pathways.

A couple of hypotheses have been suggested to explain the interaction of Ras and heterotrimeric G-protein-coupled pathways. These include, (1) the involvement of a cytoplasmic protein tyrosine kinase such as Src for the coupling of G_i to Ras activation (Van Corren et al 1993, Chen et al 1994) and (2) the release of G-protein βγ subunits to activate a tyrosine kinase or binding to proteins displaying pleckstrin homology (PH) domains (Koch et al 1994, Crespo et al 1994, Touhara et al 1994).

Transient expression of $G_{\beta\gamma}$ in COS-7 cells or other cells resulted in 2–10-fold increase in mitogen-activated protein kinase (MAPK) activity, and this increase could be blocked by antagonists of $G_{\beta\gamma}$, derived from the βγ binding site of the β-adrenergic receptor kinase 1 (BARK1). Using a dominant-negative form of Ras, it was shown that $G_{\beta\gamma}$-mediated activation of MAPK was dependent on Ras.

Elegant data from Lefkowitz's laboratory (Lutterell et al 1995, van Biesen et al 1995) have shown that $G_{\beta\gamma}$-mediated tyrosine phosphorylation of Sch precedes the activation of MAPK by G_i-coupled receptors. Sch belongs to a group of adapter molecules involved in the coupling of various protein components of signal transduction pathways (Pawson and Schlessinger 1993). Three isoforms of Sch (46, 52, and 66 kDa) have been reported, each of which contains a single SH2 domain and a proline-rich SH3 binding motif. In addition, there is a tyrosine residue which, when phosphorylated, mediates binding to other SH2-containing proteins. Expression of $G_{\beta\gamma}$ in COS cells or activation of G_i-coupled receptors such as lysophosphatidic acid and α2-adrenergic receptors resulted in 3–4 fold increase in Sch phosphorylation, and this increase was blocked by coexpression of a $G_{\beta\gamma}$-binding peptide derived from BARK1. Sch phosphorylation has also been shown to be mediated by pp60c-Src or a related member of the Src kinase family such as *lyn* tyrosine kinase (Ptasznik et al 1995).

Involvement of a second adapter protein, Grb2, in the binding of phosphorylated Sch has been demonstrated in EGF-receptor-mediated activation of Ras (Pawson and Schlessinger 1993, Downward 1992). Grb2 interacts with the Ras guanine nucleotide exchange protein Sos (Son of sevenless), which in turn interacts and activates Ras by catalyzing guanine nucleotide exchange (Downward 1992). A similar sequence of events has also been reported for G-protein-coupled receptors (van Biesen et al 1995). In essence, all the above data from the literature suggest that Src-related kinases such as Lyn are involved in the coupling of G-proteins to the Sch-Grb2-Sos-Ras-MAPK pathway. Although the exact mechanism(s) of $G_{\beta\gamma}$ regulation of Src kinase activity is (are) not known, there is a clearly defined link between heterotrimeric G-proteins and the Ras pathway.

THE RHO SUBFAMILY

The Rho proteins are part of the Ras superfamily, and in fact, the name Rho is derived from *ras homologue*. The *rho* gene was first identified in *Aplysia* (Madaule and Alex 1985) and subsequently discovered in mammals and *Saccharomyces cerevisiae* (Madaule et al 1987, Yeramian et al 1987). The mammalian Rho family includes three isoforms of *rho* (A, B, and C), two isoforms of *rac* (1 and 2), and one isoform each of *rho*G, Cdc42, and TC10. Although most of the cellular functions identified are mediated by *rho*A, *rho*B has been reported to be an early gene induced by growth factors such as EGF and PDGF, nonreceptor tyrosine kinases, DNA-damaging drugs, and UV irradiation (Jahner and Hunter 1991, Pritz et al 1995).

Rho contains the consensus amino acid sequences responsible for GDP/GTP binding and GTPase activities which are highly conserved in all small GTP-binding proteins. The target domains of Rho reside between residues 34–42; another putative target domain is located in the carboxy-terminal two-thirds of Rho. This region also contains a unique sequence, CAAL, where A is an aliphatic amino acid (Takai et al 1995). The posttranslational modifications in this region appear to be critical for Rho activity and function.

Involvement of Rho in cellular functions was identified using *Clostridium botulinum* C3 transferase, a bacterial exoenzyme that selectively ADP ribosylates and inactivates endogenous Rho in intact cells. Inhibition of cell functions by this enzyme suggested the involvement of Rho in that particular function. A number of seven-transmembrane G-protein-coupled receptors such as thrombin, lysophosphatidic acid (LPA), endothelin, and bombesin mediate the activation of Rho (Ridley and Hall 1992, Jalink et al 1994, Rankin et al 1994). Most of these receptors are coupled to multiple signal transduction pathways through different G-proteins. LPA receptors have been shown to mediate cell growth and stress fiber induction through G_i and G_q pathways, respectively (Moolenaar 1995). Further experiments by Buhl et al (1995) demonstrated that Rho activation was downstream of the G_{12} family of G-proteins. The observation that LPA-mediated fiber formation was inhibited by tyrphostin, whereas activated Rho-mediated stress fiber formation was insensitive to tyrphostin, a tyrosine kinase inhibitor, suggested that Rho was activated by a tyrosine kinase (Nobes et al 1995). In fact, heterotrimeric G-proteins have also been shown to be linked to the tyrosine kinases Lyn and Syk, which mediate the activation of MAPK (Wan et al 1996).

A number of molecules such as ect2 (Miki et al 1993), ost (Horii et al 1994), tiam-1 (Habets et al 1994), 1bc (Zheng et al 1995), and FGD1 (Pasteris et al 1994) have been shown to have GDP/GTP exchange activity for Rho. All these molecules show oncogenic activity and different tissue distribution, but the mechanism for the differential use of these exchange proteins in cells is not known.

GTPase activating proteins, or GAPs, are an important class of regulatory proteins that down-regulate Rho activity (Bogulski and McCormick 1993). The first Rho GAP was identified from human spleen, and since then, a number of other GAP proteins, including bcr, chimerin, p190, and abr, have been added to this list (Garrett et al 1991, Diekmann et al 1991, Lamarche and Hall 1994). In all nine mammalian genes, the *rotund* locus of *Drosophila melanogaster*, the *BEM2* and *BEM3* genes of *Saccharomyces cerevisiae* and a *Caenorhabditis elegans* gene have now been found to encode proteins containing a *rho*GAP domain (Lamarche and Hall 1994). The domains show about 20–40% amino acid identity and encode fairly large and multifunctional proteins. *In vitro* experiments indicate that the GAPs interact with multiple members of the Rho family, although with different affinities (Barford et al 1993). Recently, the crystal structure of an active 242-residue C-terminal fragment of human p50*rho* GAP was reported. In this study the investigators proposed that Arg-85 and Asn-194 are involved in binding G-proteins and enhancing GTPase activity (Barrett et al 1997).

It is intriguing that the number of proteins containing the *rho*GAP domain is so large. Except for the tissue-specific chimerins, the expression of these *rho*GAP proteins is also quite ubiquitous; it is therefore reasonable to speculate that they must be highly regulated in the cell so that Rho-like GTPases are not always turned off.

Another important molecule that regulates Rho activity is the guanine nucleotide dissociation inhibitor called *rho*GDI. It is a ubiquitous protein of 23-kDa molecular weight that forms 1:1 complex with all members of Rho subfamily (Hart et al 1992, Ueda et al 1990). This protein has two regulatory functions on Rho proteins: (1) it inhibits the dissociation of guanine nucleotide (GDP) bound to Rho proteins, and functions as a negative regulator and (2) by forming a complex with Rho, it inhibits Rho from binding to the membrane and the release of GDP-bound Rho from the membrane. This latter function has also been demonstrated *in vitro* in a cell-free system (Isomura et al 1991). The negative regulation of *rho*GDI has been observed by microinjection of *rho*GDI or by expressing it in various cell systems where it blocked stimulus-induced Rho effects (Kishi et al 1993).

The biochemical mechanisms of Rho action involve tyrosine phosphorylation in focal adhesions assembly, and stress fiber formation, because tyrosine kinase inhibitors (genistein and herbimycin) have been shown to inhibit stimulus-mediated cell adhesion and focal adhesion formation (Ridley and Hall 1994, Burridge et al 1992). In addition, a number of serine/threonine kinases have been implicated in Rho action, since staurosporine inhibited LPA as well as Rho-induced focal adhesion. These data also suggest that Rho-mediated actin filament assembly and focal adhesion formation may involve different effectors.

A number of putative target proteins having a homologous Rho-binding sequence have been identified using different techniques. These proteins include serine/threonine protein kinase N (PKN), rhophillin and rhotekin, although rhophillin and rhotekin have no domains for catalytic activity (Amano et al 1996 Reid et al 1996). These data suggest that these proteins may have the same binding motifs but different activities. In addition to these direct binding molecules, other molecules such as PI4-phosphate-5 kinase, PI3 kinase, and phospholipase D were shown to be activated by GTP Rho in crude membrane or cell lysates (Chong et al 1994).

In conclusion, it appears that Rho-mediated actions on nuclear signaling and changes to the actin cytoskeleton may involve separate pathways.

INTEGRIN-MEDIATED SIGNALING

There is now increasing evidence that small molecular weight G-proteins are also involved in integrating signals from yet another class of proteins, called integrins. Integrins are a family of adhesion receptors consisting of α and β subunits that can pair to form more than 20 receptors. In general, they contain a large extracellular domain formed by the α ($\sim$1000 residue) and β ($\sim$750 residue) subunits, a transmembrane domain from each subunit, and a short cytoplasmic tail, with the exception of β_4 whose cytoplasmic tail is >1000 residues in length (Hogervorst et al 1990). Integrins couple components of the extracellular matrix (ECM) with the actin cytoskeleton and regulate a variety of complex biological processes that include cell migration, tissue organization, cell growth, blood clotting, inflammation, target recognition by lymphocytes, and differentiation of many cell types (Hynes 1992).

The signals transmitted by ECM interaction with integrin receptors result in the activation of a number of signaling pathways (Richardson and Parsons 1995, Schwartz et al 1995), among which are those controlling activation of both protein tyrosine kinases (PTK) and members of the Rho family of small GTP-binding proteins. The coordinated actions of these molecules are extremely important for the formation of focal adhesions and subsequent activation of the downstream signaling cascade.

PTKs play a very important role in the assembly of focal adhesions (Schaller and Parsons 1994, Burridge et al 1988). Transformation of cells by the oncogenic *src* results in a significant increase in tyrosine phosphorylation of several focal adhesion proteins, including the focal adhesion kinase -FAK (Schaller et al 1992). FAK is a structurally distinct and highly conserved member of the nonreceptor PTKs that is expressed in all vertebrate species examined to date.

While activation of PTKs is critical for integrin-mediated signaling, activation of Rho and structurally related members of this family of proteins, namely Rac and Cdc42, has been shown to be equally important to regulating the formation of focal adhesions and actin cytoskeleton (Hall 1992). In Swiss 3T3 cells, serum or lysophosphatidic acid (LPA) cause a rapid induction in the formation of actin stress fibers and focal adhesions (Ridley and Hall 1992). This effect can be mimicked by microinjection of activated Rho in these cells (Nobes and Hall 1995). Furthermore, inhibition of Rho activity by C3 transferase, which ADP-ribosylates endogenous Rho protein, can completely abolish LPA-induced formation of focal adhesions and actin-stress fibers (Ridley and Hall 1992, Nobes and Hall 1995).

Rac and Cdc42 also play a role in the regulation of the actin cytoskeleton. In the Swiss 3T3 cells, platelet-derived growth factor (PDGF) and insulin rapidly induce formation of lamellipodia and membrane ruffling, processes that are associated with the assembly of multimolecular focal complexes. This process can be mimicked by microinjection of activated forms of Rac (Nobes et al 1995). Similarly, microinjection of Cdc42 induces formation of filopodia which are actin-based membrane projections at the cell surface.

There is indication to suggest that *rho* acts upstream of FAK, since C3 inhibition of Rho blocks both LPA-dependent and adhesion-dependent activation of focal adhesions (Chzanowska-Wodnicka and Burridge 1994). C3 also blocks bombesin-induced FAK activation in Swiss 3T3 cells (Rankin et al 1994).

The exact mechanism by which Rho and FAK coordinate to regulate the assembly of focal adhesion complexes is still speculative. While Rho appears to be required for the early events regulating integrin clustering following attachment to the ECM, FAK facilitates the assembly of protein-protein interactions (Parsons 1996). FAK knockout mice show deficiencies in mesoderm development and die by the eighth day of embryonic development. Cells recovered from such mice exhibit defects in cell migration (Ilic et al 1995). In fact, cells from *src*-deficient mice also exhibit similar defects in cell spreading (Kaplan et al 1995). Recently, the FAK carboxy-terminal domain has been shown to bind to a new member of the Rho-GTPase activating protein (GAP) referred to as Graf (GTPase regulator associated with FAK) (Parsons 1996). This observed association is indeed provocative and suggests the possibility that recruitment of Graf to the focal adhesion complex may play a role in providing a point of cross-talk between Rho and/or Cdc42 and the PTKs. An understanding of these control mechanisms will greatly aid in dissecting the pathways that control and regulate cell motility, differentiation, and survival.

ACKNOWLEDGMENTS

We thank Sue Tirri for her excellent secretarial assistance.

REFERENCES

Alblas J, van Corven EJ, Hordijk PL, Milligan G, and Moolenaar WH. 1993. Gi-mediated activation of the p21 *Ras* mitogen activated protein kinase pathway by α2 adrenergic receptors expressed in fibroblasts. J Biol Chem 268:22235–22238.

Amano M, Mukai H, Ono Y, Chihara K, Matsui T, Hamajima Y, Osawa K, Iwamatsu A, and Kaibuchi K. 1996. Identification of a putative target for *Rho* as the ser/thr kinase protein kinase N. Science 271:648–650.

Andreas C, Wagner C, and Williams JA. 1994. Low molecular weight GTP-binding proteins: molecular switches regulating diverse cellular functions. Amer J Phys G1–G14.

Barbacid MC. 1987. *Ras* genes. Ann Rev Biochem 56:779–827.

Barford ET, Zheng Y, Kuang WJ, Hart MJ, and Evans T. 1993. Cloning and expression of a human CDC42 GTPase activating protein reveals a functional SH3 domain. J Biol Chem 268:26059–26062.

Barrett T, Xiao B, Dodson EJ, Dodson G, Ludbrook SB, Nurmahomed K, Gamblin SJ, Musacchio A, Smerdon SJ, and Eccleston JF. 1997. The structure of the GTPase activating domain of p50rhoGAP. Nature, London 385(6615):458–461.

Bogoski MS, and McCormick F. 1993. Proteins regulating Ras and its relatives. Nature 366:643–654.

Bokoch GM. 1993. Biology of *Ras* proteins, members of the *Ras* superfamily of GTP-binding proteins. Biochem J 289:17–24.

Bokoch GM. 1995. Regulation of phagocytic respiratory burst by small GTP-binding proteins. Trends Cell Biol 5:109–113.

Bokoch GM and Der CJ. 1993. Emerging concepts in the ras superfamily of GTP-binding proteins. FASEB J 7:750–759.

Bollag G and McCormick F. 1991. Regulators and effectors of *Ras* proteins. Ann Rev Cell Biol 7:601–632.

Bourne HR. 1986. GTP binding proteins. One molecular machine can transduce diverse signals. Nature, London 321(6073):814–816.

Buhl AM, Johnson NL, Dhanarekara N, and Johnson GL. 1995. Gα12 and Gα13 stimulate *Rho*-dependent stress fiber formation and focal adhesion assembly. J Biol Chem 270:24631–24636.

Burgoyne RD. 1989. Small GTP binding proteins. Trends Biochem Sci 14:394–396.

Burridge K, Fath K, Kelly T, Nuckolls T, and Turner C. 1988. Focal adhesions: transmembrane junctions between the extracellular matrix and cytoskeleton. Ann Rev Cell Biol 4:487–525.

Burridge K, Turner CE, and Romer LH. 1992. Tyrosine phosphorylation of paxillin and pp125 FAK accompanies cell adhesion to extracellular matrix: a role in cytoskeletal assembly. J Cell Biol 119:893–903.

Capon DJ, Chen EY, Levinson AD, Seeburg PH, and Goeddel, DP. 1983. Complete nucleotide sequence of the T24 human bladder carcinoma oncogene and its normal homolog. Nature 302:33–37.

Chavrier P, Gorvel JP, Stelzer E, Simons K, Gruenberg J, and Zerial M. 1991. The hypervariable C-terminal domain of *rab* proteins acts as a targeting signal. Nature, London 353:769–772.

Chen Y-H, Pouyssegur J, Courtneidge SA, and Van Obbergheu-Schilling E. 1994. Activation of *Src* family kinase activity by the G-protein-coupled thrombin receptor in growth-responsive fibroblasts. J Biol Chem 269:27372–27377.

Chong LD, Traynar-Kaplan A, Bokoch GM, and Schwartz MA. 1994. The small GTP binding protein *Rho* regulates a phosphatidyl inositol 4-phosphate 5 kinase in mammalian cells. Cell 79:507–513.

Chrzanowska-Wodnicka M and Burridge K. 1994. Tyrosine phosphorylation is involved in reorganization of the actin cytoskeleton in response to serum or LPA stimulation. J Cell Sci 107:3643–3654

Crespo P, Xu N, Simonds WF, and Gutkind JS. 1994. *Ras*-dependent activation of MAP kinase pathway mediated by G-protein βγ subunits. Nature, London 369:418–420.

Diekmann D, Brill S, Garrett MD, Totty N, and Hsuan J. 1991. Bcr encodes a GTPase activating protein for p21 *Rac*. Nature 351:400–402.

Downward J. 1992. Regulatory mechanisms for *Ras* proteins. Bioessays 14:177–184.

Fritz G, Kaina B, and Aktories K. 1995. The *Ras*-related small GTP binding protein *Rho*B is an immediate early gene inducible by DNA damaging treatments. J Biol Chem 270:25172–25177.

Garrett MD, Major GN, Totty N, and Hall A. 1991. Purification and N-terminal sequence of the p21*Rho* GTPase activating protein. Biochem J 276:833–836.

Habets GGM, Scholtes EHE, Zuydgeest D, van der Kammen RA, Stam JC, Berus A, and Collard JG. 1994. Identification of an invasion-inducing gene, Tiam-1, that encodes a protein with homology to GDP-GTP exchangers for *Rho*-like proteins. Cell 77:537–549.

Hall A. 1992. *Ras*-related GTPases and the cytoskeleton. Mol Bio Cell 3:475–479.

Hart MJ, Maru Y, Leonard D, Witte ON, Evans T, and Cerione RA. 1992. A GDP dissociation inhibitor that serves as a GTPase inhibitor for the *Ras*-like protein CDC42 HS. Science 258:812–815.

Hattori S and Baba H. 1996. Heterogeneity of GTPase-activating proteins for ras in the regulation of ras signal transduction pathway. Yakugako Zasshi 116(1):21–38.

Hawes BE, van Biesen T, Koch WJ, Lutterell LM, and Lefkowitz RJ. 1995. Distinct pathways of Gi and Gq mediated mitogen activated protein kinase activation. J Biol Chem 270:17148–17153.

Hogervorst F, Kuikman I, von dem Borne AEGK, and Sonnenberg A. 1990. Cloning and sequence analysis of β_4 DNA: an integrin subunit that contains a unique 118 kD cytoplasmic domain. EMBO J 9:765–770.

Horii Y, Beiler JF, Sakaguchi K, Tachibana M, and Miki T. 1994. A novel oncogene, *ost*, encodes a small guanine nucleotide exchange factor that potentially links *Rho* and *Rac* signaling pathways. EMBO J 13:4776–4786.

Howe LR and Marshall CJ. 1993. Lysophosphatidic acid stimulates mitogen activated protein kinase via a G-protein-coupled pathway requiring p21*Ras* and P24*raf*-1. J Biol Chem 268:20717–20720.

Hunter T and Karin M. 1992. The regulation of transcription by phosphorylation. Cell 70:375–387.

Hynes RO. 1992. Integrins: versatility, modulation, and signaling in cell adhesion. Cell 69:11–25.

Ilic D, Furuta Y, Kanazawa S, Takeda N, Sobue K, Nakatsuji N, Nomura S, Fujimoto J, Okada M, Yamamoto T, and Alzawa S. 1995. Reduced cell motility and enhanced focal adhesion contact formation in cells from FAK-deficient mice. Nature, London 377:539–543.

Isomura M, Kikuchi A, Ohga N, and Takai Y. 1991. Regulation of binding of *Rho*B p20 to membranes by its specific regulatory protein, GDP dissociation inhibitor. Oncogene 6:119–124.

Jahner D and Hunter T. 1991. The *Ras*-related gene *Rho*B is an immediate early gene inducible by V-Fps, epidermal growth factor and platelet-derived growth factor in rat fibroblasts. Mol Cell Biol 11:3682–3690.

Jalink K, van Corven EJ, Hengeveld T, Moril N, Narumiya S, and Moolenaar WH. 1994. Inhibition of lysophosphatidate- and thrombin-induced retraction and neuronal cell rounding by ADP ribosylation of the small GTP binding protein *Rho*. J Cell Biol 126:801–810.

Kaplan KB, Swedlow JR, Morgan DO, and Varmus HE. 1995. C-*Src* enhances the spreading of *scr-l-* fibroblasts on fibronectin by a kinase-independent mechanism. Genes Dev 9:1505–1517.

Khosravi-far R, Kato K and Der CJ. 1992. Protein prenylation: key to *Ras* function and cancer termination? Cell Growth Differ 3:461–469.

Kishi K, Sasaki T, Kuroda S, Ito T, and Takai T. 1993. Regulation of cytoplasmic division of Xenopus embryo by *Rho* p21 and its inhibitory GDP/GTP exchange proteins (*Rho* GDI). J Cell Biol 120:1187–1195.

Koch WJ, Hawes BE, Allen LF, and Lefkowitz, RJ. 1994. Direct evidence that Gi-coupled receptor stimulation of mitogen activated protein kinase is mediated by G$\beta\gamma$ activation of p21 *Ras*. Proc Natl Acad Sci USA 91:12706–12710.

Lamarche N and Hall A. 1994. GAPs for rho-related GTPases. Trends Genet 10(12):436–440.

LaMorte VJ, Kennedy ED, Collins LR, Goldstein D, Harootuniak AT, Brown JH, and Feramisco JR. 1993. A requirement for *Ras* protein function in thrombin-stimulated mitogenesis in astrocytoma cells. J Biol Chem 268:19411–19415.

Leavers SJ, Paterson HF, and Marshall CJ. 1994. Requirement for *Ras* in *Raf* activation is overcome by targeting Raf to the plasma membrane. Nature, London 369:411–414.

Lowy DR and Willumsen BM. 1993. Function and regulation of *Ras*. Ann Rev Biochem 62:851–891.

Luttrell LM, van Biesen T, Hawes BE, Koch WJ, Touhara K, and Lefkowitz RJ. 1995. Gβγ subunits mediate mitogen-activated protein kinase activation by the tyrosine kinase insulin-like growth factor 1 receptor. J Biol Chem 270:16495–16498.

Madaule P and Axel R. 1985. A novel ras-related gene family. Cell 41(1):31–40.

Madaule P, Axel R, and Myers AM. 1987. Characterization of two members of the rho gene family from the yeast Saccharomyces cerevisiae. Proc Natl Acad Sci USA 84(3):779–83.

Marshall CJ. 1993. Protein prenylation: a mediator of protein-protein interactions. Science 259:1865–1866.

Marshall MS. 1995. *Ras* target proteins in eukaryotic cells. FASEB J 9:1311–1318.

Miki T, Smith CL, Long JE, Eva A, and Fleming TP. 1993. Oncogene ect2 is related to regulators of small GTP binding proteins. Nature 362:462–465.

Milburn MV, Tong L, deVos AM, Brunger A, Yamaizumi Z, Mishimura S, and Kim SH. 1990. Molecular switch for signal transduction: structural differences between active and inactive forms of protooncogenic *Ras* proteins. Science 247:939–945.

Moolenaar WH. 1995. Lysophosphatidic acid signaling. Curr Opin Cell Biol 7:203–210.

Mulcahy LS, Smith MR, and Stacey DW. 1985. Requirements for *Ras* proto oncogene function during serum-stimulated growth of NIH 3T3 cells. Nature, London 313:241–243.

Nobes CD and Hall A. 1995. *Rho, Rac*, and Cdc42 GTPases regulate the assembly of multimolecular focal complexes associated with actin stress fibers, lamellipodia, and filopodia. Cell 81:53–62.

Nobes C, Hawkins P, Stephens L, and Hall A. 1995. Activation of the small GTP binding proteins *Rho* and *Rac* by growth factor receptors. J Cell Sci 108:225–233.

Nuoffer C and Balch WE. 1994. GTPases: multifunctional molecular switches regulating vesicular traffic. Annu Rev Biochem 63:949–990.

Pai EF, Kabsch W, Krengel U, Holmes K, John J, and Wittinghofer A. 1989. Structure of the guanine-nucleotide binding domain of the Ha-*Ras* oncogene product p21 in the triphosphate formation. Nature, London 341:209–214.

Pai EF, Krengel U, Petsko GA, Goody RS, Kabsch W, and Wittinghofer A. 1990. Refined crystal structure of the triphosphate conformation of H-*Ras* p21 at 1.35 Å resolution: implications for mechanism of GTP-hydrolysis. EMBO J 9:2351–2359.

Parsons TJ. 1996. Integrin-mediated signalling: regulation by protein tyrosine kinases and small GTP-binding proteins. Curr Opin Cell Biol 8:146–152.

Pasteris N-G, Cadle A, Logie LL, Portious MEM, Schwartz CE, Stevenson RE, Glover TW, Wilroy RS, and Gorski JL. 1994. Isolation and characterization of the faciogenital dysplasia (Aarskog-Scott Syndrome) gene: a putative *Rho/Rac* guanine nucleotide exchange factor. Cell 79:669–678.

Pawson T and Schlessinger J. 1993. SH2 and SH3 domains. Curr Opin Cell Biol 3:434–442.

Polakis P and McCormick F. 1993. Structural requirements for the interaction of p21 *Ras* with GAP, exchange factors, and its biological effector target. J Biol Chem 268:9157–9160.

Pryer NK, Wuestenhube LJ, and Schekman R. 1992. Vesicle-mediated protein sorting. Ann Rev Biochem 61:471–516.

Ptasznik A, Traynor-Kaplan A, and Bokoch GM. 1995. G-protein-coupled chemoattractant receptors regulate lyn tyrosine kinase-*Sch* adapter protein signaling complexes. J Biol Chem 270:19969–19973.

Quilliam LA, Lacal JC, and Bokoch GM. Identification of rho as a substrate for botulinum toxin C3-catalyzed ADP-ribosylation. FEBS Lett. 247:221–226

Rankin S, Morii N, Narumiya S, and Rozengart E. 1994. Botulinum C3 exoenzyme blocks tyrosine phosphorylation of p125^{FAK} and paxillin induced by bombesin and endothelin. FEBS Lett 354:315–319.

Reid T, Furuyashiki T, Ishizaki T, Watanabe G, Watanabe N, Fujiisawa K, Horii N, Madule P, and Narumiya S. 1996. Rhotekin, a new putative target for *Rho* bearing homology to a ser/thr kinase, PKN, and Rhodophillin in the *Rho* binding domain. J Biol Chem 271:13556–13560.

Richardson A and Parsons TJ. 1995. Signal transduction through integrins: a central role for focal adhesion kinase? Bioessays. 17:229–236.

Ridley AJ and Hall A. 1992. The small GTP binding protein *Rho* regulates the assembly of focal adhesions and actin stress fibers in response to growth factors. Cell 70:389–399.

Ridley AJ and Hall A. 1994. Signal transduction pathways regulating *Rho*-mediated stress fiber formation. Requirement for a tyrosine kinase. EMBO J 13:2600–2610.

Rodbell M. 1992. The role of GTP-binding proteins in signal transduction: from the sublimely simple to the conceptually complex. Curr Top Cell Regul 32:1–47.

Rodbell M. 1995a. The complex structure and function of G-proteins in cellular communication. Bull Mem Acd R Med Belg 150(7–9):316–319.

Rodbell M. 1995b. Nobel Lecture: Signal transduction: evolution of an idea. BioSci Rep 15(3):117–133.

Rodbell M. 1996. G-proteins: out of the cytoskeletal closet. Mt Sinai J Med 63(5–6):381–386.

Santos E and Nebreda AR. 1989. Structural and functional properties of *Ras* proteins. FASEB J 3:2151–2163.

Schafer WR and Rhine J. 1992. Protein prenylation: genes, enzymes, targets and functions. Ann Rev Genet 30:209–237.

Schaller MD, Borgman CA, Cobb BS, Vines RR, Reynolds AB, and Parsons TJ. 1992.

pp125^{FAK}, a structurally distinctive protein-tyrosine kinase associated with focal adhesions. Proc Natl Acad Sci USA 89:5192–5196.

Schaller MD and Parsons TJ. 1994. Focal adhesion kinase and associated proteins. Curr Opin Cell Biol 6:705–710.

Schlichting I, Almo SC, Rapp G, Wilson K, Lentfer A, Wittinghofer A, Kabsch W, Pai EF, Petsko GA, and Goody RS. 1990. Time resolved x-ray crystallographic study of the conformational change in Ha-*Ras* p21 on GTP-hydrolysis. Nature, London 345:309–314.

Schmidt HD, Wagner P, Pfaff E, and Gallwitz D. 1986. The *Ras* related YPT1 gene product in yeast. A GTP binding protein that might be involved in microtubule organization. Cell 47:401–412.

Schwartz MA, Schaller MD, and Ginsberg MH. 1995. Integrins: emerging paradigms of signal transduction. Annu Rev Cell Dev Biol 11:549–599.

Sigal Ss, Gibbs JB, D'Alanzo JS, and Scholnik EM. 1986. Identification of effector residues and a neutralizing epitope of Ha-*Ras* encoded p21. Proc Natl Acad Sci USA 83:4725–4729.

Stang S, Bottorf D, and Stone JC. 1996. Ras effector loop mutations that dissociate p120GAP and neurofibromin interactions. Mol Carcinog 15(1):64–69.

Stryer L. 1986. Cyclic GMP cascade of vision. Annu Rev Neurosci 9:87–119.

Szeberenyi J, Cai H, and Cooper GM. 1990. Effect of a dominant inhibitory Ha-*Ras* mutation on neuronal differentiation of PC12 cells. Mol Cell Biol 10:5324–5332.

Takai Y, Sasaki T, Tanaka K, and Nakanishi H. 1995. Rho as a regulator of the cytoskeleton. Trends in Biochem Sci 20(6):227–231.

Thornburn J and Thornburn A. 1994. The tyrosine kinase inhibitor, genestein, prevents α-adrenergic-induced cardiac muscle hypertrophy by inhibiting activation of the *Ras*-MAP kinase signaling pathway. Biochem Biophys Res Commun 202:1586–1591.

Touhara K, Inglese J, Pitcher JA, Shaw G, and Lefkowitz RJ. 1994. Binding of G-protein βγ subunits to pleckstrin homology domains. J Biol Chem 269:10217–10220.

Ueda T, Kikuchi A, Ohga N, Yamamoto J, and Takai Y. 1990. Purification and characterization from bovine brain cytosol of a novel regulatory protein inhibiting the dissociation of GDP from and the subsequent binding of GTP to *Rho*B p20, a *Ras* p21-like GTP binding protein. J Biol Chem 265:9373–9380.

van Biesen T, Hawes BE, Lutterell DK, Krueger KM, Touhara K, Porfiri E, Kakaue K, Lutterell LM, and Lefkowitz RJ. 1995. Receptor tyrosine kinase and Gβγ-mediated MAP kinase activation by a common signaling pathway. Nature, London 376:781–784.

Van Corren EJ, Hordijk PL, Medema RH, Bos JL, and Moolenaar WH. 1993. Pertussis toxin sensitive activation of p21 *Ras* by G-protein-coupled receptor agonists in fibroblasts. Proc Natl Acad Sci USA 90:1257–1261.

Wagner AC and Williams JA. 1994. Low molecular weight GTP binding proteins: molecular switches regulating diverse cellular functions. Am J Physiol 266:G1–G14.

Wan Y, Kurosaki T, and Huang X-Y. 1996. Tyrosine kinases in activation of the MAP kinase cascade by G-protein-coupled receptors. Nature 380:541–544.

Willumsen BM, Papageorge AG, Kung HF, Bekesi E, Robins T, Johnson M, Vass WC, and Lowy DL. 1986. Mutational analysis of a *Ras* catalytic domain. Mol Cell Biol 6:2646–2654.

Winitz S, Russell M, Qian H, Gardner A, Dwyer L, and Johnson GL. 1993. Involvement of *Ras* and *raf* in the Gi-coupled acetylcholine muscarinic m2 receptor activation of mitogen activated protein kinase and MAD kinase. J Biol Chem 268:19196–19199.

Yamamoto K, Kondo J, Hishida T, Teranishi Y, and Takai Y. 1988. Purification and characterization of a GTP binding protein with a molecular weight of 20,000 dalton in bovine membrane. Identification of the *Rho* gene product. J Biol Chem 263:9926–9932.

Yamamoto T, Matsui T, Nakafuku M, Iwamatsu A, and Kaibuchi K. 1995. A novel GTPase-activating protein for ras. J Biol Chem 270(51):30557–30561.

Yeramian P, Chardin P, Madaule P, and Tavitian A. 1987. Nucleotide sequence of human rho cDNA clone 12. Nucleic Acids Res 15(4):1869.

Zheng Y, Olson MF, Hall A, Cerione RA, and Toksoz D. 1995. Direct involvement of the small GTP binding protein *Rho* in 1bc oncogene function. J Biol Chem 270:9031–9034.

Part III

SECOND MESSENGERS

4

Cyclic Nucleotides: Synthesis by Adenylyl and Guanylyl Cyclases

AKIO YAMAZAKI

Cyclic-3′,5′-adenosine monophosphate (cAMP) and cyclic-3′,5′-guanosine monophosphate (cGMP) are second messengers in intracellular signaling cascades. The intracellular levels of these cyclic nucleotides are regulated by two enzyme systems. Adenylyl cyclase and guanylyl cyclase synthesize cAMP and cGMP, respectively, and cyclic nucleotide phosphodiesterase hydrolyzes these cyclic nucleotides. Each cyclase system synthesizes only one of these cyclic nucleotides; however, various kinds of phosphodiesterases hydrolyze both cAMP and cGMP (see chapter on cyclic nucleotide phosphodiesterases). Complete details of the functional relationship between cyclase systems and phosphodiesterase systems in most cells remain unclear. However, it is evident that a wide variety of exogenous stimuli, such as hormones, neurotransmitters, and physical and chemical signals, controls the intracellular level of these cyclic nucleotides by regulating these enzyme systems directly or indirectly. The physiological effects of these cyclic nucleotides are exerted through binding to a number of various proteins, including cyclic nucleotide-dependent protein kinases, cyclic nucleotide-gated channels, and cyclic nucleotide-regulated (or-bound) phosphodiesterases (see these chapters). Thus, the phosphorylated state of various proteins and the concentrations of ions and cyclic nucleotides in cells are controlled by extracellular signals. The presence of these cyclic nucleotide-binding proteins indicates the essential roles of these cyclic nucleotides in the organization of cellular functions. The basic concepts of adenylyl and guanylyl cyclase functions are described in this chapter.

Introduction to Cellular Signal Transduction
A. Sitaramayya, Editor
©1999 Birkhäuser Boston

I. ADENYLYL CYCLASES

Adenylyl cyclases [ATP pyrophosphate-lyase (cyclizing), EC 4.6.1.1] convert ATP into cAMP and pyrophosphate. It has been widely recognized that adenylyl cyclases are membrane-bound in most tissues and that the soluble forms are found only in a few cell types (for example, some bacterial and mammalian testis enzymes). Membrane-bound enzymes may be classified into three groups: (1) Adenylyl cyclases bound peripherally to membranes. Enzymes from *E. coli* and *Saccharomyces cerevisiae* do not have a membrane-spanning domain. (2) Adenylyl cyclase with one membrane-spanning domain. An enzyme from slime mold, *Dictyostelium discoideum*, has one membrane-spanning domain, like membrane-bound guanylyl cyclase. (3) Adenylyl cyclases with two sets of transmembrane domains. The mammalian membrane-bound enzymes have two sets of transmembrane domains. These mammalian membrane-bound enzymes are expressed in almost all cells, but in small amounts, and the enzymes are labile and difficult to manipulate in detergent-containing solutions. These characteristics have hampered purification and characterization of these mammalian adenylyl cyclases. However, recent studies have revealed complex properties of mammalian membrane-bound adenylyl cyclases.

A. Structure of Mammalian Membrane-bound Adenylyl Cyclase

1. COMMON STRUCTURE.   Adenylyl cyclase from mammalian tissues was first purified from rabbit myocardium by forskolin-affinity column chromatography (Pfeuffer and Metzger 1982). Using similar affinity column chromatography, Gilman and coworkers purified calmodulin-sensitive adenylyl cyclase from bovine brain and isolated cDNA encoding the full-length protein (Krupinski et al 1989). Then, the sequence of this brain adenylyl cyclase was used to identify cDNAs encoding other mammalian adenylyl cyclases (Bakalyar and Reed 1990, Feinstein et al 1991, Tang and Gilman 1992, Iyengar 1993). Analysis of amino acid sequences of these adenylyl cyclases indicates that these enzymes have a common structure. These adenylyl cyclases are glycoproteins with a molecular weight of roughly 120 kDa and an unexpected topology within the membranes, as depicted in Figure 4-1. These enzymes contain a short and variable amino-terminus (N), followed by two repeats of a unit composed of six transmembrane α-helical spans (M_1 and M_2), a 350 amino acid loop (C_{1a} and C_{1b}) between the first and second unit of the transmembrane spans, and a 250–300 amino acid tail (C_{2a} and C_{2b}) after the transmembrane domain. Although these two transmembrane spans (M_1 and M_2) share some similarity with each other, there is little sequence homology among the various enzymes. The most highly conserved sequences are within the large cytoplasmic loop (C_{1a}) and tail (C_{2a}) (50~90%). Consensus sites for N-linked glycosylation are always present in

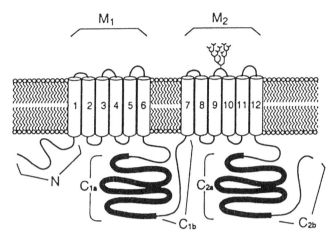

Figure 4-1. Structure of membrane-bound adenylyl cyclases. N,N-terminal of the cyclase. Cylinders (1–12) represent membrane-spanning regions. Boldface lines indicate regions of high amino acid similarity among all members of the family. Sites for N-linked glycosylation are present in at least one domain in the M_2 region. Reproduced with permission from R. Taussig and A.G. Gilman. 1995. Mammalian membrane-bound adenylyl cyclases. J Biol Chem 270:1–4.

at least one extracellular domain associated with the second set of six transmembrane spans. Adenylyl cyclases may serve as channels and transporters. The proposed topography of these adenylyl cyclases closely resembles the structure proposed for molecules such as the P-glycoprotein and the cystic fibrosis transmembrane conductance-regulator chloride channels (Riordan et al 1989). However, there is no additional evidence to support this prediction.

Cloned mammalian adenylyl cyclases do not have any obvious signature sequences for nucleotide binding. However, regions in both the central cytoplasmic loop (C_1) and the carboxyl-terminal tail (C_2) appear to be similar to a single domain of guanylyl cyclase (Chinkers and Garbers 1991). Many residues within these regions seem to be conserved among the members of both adenylyl and guanylyl cyclases, suggesting that these domains function as nucleotide binding sites for catalysis. This conclusion is supported by data that point mutation in either the C_{1a} or C_{2a} domain of the adenylyl cyclases impairs enzymatic activity (Wu et al 1993). It is notable that expression of either domain does not yield catalytic activity. Coexpression of these domains results in the expression of the enzymatic activity, although less activity than in the wild-type, but the activity can be regulated by G-protein subunits or calmodulin as effectively as in the wild-type of the adenylyl cyclase (Tang et al 1991). Interactions between these two cytoplasmic domains are required for the catalytic activity, and specific protein holding is needed for the interaction.

Functions of the amino-terminal region (N) of the adenylyl cyclases remain unclear. Roles of the transmembrane domains (M_1 and M_2) are also not clear. However, because coexpression of the C_1 and C_2-domains does not result in full restoration of activity, it is speculated that a function of the transmembrane domains may provide optional molecular interactions between the C_1 and C_2 domains (Iyengar 1993). Forskolin, a lipid-soluble diterpen and an activator of all types of these mammalian adenylyl cyclases, may alter fluidity of the liquid microenvironment around a portion of the transmembrane regions. However, the roles of these transmembrane regions may not be limited to orienting the enzyme in membranes. A recombinant adenylyl cyclase that is constructed by mostly C_1 and C_2 domains is stimulated by forskolin (Tang and Gilman 1995).

2. ISOFORMS OF MAMMALIAN ADENYLYL CYCLASE.   Recent studies have demonstrated a great diversity of adenylyl cyclases in mammalian tissues (Krupinski et al 1989, Bakalyer and Reed 1990, Feinstein et al 1991, Tang and Gilman 1992, Iyengar 1993, Taussig and Gilman 1995). Knowledge of the molecular diversity in this family will probably be extended; however, analysis of sequences of the cloned adenylyl cyclases permits some grouping, as summarized in Figure 4-2. More than eight isoforms of mammalian adenylyl cyclase have been identified. The overall similarity between these

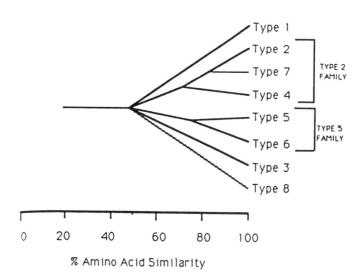

Figure 4-2. Homology of mammalian adenylyl cyclase. Reproduced with permission from R. Iyengar. 1993. Molecular and functional diversity of mammalian Gs-stimulated adenylyl cyclases. FASEB 7:768–775.

adenylyl cyclases is about 50%. The type 2, 4, and 7 isoforms show close homology to each other (type 2 family). Close relationship is also seen between type 5 and type 6 (type 5 family). Analysis of their mRNA expression indicates that all isoforms are present in brain, but distribution of these isoforms in other tissues appears to be isoform-specific. Type 1 adenylyl cyclase is present mainly in areas that have been implicated in processes of learning and memory (Xia et al 1991). In the type 2 family, the type 2 isoform is found in brain and lung, while type 4 and 7 are more widely distributed (liver, brain, heart, kidney, lung, and testis). The expression of type 2 mRNA in brain is not restricted to any particular area (Mons and Cooper 1993). Type 3 adenylyl cyclase is abundant in olfactory neuroepithelium (Bakalyar and Reed 1990). The type 5 and 6 isoforms are highly expressed in brain and heart, but these isoforms are also expressed widely at a low level. In brain, type 5 is restricted to the caudate putamen, the nucleus accumbens, and the olfactory tubercle. Type 8 adenylyl cyclase appears to be present only in brain. It should be emphasized that the cellular and subcellular localization of these isoforms was studied using polymerase chain reaction (PCR)-amplified mRNA, but not with isoform-specific antibodies. Thus, it is not clear how much these adenylyl cyclase proteins are expressed in these regions.

B. Regulation of Mammalian Membrane-bound Adenylyl Cyclases

A variety of hormones, neurotransmitters, and olfactants regulates mammalian membrane-bound adenylyl cyclases. The regulatory mechanism of these adenylyl cyclases is typical G-protein-dependent signal transduction. The G-protein-dependent signal transduction mechanism is composed of three components: receptor, G-protein, and effector. Agonist binding to the receptor leads to conformational change of the receptor that promotes GDP/GTP exchange on the α subunit of the G-protein and dissociation of the GTP-bound α subunit from the $\beta\gamma$ subunits of the G-protein. The GTP-bound α subunit and/or the $\beta\gamma$ subunits then associate with effector proteins and regulate their functions (see chapter on G-proteins). The molecular mechanism of the activation of adenylyl cyclases is not clear. All of the adenylyl cyclases are also activated by forskolin (Seamon et al 1981). The stimulation of activity by forskolin and $G_{s\alpha}$, the α subunit of the stimulatory G-protein, is highly synergistic in many cases (types 2, 4, and 6). All family members are also inhibited by certain adenosine analogs termed P-site inhibitors (Londos and Wolff 1977). The physiological significance of P-site inhibition is not clear. However, it is speculated that some relevant compounds appear to be present *in vivo* to affect the enzyme activity (Bushfield et al 1990). All adenylyl cyclases are inhibited by high (100–1000 μM) concentration of Ca^{2+} as a result of competition for Mg^{2+}. In addition, the adenylyl cyclase isoforms are further regulated in type-specific patterns, especially by the

G-protein $\beta\gamma$ subunits and Ca^{2+}/calmodulin. Moreover, these adenylyl cyclases appear to be modified by several kinases in a type-specific manner.

1. REGULATION BY G-PROTEIN SUBUNITS. Two functionally different G-proteins, G_s, the stimulatory G-protein, and G_i, the inhibitory G-protein, are involved in adenylyl cyclase regulation (Table 4-1). The GTP-bound $G_{s\alpha}$ associates with adenylyl cyclase and stimulates the rate of cAMP synthesis. In contrast, the GTP-bound $G_{i\alpha}$ inhibits the catalytic activity. For example, coexpression of activated $G_{i\alpha}$ with type 2, 3, and 6 adenylyl cyclases results in the inhibition of the expressed adenylyl cyclase activity. It appears likely that all the adenylyl cyclases would be subjected to inhibition by $G_{i\alpha}$; however, it is not clear if the $G_{i\alpha}$ inhibition is caused by direct interaction with these adenylyl cyclases. Various isoforms of the adenylyl cyclases are also regulated differently by $\beta\gamma$ subunits of G-protein (Table 4-1). Type 1 adenylyl cyclase is directly inhibited by $\beta\gamma$ subunits. However, type 2 and 4 adenylyl cyclases are stimulated by $\beta\gamma$ subunits. This stimulation is highly conditional. The $\beta\gamma$ subunits alone have little or no stimulatory effects on these adenylyl cyclases; however, in the presence of $G_{s\alpha}$, the magnitude of stimulation of the activity is substantial. Type 3 enzyme activity is not directly affected by the $\beta\gamma$ subunits. The activity of both type 5 and 6 isoforms is also not susceptible to the direct effect of the $\beta\gamma$ subunits.

2. REGULATION BY CA^{2+}. It has been known for more than two decades that some adenylyl cyclases are regulated by Ca^{2+}/calmodulin (Brostrom et al 1975, Cheung et al 1975). Recent studies have demonstrated a great diversity of Ca^{2+} -regulated adenylyl cyclases in mammalian tissues (Table 4-1). Type 1, 3, and 8 enzymes are stimulated by Ca^{2+}/calmodulin. However, types 2 and 4 are not sensitive to Ca^{2+}/calmodulin. Interestingly, types 5 and 6 appear to be inhibited by low (μM) concentrations of free Ca^{2+}. This effect

Table 4-1. Regulation of mammalian membrane-bound adenylyl cyclases

Regulator	Mode of Regulation	Adenylyl Cyclase Types
$G_{s\alpha}$	stimulation	1–6
$G_{i\alpha}$	inhibition	1–6
$G\beta\gamma$	stimulation (in the presence of $G_{\alpha s}$)	2,4
	inhibition	1
Forskolin	stimulation	1–6
Ca^{2+}/calmodulin	stimulation	1, 3, 8
Ca^{2+}	inhibition	5,6
Adenosine	inhibition	1–6
Protein kinase C	stimulation	2, 5
Protein kinase A	inhibition	6

is independent of calmodulin. None of the other forms is inhibited by low (μM) concentration of Ca^{2+}. Several studies suggest that the binding of calmodulin to the C_{1b} domain (Figure 4-1) is a mechanism of the Ca^{2+}/calmodulin regulation of adenylyl cyclases. A peptide corresponding to residues 495–522 of the C_{1b} domain of type 1 adenylyl cyclase binds calmodulin and inhibits calmodulin-stimulated enzymatic activity (Vorherr et al 1993). Moreover, point mutations in this domain also suppress the activation of adenylyl cyclase by calmodulin (Wu et al 1993).

3. REGULATION BY PHOSPHORYLATION. Many studies have shown that cAMP production is regulated by protein kinase C stimulation. Biochemical analysis has shown that adenylyl cyclase is a direct target for phosphorylation by protein kinase C (Yoshimasa et al 1987). Type 2 adenylyl cyclase has been reported to be phosphorylated and stimulated by protein kinase C (Jacobowitz et al 1993, Yoshimura and Cooper 1993). The phosphorylation also alters the adenylyl cyclase 2 responsiveness to G-protein regulation (Zimmermann and Taussig 1996). Type 5 also demonstrates a marked increase in enzymatic activity by phosphorylation by protein kinase C (Kawabe et al 1994). This effect is specific for the α and δ isoforms of protein kinase C. Type 7 enzyme may also be phosphorylated by protein kinase C. It is also possible that adenylyl cyclase activity is modified by feedback regulation by cAMP-dependent protein kinase (protein kinase A) (Yoshimasa et al 1987). Examination of the sequence of the cloned adenylyl cyclases shows that all of the isoforms, except the type 4, contain one or two putative protein kinase A phosphorylation sites. However, the positions of the putative kinase A phosphorylation sites are not conserved among these adenylyl cyclases. This may suggest that adenylyl cyclase isoforms may be differentially regulated by protein-kinase-A-dependent phosphorylation. More extensive study about phosphorylation of adenylyl cyclases is required to reveal the regulatory mechanisms of adenylyl cyclases by integration and cross-talk with other signal transduction mechanisms.

4. ROUTES OF SIGNALS TO ADENYLYL CYCLASES. Net cellular activity of adenylyl cyclase reflects the effects of a variety of regulatory processes functioning simultaneously. The level of cAMP in cells also depends on the specific types of adenylyl cyclase present and their relative ratios in cells. As described above, all cloned mammalian adenylyl cyclases are stimulated by receptors that activate G_s. Receptors coupled to G_i also appear to inhibit all adenylyl cyclases through activation of $G_{i\alpha}$. However, the $\beta\gamma$ subunits of G-protein regulate adenylyl cyclase isoforms differently. Furthermore, Ca^{2+}/calmodulin stimulates specific isoforms of these adenylyl cyclases. In addition, several protein kinases appear to be involved in the regulation of specific adenylyl cyclases. The regulation of adenylyl cyclases by Ca^{2+} is connected with the activation of another G-protein, G_q, family. The $\beta\gamma$-de-

pendent regulation of adenylyl cyclases is connected with the activation of other G-proteins: the G_i, G_o, and G_z families. The regulation of adenylyl cyclases by protein kinase C is also connected to the regulatory mechanism of protein kinase C, including various growth factors and the pathways for the activation of the G_q subfamily. Thus, the adenylyl cyclase systems are involved in a very large number of signal transduction systems.

II. GUANYLYL CYCLASES

The cGMP metabolism is related to smooth muscle relaxation, natriuresis, sperm chemotaxis, retinal photoreceptor function, and other biological events. Many hormones and other agents cause, directly or indirectly, elevation of the cGMP level through inhibition of cyclic nucleotide phosphodiesterases, activation of guanylyl cyclases, or both. Guanylyl cyclases (GTP pyrophosphate-lyase [cyclizing], EC 4.6. 1.2) convert GTP into cGMP and pyrophosphate. Guanylyl cyclases are found associated with membranes and cytoskeletal elements as well as in the cytosol. These guanylyl cyclases coexist in the same tissue and cell types, in many cases. As described above, mammalian membrane-bound adenylyl cyclases are regulated by agonist binding to receptors through activation of G-proteins. Unlike adenylyl cyclases, almost all guanylyl cyclases represent complete signal-transducing units that do not require additional components. The guanylyl cyclase serves as a signal receptor, a signal transducer, and an effector, all in one molecule. All membrane-bound guanylyl cyclases, except guanylyl cyclases regulated by Ca^{2+}/Ca^{2+}-binding proteins, appear to be receptors for extracellular peptides. These membrane-bound guanylyl cyclases also contain domains for the transduction of peptide signals from their extracellular receptor to their intracellular catalytic domain. Thus, peptide binding to the extracellular receptor directly stimulates the catalytic activity of guanylyl cyclases. The cytoplasmic forms of guanylyl cyclase are heme-containing heterodimers that are directly activated by signals such as nitric oxide (NO) and related nitrovasodilatory compounds. NO is a reactive and toxic, free radical gas and rapidly diffuses across cell membranes. NO presumably reacts with the heme prosthetic group of the soluble guanylyl cyclase. Thus, NO produced extrcellulally and/or intracellulally stimulates catalytic activity of the soluble guanylyl cyclases.

A. Structure of Guanylyl Cyclases

The membrane-bound form of guanylyl cyclase was first purified from sea urchin spermatozoa (Radany et al 1983). Partial amino acid sequencing of the sea urchin guanylyl cyclase led to the cloning and sequencing of cDNA of the enzyme (Singh et al 1988, Thorpe and Garbers 1989). cDNAs encoding other membrane forms of guanylyl cyclase were also cloned and sequenced by using cDNA of the sea urchin guanylyl cyclase (Chinkers et al 1989, Lowe et al 1989). Amino acid sequences of these cDNA clones show that their overall topology resembles receptor tyrosine kinases and phosphatases (Chinkers and Garbers 1991, Garbers and Lowe 1994, Fulle and Garbers 1994). Antibodies specific for sea urchin guanylyl cyclase precipitate mammalian membrane forms of guanylyl cyclases but not soluble forms, suggesting that the sea urchin and mammalian membrane forms of guanylyl cyclase share regions of identity, but that the soluble forms of guanylyl cyclases do not have such regions. The soluble forms of guanylyl cyclases were purified from various tissues (Kamisaki et al 1986, Humbert et al 1990). They exist as heterodimers (α and β), and these two subunits have been cloned and sequenced.

1. MEMBRANE GUANYLYL CYCLASES.   Sequences of cDNA clones encoding various membrane guanylyl cyclases have been obtained as described. Hydropathic analysis of the deduced amino acid sequences indicates that all of the membrane guanylyl cyclases contain a single transmembrane domain of about 20 hydrophobic amino acids (Figure 4-3). The extracellular domain (about 500 amino acids) contains a unique sequence that does not match any known sequence. The domain has been shown to be required for ligand binding by cross-linking of radioactive peptides. The intracellular region, which represents about half of the protein (about 500 amino acids), contains two different domains. A domain (about 300 amino acids) located closer to the transmembrane domain shows homology to the catalytic domain of protein kinases. The transmembrane domain and the so-called protein-kinase-like domain function for the signal transduction from the extracellular domain to the carboxyl-terminal catalytic domain. The kinase-like domain appears to be regulated by binding of ATP. The kinase-like domain is not present in either the soluble form of guanylyl cyclases or the mammalian membrane-bound adenylyl cyclases. The carboxyl-terminal domain is highly homologous to that of the α and β subunits of soluble guanylyl cyclases (Nakane et al 1990) and to two regions (C_{1a} and C_{2a}) of mammalian membrane-bound adenylyl cyclases (Tang and Gilman 1992, Iyenger 1993, Taussig and Gilman 1995), suggesting that these regions have a common function: nucleotide binding for catalytic activity. Proteolytic fragmentation and various mutations of the native enzyme are also used for the direct demonstration of cyclase activity with the carboxyl region.

2. SOLUBLE GUANYLYL CYCLASES.   The soluble form of guanylyl cyclases is a heterodimer of a large (α) and a small (β) subunit (Figure 4-3). The α and β subunits are linked by an interchain disulfide bond. cDNA sequences for three α ($\alpha1$–$\alpha3$) and three β ($\beta1$–$\beta3$) subunits have been reported (Nakane et al 1990, Giuili et al 1992). The $\alpha1$ and $\beta1$ subunits have been cloned from rat and bovine lungs. These subunits have also been detected in

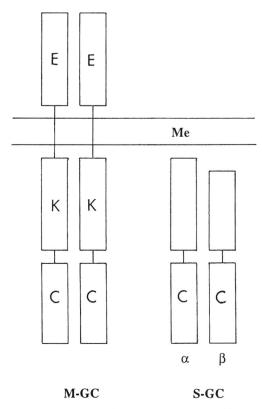

M-GC   **S-GC**

Figure 4-3. Structures of guanylyl cyclases. Membrane-bound guanylyl cyclases (M-GC). E, extracellular domain; K, kinase-like domain; and C, catalytic domain. It is speculated that membrane-bound guanylyl cyclases dimerize for the expression of the enzyme activity. Me, membranes. Soluble guanylyl cyclases (S-GC). C, catalytic domain. Soluble guanylyl cyclase has two subunit α and β.

liver, heart, kidney, cerebrum, and cerebellum. α2 has been cloned from fetal human brain, and β2 has been cloned from rat kidney. α3 and β3 were cloned from adult human frontal lobe. The α1 and α3 subunits have 80% homology at the protein level, while α2 subunits have only 34% homology to α1. β1 and β3 share 99% homology at the amino acid level; however, β2 has less than 40% homology with β1 and β3. The carboxyl domains of the α and β subunits share approximately 45% amino acid sequence identity and also have an amino acid sequence similar to those found in the carboxyl catalytic region of the membrane guanylyl cyclases, suggesting that the carboxyl domain functions as a catalytic site. For the expression of enzymatic activity, both α and β subunits appear to be required (Nakane et al 1990). No report has been published to show the active form of homodimers. Moreover, reconstitution of the guanylyl cyclase activity with known α or β subunits has not yet been accomplished.

A new type of guanylyl cyclase has been reported (Kojima et al 1995). The guanylyl cyclase is a soluble protein but contains a kinase-like domain in its N-terminal. The kinase-like domain is shorter than those of known particulate guanylyl cyclases. Analysis of cDNA suggests that the guanylyl cyclase is a gene product different from known membrane and soluble guanylyl cyclases. Northern blot analysis indicates that the guanylyl cyclase is expressed in kidney, lung, and skeletal muscle but not in cerebrum, cerebellum, heart, liver, stomach, or testis.

B. Regulation of Guanylyl Cyclases

1. MEMBRANE-BOUND GUANYLYL CYCLASES.   In many cases, membrane-bound guanylyl cyclases serve as signal receptors, signal transducers, and effect or enzymes. For these guanylyl cyclases, another protein is not involved as an essential component for the expression of the enzyme activity. Thus, these guanylyl cyclases are regulated not only as effectors but also as receptors and transducers. The extracellular domain of these guanylyl cyclases functions as a receptor. These guanylyl cyclases are activated by binding of regulatory peptides to their extracellular domains. Guanylyl cyclases GC-A and GC-B are stimulated by binding of natriuretic peptides. GC-C is a receptor for bacterial heat-stable enterotoxin/guanylin. In addition, guanylyl cyclases from sea urchin sperm plasma membrane are activated by binding of sea urchin egg peptides. It is interesting that both homologous desensitization and heterologous desensitization, which are seen in the regulation of G-protein-coupled receptors, have also been detected in these guanylyl cyclases. The membrane-spanning domain and the kinase-like domain of guanylyl cyclase appear to function for transduction of the signal to the catalytic domains. Although regulation of the membrane-spanning domain is unknown, binding of ATP to the kinase-like domain is involved in the regulation of membrane guanylyl cyclase activity. Activity of some membrane-bound guanylyl cyclases is also regulated by intracellular protein regulators. These include retinal photoreceptor guanylyl cyclases and cyclases from *Tetrahymena* (Kakiuchi et al 1983) and *Paramecium* (Klumpp et al 1987). Retinal guanylyl cyclases are regulated by Ca^{2+} through Ca^{2+}-binding proteins without involvement of the extracellular domain. Although the topography of the retinal guanylyl cyclases is very similar to that of the peptide-regulated guanylyl cyclases, involvement of extracellular peptides in their regulation has never been reported. The *Tetrahymena* and *Paramecium* enzymes are activated by calmodulin in the presence of µM of calcium.

2. Peptide Receptor Guanylyl Cyclases.

a. GC-A and GC-B. The family of natriuretic peptides (ANP, atrial natriuretic peptide; BNP, brain natriuretic peptide; and CNP, C-type natriuretic peptide) is synthesized and released into blood by the cardiac atria. Amino acid sequences of these peptides are summarized in Figure 4-4. ANP is highly conserved across various species, but BNP shows considerable diversity across species. These peptides exert various receptor-mediated actions in kidney, adrenal gland, and vasculature, that are important in fluid homeostasis and cardiovascular regulation. (Rosenzweig et al 1991). These peptides are also found in the central nervous system and involved in the control of arterial pressure, vasopressin release, and drinking behavior. These peptides bind to two different size-classes of receptors, which were identified by affinity labeling as a ~140-kDa protein and a ~65-kDa protein (Maack 1992). The ~65-kDa protein forms disulfide-linked homodimers. The partially purified ~140-kDa protein has guanylyl cyclase activity (Paul et al 1987), whereas the ~65-kDa receptor is apparently not coupled to guanylyl cyclases. The sequencing of cDNAs encoding the ~140-kDa proteins reveals

---

ANP   S-L -R-R-S-S-**C-F-G**-G-R-M-**D-R-I-G**-A-Q-**S-G-L-G-C**-N-S-F-R- -Y

BNP   S-P-K-T-M-R-D-S-**G-C-F-G**-R-R-L-**D-R-I-G**- S-L-**S-G-L-G-C**-N-V-L-R-R-Y

CNP   G-L-S-K-**G-C-F-G**-L-K-L-**D-R-I-G**-S-M-**S-G-L-G-C**

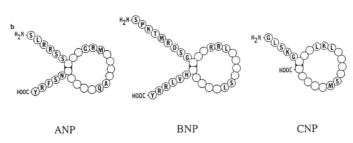

ANP   BNP   CNP

Figure 4-4. Amino acid sequence of porcine atrial natriuretic peptides: ANP, BNP, and CNP. Bold letters indicate conserved residues among three peptides. Open circles indicate identical residues among three peptides. Intermolecular disulfide linkage is formed between two cysteine residues in each peptide. Reproduced with permission from T. Sudoh, N. Minamino, K. Kangawa, and H. Matsuo. 1990. C-type natriuretic peptide (CNP): A new member of natriuretic peptide family identified in porcine brain. Biochem Biophys Res Commun 168:863–870.

that the ~140-kDa proteins have an intracellular domain exhibiting similarity to the catalytic domains of bovine soluble and sea urchin membrane-bound guanylyl cyclases. These data indicate that the atrial natriuretic peptide receptors are actually membrane-bound guanylyl cyclases. The ~65-kDa receptor is considered to function as an ANP clearance-storage binding protein (Fuller et al 1988).

When DNA corresponding to the sea urchin sperm guanylyl cyclase is used as a probe to screen mammalian cDNA libraries, two different cDNA clones, GC-A (Fulle and Garbers 1994) and GC-B (Schultz et al 1989), are isolated. The sequence homology between GC-A and GC-B is highest in the catalytic domain (91%) and lowest in the extracellular domain (43%). Activities of GC-A and GC-B, especially GC-A, expressed in COS cells are stimulated by ANP and BNP in a dose-dependent manner. The radiolabeled peptides also cross-link to these guanylyl cyclases, indicating that these guanylyl cyclases are atrial natriuretic peptide receptor enzymes. Subsequent studies have shown that CNP is a very potent and selective stimulator of GC-B (Koller et al 1991). CNP does not bind to GC-A.

ATP is obligatory in peptide signaling of GC-A, but ATP by itself has no effect on the cyclase activity. The ATP-mediated activation is driven by the direct binding of ATP to the cyclase kinase-like domain (Chinkers and Garbers 1989). Moreover, binding of ATP inhibits the ANP binding to GC-A by converting the high-affinity receptor form to the low-affinity form. This is a mechanism of desensitization of GC-A. These phenomena are also detected in the presence of ATPγS, a nonhydrolyzable analog of ATP, indicating that phosphorylation is not involved. Therefore, the mechanism for the desensitization of GC-A is totally different from desensitization of the receptor coupled to G-protein. G-protein-coupled receptors are desensitized by specific protein-kinase-dependent phosphorylation.

b. GC-C. It is known that the heat-stable enterotoxins (STs) secreted by pathogenic bacteria elicit diarrhea by a cGMP/guanylyl-cyclase-dependent mechanism (Field et al 1978). These bacterial peptides (Figure 4-5) appear to act on a specific intestinal receptor to cause a higher intracellular cGMP level. STs activate GC-C, a third form of mammalian membrane-bound guanylyl cyclase (de Sauvage et al 1991). Neither ANP nor CNP stimulates GC-C activity. An endogenous peptide named guanylin (Forte and Currie 1995) also activates GC-C. Guanylin and STs similarly activate GC-C, indicating that guanylin is an agonist bound to GC-C. A similar peptide, uroguanylin, is also isolated from human urine (Hamra et al 1993). Uroguanylin has two additional acidic amino acids that are not present in guanylin (Figure 4-5). Synthetic uroguanylin is more potent than guanylin for the activation of GC-C.

GC-C binds ^{125}I-labeled ST with high affinity when cDNA encoding GC-C is expressed in cells. GC-C is structurally related to other membrane

```
                        1           5           10          15

Guanylin                P  G  T  C  E  I  C  A  Y  A  A  C  T  G  C

Uroguanylin             N  D  D  C  E  L  C  V  N  V  A  C  T  G  C  L

E.coli ST         N  S  S  N  Y  C  C  E  L  C  C  N  P  A  C  T  G  C  Y

V. cholerae-01 ST    L  I  D  C  C  E  I  C  C  N  P  A  C  F  G  C  L  N
```

Figure 4-5. Structures of bacterial STs, human guanylin, and human uroguanylin. Each amino acid is shown by single letter abbreviation. Bold letters indicate conserved amino acids.

guanylyl cyclases such as sea urchin sperm guanylyl cyclase, GC-A, and GC-B (Schultz et al 1990). GC-C contains an extracellular ligand-binding domain (410-residue NH_2-terminal region), a single transmembrane region of 21 hydrophobic amino acids, and an intracellular domain. The extracellular domain of GC-C has a sequence identity of 10% when compared to other guanylyl cyclases. The intracellular region has a kinase-like domain and a catalytic domain. ATP's effects on the activity of GC-C are very similar to its effects on GC-A. It is notable that GC-C has an extended carboxyl-terminal tail (about 70 amino acids) rich in uncharged polar amino acids. This structure may be responsible for its relative resistance to solubilization with nonionic detergents, perhaps because it anchors the enzyme to cytoskeletal structures. Localization of GC-C and guanylin indicates that enterocytes may be the major cells responsible for net secretion of salt and water, and that the secretion is physically controlled by guanylin through GC-C activation and disrupted by bacterial STs that cause diarrhea.

c. Guanylyl cyclase from sea urchin sperm. The sea urchin egg peptides, speract (GFDLNGGGVG) and resact (CVTGAPGCVGGGRL-NH_2), bind to spermatozoa or to spermatozoa membranes in a species-specific manner and cause a number of biochemical responses, including elevation of cGMP levels (Garbers et al 1989). Studies about guanylyl cyclases from spermatozoan membrane have made important contributions to the understanding of membrane-bound guanylyl cyclases. The guanylyl cyclase that is activated by resact was first purified as a membrane-bound form of guanylyl cyclase. The guanylyl cyclases from two sea urchin species were initially cloned and sequenced as membrane-bound guanylyl cyclases. These two sea urchin clones of guanylyl cyclases were used to isolate guanylyl cyclases from various mammalian tissues. The clones were also used to define the overall topology of membrane-bound guanylyl cyclase, depicted in Figure 4-3. Moreover, the interaction between membrane-bound guanylyl cyclase and its specific peptide was also first revealed in the sea urchin sperm guanylyl cyclase system. An analog of resact was covalently coupled to spermatozoan guanylyl cyclase with high specificity. The interaction led to the hypothesis that a peptide binds to an extracellular, peptide-specific cell surface receptor domain, and then guanylyl cyclase is activated. It is clear now that this kind of mechanism is involved in membrane-bound guanylyl cyclases such as GC-A, GC-B, and GC-C.

The sea urchin guanylyl cyclase is also regulated by phosphorylation (Garbers 1989). Under normal conditions the guanylyl cyclase contains up to 17 mol phosphate/mol enzyme. However, after the treatment of sperm cells with the peptide, the phosphorylation level is reduced to less than 2 mol/mol enzyme. Interestingly, the peptide resact activates guanylyl cyclase before dephosphorylation, but after dephosphorylation the enzyme activity of the sea urchin guanylyl cyclase is reduced by approximately 80%. This suggests that dephosphorylation of the sperm guanylyl cyclase is a desensitization step.

3. Ca^{2+}/CALCIUM-BINDING PROTEIN-REGULATED GUANYLYL CYCLASES. Guanylyl cyclases from *Tetrahymena* and *Paramecium* have not been characterized. In contrast, retinal photoreceptor guanylyl cyclases have been investigated extensively. Visual signal transduction takes place in the outer segments of retinal rod and cone photoreceptors (Pugh and Lamb 1993, Kontalos and Yau 1996, Helmreich anf Hofmann 1996). In darkness, the Ca^{2+} influx into the outer segment through cGMP-gated channels is balanced by an efflux through a Na^+/Ca^{2+} exchanger. Illumination of rhodopsin activates a biochemical cascade that leads to stimulation of cGMP phosphodiesterase. The reduction of free cGMP releases cGMP from the cGMP-gated cation channels, blocks Na^+ and Ca^{2+} influx, and allows a Na^+/Ca^{2+} exchanger to decrease the concentration of free Ca^{2+}. This decrease in the Ca^{2+} concentration triggers a negative feedback of various phototransduction pathways to produce light adaptation. One of the pathways is activation of membrane-bound guanylyl cyclase in the outer segments.

The photoreceptor guanylyl cyclase has been purified from frog, toad, and bovine photoreceptor rod outer segments and identified as a 115-kDa protein (Hayashi and Yamazaki 1991, Koch 1991). The structure of the membrane guanylyl cyclase deduced from cloned cDNA sequences indicates that these enzymes are members of the membrane-bound guanylyl cyclases depicted in Figure 4-3. In the human retinal cDNA library, two photoreceptor guanylyl cyclases have been cloned, RetGC-1 (Shyjan et al 1992) and RetGC-2 (Lowe et al 1995). Homologs of RetGC-1 have also been cloned from bovine (ROS-GC) (Goraczniak et al 1994) and rat retinal cDNA libraries (Yang et al 1995). RetGC-1 is found predominantly in the photoreceptor layer (Liu et al 1994). In particular, cone outer segments are more densely labeled with a specific antibody than rod outer segments. RetGC-1 is localized exclusively in the membrane-rich domains and associated with the marginal region of the disc membranes and/or the plasma membranes. Electron microscopic immunocytochemistry also reveals that RetGC-1 (Liu et al 1994) and ROS-GC (Cooper et al 1996) are also present in the plexiform layers, suggesting that these photoreceptor guanylyl cyclases may play a role in transduction processes of retinal synapses.

Two Ca^{2+}-binding proteins, GCAP-1 (Subbaraya et al 1994) and GCAP-2 (Dizhoor et al 1995), have been shown to be involved in the activation of photoreceptor guanylyl cyclases. They are distinct gene products. The GCAPs (molecular weight ~23 kDa) are members of the calmodulin family and contain three EF hands functional for Ca^{2+}-binding. The GCAPs have been shown in photoreceptors; however, their precise subcellular distribution has not been determined. Biochemical experiments have shown that the photoreceptor guanylyl cyclases are activated by either GCAP, and that the activator is inhibited by Ca^{2+} with an EC50 for Ca^{2+} of about 200 nM. Various deletion mutants of RetGC-1 and ROS-GC suggest that GCAPs regulate guanylyl cyclases by binding to the intracellular domain rather than the extracellular domain (Laura et al 1996, Duda et al 1996). Function of the extracellular domain remains unclear. As described, the activation of photoreceptor guanylyl cyclases is not dependent on regulatory peptides. Natriuretic peptides do not activate photoreceptor guanylyl cyclases. These results may suggest that photoreceptor guanylyl cyclases are regulated by a mechanism distinct from the mechanism that regulates peptide-dependent guanylyl cyclases such as GC-A, GC-B, and GC-C.

As described above, the photoreceptor guanylyl cyclases appear to be negatively regulated by Ca^{2+}. When the Ca^{2+} concentration is reduced, the enzyme is activated by GCAPs. However, the photoreceptor guanylyl cyclases are also positively regulated by Ca^{2+} (Pozdnyakov et al 1995). When the Ca^{2+} concentration is increased to 2~5 μM, the enzyme is found to be activated up to 25-fold. The Ca^{2+}-dependent activator appears to be S100-like proteins (Margulis et al 1996). Guanylyl cyclases in retinal synapses may be regulated in this manner. The outer segments of photoreceptors may also contain soluble (Koch et al 1994) and ANP- and CNP-activated guanylyl cyclases (Ahmed and Barnstable 1993, Duda et al 1993). NO synthesis is also reported in retinal photoreceptor cells (Yoshida et al 1995). ANP is also found in rat retina (Stone and Glembotski 1986). However, the relationships between photoreceptor guanylyl cyclases and these enzymes and the functions of these enzymes in phototransduction have never been reported.

The photoreceptor guanylyl cyclase is also regulated by ATP (Sitaramayya et al 1991, Gorczyca et al 1994). The mechanism of action of ATP appears to be independent of hydrolysis, and the binding of ATP to the kinase-like domain is suspected to be the activation mechanism. In addition, the photoreceptor guanylyl cyclase is reported to be phosphorylated by protein kinase C (Wolbing and Schnetkamp 1995). The guanylyl cyclase activity seems to be increased by the phosphorylation, and the increase is inhibited by inhibitors of protein kinase C. Moreover, the photoreceptor guanylyl cyclase is reported to have protein kinase activity (Aparicio and Applebury 1996). The kinase activity is unaffected by Ca^{2+}, cyclic nucleotides, and other protein kinase activators. The effect of the phosphorylation on the enzyme activity is not reported.

4. SOLUBLE GUANYLYL CYCLASES.   Soluble guanylyl cyclase is activated by NO (Waldman and Murad 1987). Soluble guanylyl cyclase is also regulated by free radicals such as hydroxyl radicals and carbon monoxide. Activation of the enzyme by NO regulates the concentration of cGMP in cells, and the level of cGMP affects many physiological events, including endothelium-dependent relaxation, platelet aggregation, and neuronal communication. Soluble guanylyl cyclase contains one ferro-protoporphyrin IX per monomer (Gerzer et al 1981). The ferrous heme appears to be a receptor for NO, because NO exhibits high affinity for ferrous heme (Traylor and Sharma 1992), and a heme-deficient mutant of soluble guanylyl cyclase is insensitive to NO (Wedel et al 1994). NO activates the enzyme through binding to the prosthetic heme group, resulting in formation of a ferrous nitrosyl heme complex and consequent change in the enzyme conformation (Wedel et al 1994). Dissociation of NO from the heme may be a turnoff mechanism of activation. Other activators include arachidonic acid, protoporphyrins, and polyunsaturated fatty acids (Ignarro et al 1984). Inhibitors are methylene blue and metalloporphyrins (Ignarro et al 1984). An endogeneous protein inhibitor (149 kDa) is also reported in bovine lung (Kim and Burstyn 1994). This inhibitor protein appears to regulate the enzyme activity as an allosteric regulator of the soluble guanylyl cyclase.

REFERENCES

Ahmed I and Barnstable CH. 1993. Differential laminar expression of particulate and soluble guanylate cyclase genes in rat retina. Exp Eye Res 56:51–62.
Aparicio JG and Applebury ML. 1996. The photoreceptor guanylate cyclase is an autophosphorylating protein kinase. J Biol Chem 271:27083–27089.
Bakalyar HA and Reed RR. 1990. Identification of a special adenylyl cyclase that may mediate odorant detection. Science 250:1403–1406.
Brostrom CO, Huang YC, Breckenridge BM, and Wolff DJ. 1975. Identification of a calcium-binding protein as a calcium-dependent regulator of brain adenylate cyclase. Proc Natl Acad Sci USA 72:64–68.
Bushfield M, Shoshani I, and Johnson RA. 1990. Tissue levels, source, and regulation

of 3'-AMP: An intracellular inhibitor of adenylyl cyclases. Mol Pharmacol 38:848–853.

Cheung WY, Bradham LS, Lynch TJ, Lin YM, and Tallant EA. 1975. Protein activator of 3':5'-nucleotide phosphodiesterase of bovine or rat brain also activates its adenylate cyclase. Biochem Biophys Res Commun 66:1055–1062.

Chinkers M and Garbers DL. 1989. The protein kinase domain of the ANP receptor is required for signaling. Science 245:1392–1394.

Chinkers M and Garbers DL. 1991. Signal transduction by guanylyl cyclases. Annu Rev Biochem 60:553–575.

Chinkers M, Garbers DL, Chang MS, Lowe DG, Chin H, Goeddel DV, and Schultz S. 1989. A membrane form of guanylate cyclase is an atrial natriuretic peptide receptor. Nature 338:78–83.

Cooper N, Liu L, Yoshida A, Pozduyakov N, Margulis A, and Sitaramayya A. 1996. The bovine rod outer segment guanylate cyclase, ROS-GC, is present in both outer segment and synaptic layers of the retina. J Mol Neurosci 6:211–222.

de Sauvage FJ, Camerato TR, and Goeddel DV. 1991. Primary structure and functional expression of the human receptor for *Escherichie coli* heat-stable enterotoxin. J Biol Chem 266:17912–17918.

Dizhoor AM, Olshevskaya EV, Henzel WJ, Wong SC, Stults JT, Ankoudinova I, and Hurley JB. 1995. Cloning, sequencing, and expression of a 24-kDa Ca^{2+}-binding protein activating photoreceptor guanylyl cyclase. J Biol Chem 270:25200–25206.

Duda T. Goraczniak RM, Sitaramayya A, and Sharma RK. 1993. Cloning and expression of an ATP-regulated human retina C-type natriuretic factor receptor guanylate cyclase. Biochemistry 32:1391–1395.

Duda T. Goraczniak RM, Surgucheva I, Rudnicka-Nawrot M, Gorczyca WA, Palczewski K, Sitaramayya A, Baehr W, and Sharma RK. 1996. Calcium modulation of bovine photoreceptor guanylate cyclase. Biochemistry 35:8478–8482.

Feinstein PG, Schrader KA, Bakalyar HA, Tang WJ, Krupinski J, Gilman AG, and Reed RR. 1991. Molecular cloning and characterization of a Ca^{2+}/calmodulin-insensitive adenylyl cyclase from rat brain. Proc Natl Acad Sci USA 88:10173–10177.

Field M, Graf Jr LH, Larid WJ, and Smith PL. 1978. Heat-stable enterotoxin of *Escherichia coli*: in vivo effects on guanylate cyclase activity, cGMP concentration, and ion transport in small intestine. Proc Natl Acad Sci USA 75:2800–2804.

Forte LR and Currie MG. 1995. Guanylin: a peptide regulator of epithelial transport. FASEB J 9:643–650.

Fulle HJ and Garbers DL. 1994. Guanylyl cyclases: a family of receptor-linked enzymes. Cell Biochem Fune 12:157–165.

Fuller F, Porter JG, Arfsten AE, Miller J, Schilling JW, Scarborough RM, Lewicki JA, and Schenk DB. 1988. Atrial natriuretic peptide clearance receptor: Complete sequence and functional expression of cDNA clones. J Biol Chem 263:9395–9401.

Garbers DL. 1989. Molecular basis of fertilization. Annu Rev Biochem 58:719–742.

Garbers DL and Lowe DG. 1994. Guanylyl cyclase receptors. J Biol Chem 269:30741–30744.

Gerzer R, Bohme E, Hofmann F, and Schultz G. 1981. Soluble guanylate cyclase purified from bovine lung contains heme and copper. FEBS Lett 132:71–74.

Giuili G, Scholl U, Bulle F, and Guelleaen G. 1992. Molecular cloning of the cDNAs coding for the two subunits of soluble guanylyl cyclase from human brain. FEBS Lett 304:83–84.

Goraczniak RM, Duda T, Sitaramayya A, and Sharma RK. 1994. Structural and functional characterization of the rod outer segment membrane guanylate cyclase. Biochem J 302:455–461.

Gorezyca WA, Van Hooser JP, and Palczewski K. 1994. Nucleotide inhibitors and activators of retinal guanylyl cyclase. Biochemistry 33:3217–3222.

Hamra FK, Forte LR, Eber SL, Pidhorodeckyj NV, Krause WJ, Freeman RH, Chin DT, Tompkins JA. Fok KF, Smith CE, Duffin KL. Siegel NR, and Currie MG. 1993. Uroguanylin: Structure and activity of a second endogenous peptide that stimulates intestinal guanylate cyclase. Proc Natl Acad Sci USA 90:10464–10468.

Hayashi F and Yamazaki A. 1991. Polymorphism in purified guanylate cyclase from vertebrate rod photoreceptors. Proc Natl Acad Sci USA 88:4746–4750.

Helmreich EJM and Hofmann KP. 1996. Structure and function of proteins in G-protein-coupled signal transfer. Biochem Biophys Acta 1286:285–322.

Humbert P, Niroomand F, Fischer G, Mayer B, Koesling D, Hinsch KD, Gausepohl H, Frank R. Schultz G, and Bohme E. 1990. Purification of soluble guanylyl cyclase from bovine lung by a new immunoaffinity chromatographic method. Eur J Biochem 190:273–278.

Ignarro LJ, Wood KS, and Wolin MS. 1984. Regulation of purified soluble guanylate cyclase by prophyrins and metalloporphyrins. Adv Cyclic Neucleotide Protein Phosphorylation Res 17:267–274.

Iyengar R. 1993. Molecular and functional diversity of mammalian Gs-stimulated adenylyl cyclases. FASEB J 7:768–775.

Jacobowitz O, Chen J, Premont RT, and Iyengar R. 1993. Stimulation of specific types of Gs-stimulated adenylyl cyclases by phorbol ester treatment. J Biol Chem 268:3829–3832.

Kakiuchi S, Sobue K, Yamazaki R, Nagao S, Umaki S, Nozawa Y, Yazawa M, and Yagi K. 1983. Ca^{2+}-dependent molecular proteins from *Tetrahymena pyriformis*, sea anemone, and scallop and guanylate cyclase activation. J Biol Chem 256:12455–12459.

Kamisaki Y, Saheki S, Nakane M, Palmieri JA, Kuno T, Chang BY, Waldman SA, and Murad F. 1986. Soluble guanylate cyclase from rat lung exists as a heterodimer. J Biol Chem 261:7236–7241.

Kawabe J, Iwami G, Ebina T, Ohno S, Katada T, Ueda Y, Homcy CJ, and Ishikawa Y. 1994. Differential activation of adenylyl cyclase by protein kinase C isoenzyme J Biol Chem 269:16554–16558.

Kim TD and Burstyn JN. 1994. Identification and partial purification of an endogenous inhibitor of soluble guanylyl cyclase from bovine lung. J Biol Chem 269:11540–15545.

Klumpp S, Guerini D, Krebs J, and Schultz JE. 1987. Effect of tryptic calmodulin fragments on guanylate cyclase activity from *Paramecium tetraurelia*. Biochem Biophys Res Commun 142:857–864.

Koch KW. 1991. Purification and identification of photoreceptor guanylate cyclase. J Biol Chem 266:8634–8637.

Koch KW, Lambrecht HG, Haberecht M. Redburn D, and Schmidt HH. 1994. Functional coupling of a Ca^{2+}/calmodulin-dependent nitric oxide synthese and a soluble guanylyl cyclase in vertebrate photoreceptor cells. EMBO J 13:3312–3320.

Kojima M, Hisaki K, Matsuo H, and Kangawa K. 1995. A new type soluble guanylyl cyclase, which contains a kinase-like domain: its structure and expression. Biochem Biophys Res Commun 217:993–1000.

Koller KJ, Lowe DG, Bennett GL, Minamino N, Kangawa N, Matsuo H, and Goeddel DV. 1991. Selective activation of the B natriuretic peptide receptor by C-type natriuretic peptide. Science 252:120–123.

Kontalos Y and Yau KW. 1996. Regulation of sensitivity in vertebrate rod photoreceptors by calcium. Trends Neurosci 19:73–81.

Krupinski J, Coussen F, Bakalyar HA, Tang WJ, Feinstein PG, Orth K, Slaughter C, Reed RR, and Gilman AG. 1989. Adenylyl cyclase amino acid sequence: possible channel or transporter-like structure. Science 244:1558–1564.

Laura RP, Dizhoor AM, and Hurley JB. 1996. The membrane guanylyl cyclase, Retinal guanylyl cyclase-l, is activated through its intercellular domain. J Biol Chem 271:11646–11651.

Liu XR, Seno K, Nishizawa Y, Hayashi F, Yamazaki A, Matsumoto H, Wakabayashi T, and Usukura J. 1994. Ultrastructural localization of retinal guanylate cyclase in human and monkey retinas. Exp Eye Res 59:761–768.

Londos C and Wolff J. 1977. Two distinct adenosine-sensitive sites on adenylate cyclase. Proc Natl Acad Sci USA 74:5482–5486.

Lowe DG, Dizhoor AM, Liu K, Gu Q, Spencer M, Laura R, Lu L, and Hurley JB. 1995. Cloning and expression of a second photoreceptor-specific membrane retina guanylyl cyclase (RetGC), RetGC-2. Proc Natl Acad Sci USA 92:5525–5539.

Lowe DG, Chang MS, Hellmiss R, Singh S, Chen E, Garbers DL, and Goeddel DV. 1989. Human atrial natriuretic peptide receptor defines a new paradigm for second messenger signal transduction. EMBO J 8:1377–1384.

Maack T. 1992. Receptors of atrial natriuretic factor. Annu Rev Physiol 54:11–27.

Margulis A, Pozdnyakov N, and Sitaramayya A. 1996. Activation of bovine photoreceptor guanylate cyclase by S100 proteins. Biochem Biophys Res Commun 218:243–247.

Mons N and Cooper DMF. 1993. Adenylate cyclases: critical foci in neuronal signaling. Mol Brain Res 22:236–244.

Nakane M, Arai K, Saheki S, Kuno T, Bucchler W, and Murad F. 1990. Molecular cloning and expression of cDNAs coding for soluble guanylate cyclase from rat lung. J Biol Chem 265:16841–16845.

Paul AK, Marala RB, Jaiswal RK, and Sharma RK. 1987. Coexistence of guanylate cyclase and atrial natriuretic factor receptor in a 180-kD protein. Science 235:1224–1226.

Pfeuffer T and Metzger H. 1982. 7-O-hemisuccinyl-deacetyl forskolin-Sepharose: a novel affinity support for purification of adenylate cyclase. FEBS Lett 146:369–375.

Pozdnyakov N, Yoshida A, Cooper NGF, Margulis A, Duda T, Sharma RK, and Sitaramayya A. 1995. A novel calcium-dependent activator of retinal rod outer segment membrane guanylate cyclase. Biochemistry 34:14279–14283.

Pugh EN Jr and Lamb TD. 1993. Amplification and kinetics of the activation steps in phototransduction. Biochem Biophys Acta 1141:111–149.

Radany EW, Gerzer R, and Garbers DL. 1983. Purification and characterization of particulate guanylate cyclase from sea urchin spermatozoa. J Biol Chem 258:8346–8351.

Riordan JR, Rommens JM, Kerem, BS, Alon N, Rozmahel R, Grzelczak Z, Zielenski J, Lok S, Plavsic N, Chou JL, Drumm ML, Iannuzzi MC, Collins FS, and Tsui LC. 1989. Identification of the cystic fibrosis gene: Cloning and characterization of complementary DNA. Science 245:1066–1073.

Rosenzweig A and Seidman CE. 1991. Atrial natriuretic factor and related peptide hormones. Ann Rev Biochem 60:229–255.

Seamon KB, Padgett W, and Daly JW. 1981. Forskolin: unique diterpene activator of adenylate cyclase in membrane and in intact cells. Proc Natl Acad Sci USA 78:3363–3367.

Schultz S, Singh S, Bellet RA, Singh G. Tubb DJ, Chin H, and Garbers DL. 1989. The primary structure of a plasma membrane guanylate cyclase demonstrates diversity within this new receptor family. Cell 58:1155–1162.

Schultz S, Green CK, Yuen PST, and Garbers DL. 1990. Guanylyl cyclase is a heat-stable enterotoxin receptor. Cell 63:941–948.

Shyjan AW, de Sauvage FJ, Gillett NA, Goeddel DV, and Lowe DG. 1992. Molecular cloning of a retina-specific membrane guanylyl cyclase. Neuron 9:727–737.

Singh S, Lowe DG, Thorpe DS, Rodriguez H, Kuang WJ, Dangott LJ, Chinkers M, Goeddel DV, and Garbers DL. 1988. Membrane guanylate cyclase: A putative cell surface receptor with homology to protein kinases. Nature 334:708–712.

Sitaramayya A, Marala RB, Hakki S, and Sharma RK. 1991. Interactions of nucleotide analogues with rod outer segment guanylate cyclase. Biochemistry 30:6742–6747.

Stone RA and Glembotski CC. 1986. Immunoactive atrial natriuretic peptide in the rat eye: Molecular forms in anterior uvea and retina. Biochem Biophys Res Commun 134:1022–1028.

Subbaraya I, Ruiz CC, Helekar BS, Zhao X, Gorezyca WA, Pettenati MJ, Rao PN, Palczewski K, and Baehr W. 1994. Molecular characterization of human and mouse photoreceptor guanylate cyclase-activating protein (GCAP) and chromosomal localization of the human gene. J Biol Chem 269:31080–31089.

Tang WJ and Gilman AG. 1992. Adenylyl cyclases. Cell 70:869–872.

Tang WJ and Gilman AG. 1995. Construction of a soluble adenylyl cyclase activated by Gsα and forskolin. Science 268:1769–1772.

Tang WJ, Krupinski J, and Gilman AG. 1991. Expression and characterization of calmodulin activated adenylyl cyclase. J Biol Chem 266:8595–8603.

Taussig R and Gilman AG. 1995. Mammalian membrane-bound adenylyl cyclases. J Biol Chem 270:1–4.

Thorpe DS and Garbers DL. 1989. The membrane form of guanylate cyclase. Homology with a subunit of the cytoplasmic form of the enzyme. J Biol Chem 264:6545–6549.

Traylor TG and Sharma VS. 1992. Why NO? Biochemistry 31:2847–2849.

Vorherr T, Knopfel L, Hofmann F, Mollner S, Pfeuffer T, and Carafoli E. 1993. The calmodulin binding domain of nitric oxide synthase and adenylyl cyclase. Biochemistry 32:6081–6088.

Waldman SA and Murad F. 1987. Cyclic GMP synthesis and function. Pharmacol Rev 39:163–196.

Wedel B, Humbert P, Harteneck C, Foerster J, Malkewitz J, Bohme E, Schultz G, and Koesling D. 1994. Mutation of His-105 in the beta 1 subunit yields a nitric oxide-insensitive form of soluble guanylyl cyclase. Proc Natl Acad Sci USA 91:2592–2596.

Wolbing G and Schnetkamp PPM. 1995. Activation by PKC of the Ca^{2+}-sensitive guanylyl cyclase in bovine retinal rod outer segments measured with an optical assay. Biochemistry 34:4689–4695.

Wu Z, Wong ST, and Storm DR. 1993. Modification of the calcium and calmodulin sensitivity of the Type I adenylyl cyclase by mutagenesis of its calmodulin binding domain. J Biol Chem 168:23766–23768.

Xia Z, Refsdal CD, Merchant KM, Dorsa DM, and Storm DR. 1991. Distribution of mRNA for the calmodulin-sensitive adenylate cyclase in rat brain: Expression in areas associated with learning and memory. Neuron 6:431–443.

Yang RB, Footer DC, Garbers DL, and Fulle HJ. 1995. Two membrane forms of guanylyl cyclase found in the eye. Proc Natl Acad Sci USA 92:602–606.

Yoshida A, Pozdnyakov N, Dang L, Orselli SM, Reddy VN, and Sitaramayya A. 1995. Nitric oxide synthesis in retinal photoreceptor cells. Vis Neurosci 12: 493–500.

Yoshimasa T, Sibley DR, Bouvier M, Lefkowitz RJ, and Caron MG. 1987. Cross-talk between cellular signaling pathways suggested by phorbolester induced adenylate cyclase phosphorylation. Nature 327:67–70.

Yoshimura M and Cooper DMF. 1993. Type specific stimulation of adenylyl cyclase by protein kinase C. J Biol Chem 268:4604–4607.

Zimmermann G and Taussig R. 1996. Protein kinase C alters the responsiveness of adenylyl cyclases to G-protein α and βγ subunits. J Biol Chem 271:27161–27166.

5

Phospholipases: Generation of Lipid-Derived Second Messengers

MARY F. ROBERTS

GENERAL PROPERTIES OF PHOSPHOLIPASES

Phospholipases are lipolytic enzymes that play key roles in signal transduction by generating both lipid and in some cases soluble second messengers. Their catalytic properties are often exquisitely controlled by phosphorylation, interaction with other proteins (e.g., GTP-binding proteins), as well as interactions with other lipids. Most of these proteins fall into the category of peripheral membrane proteins. An important aspect of these enzymes is that while they are in general water-soluble, they carry out their catalysis at an interface. This complicates kinetics because the dimensionality of the reaction has changed from a single phase (bulk solution) to two phases (the phospholipid aggregate interface as well as the bulk solution). Often there is a separate binding domain/site for the interface as well as for the active site. Such multiple functional sites can be modular or incorporated into a single area of the protein. A recurring theme is that there are mechanisms that enhance protein association with the interface. This in turn increases the local concentration of substrate (and in some cases chemically modifies the protein) such that catalytic efficiency is increased. These secondary sites can serve to regulate enzyme activity by controlling access to substrate. However, there are often other modes of regulation that intertwine phospholipases with other signal transduction proteins.

Phospholipase designations indicate the phospholipid ester bond being hydrolyzed (Figure 5-1). Phospholipase A_2 (PLA$_2$) removes the *sn*-2 fatty acyl bond, liberating a fatty acid and a lysophospholipid (1-acyl-glycero-

Introduction to Cellular Signal Transduction
A. Sitaramayya, Editor
©1999 Birkhäuser Boston

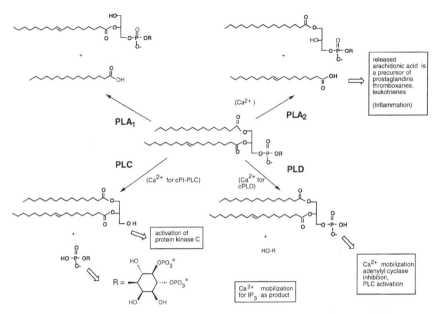

Figure 5-1. Cleavage reactions catalyzed by phospholipase A_1, A_2, C, and D and identification of lipid second messengers produced by phospholipases.

phospholipid); phospholipase A_1 (PLA_1) activities often hydrolyze both fatty acyl chains, with some preference for the *sn*-1 chain. Phospholipase C (PLC) cleaves at the phosphodiester linkage to generate diacylglycerol, which is membrane-localized, and a soluble phosphate monoester. The PLC enzyme specific for phosphatidylinositol (PI) and phosphorylated derivatives (PIP_x), denoted PI-PLC, has a key role in the PI pathway of signal transduction (Nishizuka 1992) since its lipophilic product, diacylglycerol (DAG), is a potent activator of protein kinase C (PKC), and its water-soluble product, phosphorylated inositol phosphate, is also a second messenger. Phospholipase D (PLD) hydrolyzes the other phosphodiester linkage to generate an amphiphilic product (phosphatidic acid) with reported second messenger properties, as well as a water-soluble alcohol (e.g., choline from phosphatidylcholine).

In understanding the role of these surface-active esterases in signal transduction, there are parameters that must be defined for each phospholipase. (1) What are the interactions of a phospholipase with an individual lipid molecule that transiently link the enzyme to the bilayer surface? (2) Is a cofactor such as Ca^{2+} (or another metal ion) that is highly regulated in the cell necessary for activity, and which protein residues are involved in catalysis? (3) What factors control hydrolysis, ensuring specific and localized activity of these enzymes to prevent wholesale destruction of the cell membranes? (4) How do lipophilic products, related lipids, and protein modulators affect enzyme activity—by direct interactions with the protein, or by modulation of the properties of the lipid interface? As background for discussing specific phospholipases, an understanding of the physical behavior of lipids, general strategies for analyzing interfacial kinetics, and a review of different assay systems are needed.

Physical Properties of Phospholipid Substrates and Lipophilic Products

Diacylphospholipids in bilayers have a variety of motions that can affect membrane properties (Figure 5-2): lateral diffusion in the plane of the membrane, vertical and rotational motions, and transbilayer exchange or "flip-flop." In general all but flip-flop are relatively rapid motions and ensure that a protein anchored at the bilayer can usually sample all phospholipids in one leaflet of the membrane via translational diffusion. The very slow rate of phospholipid transbilayer diffusion in membranes ensures membrane lipid asymmetry unless a catalytic protein ("flippase") is available. It also means that a water-soluble phospholipase can hydrolyze phospholipids in the leaflet of the bilayer to which it is exposed, but not on the other leaflet unless (1) the protein is transported across the membrane, or

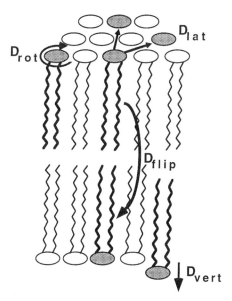

Figure 5-2. Modes of phospholipid motions that might affect phospholipase action: D_{lat}, lateral diffusion in the plane of the bilayer; D_{flip}, transverse or "flip-flop" diffusion across the bilayer; D_{rot}, rotational diffusion; D_{vert}, vertical displacements.

(2) a mechanism exists to enhance substrate flip-flop (e.g., localized membrane defects or partial fusion, etc.). Anionic diacylphospholipids (PS, PG, PA) interact strongly with metal ions (e.g., Ca^{2+}) or other cationic moieties (peptides, amino compounds, etc.), and this can have a strong influence on the binding of extrinsic membrane proteins. They can undergo partial phase separation induced by ions or other proteins binding. Limiting long-range lateral diffusion in this fashion might be important for phospholipase action.

The lipophilic products of phospholipase action can behave like the parent phospholipids or in some cases act as bilayer perturbants. Phosphatidic acid (PA) is a phospholipase D product generated in the early phases of cell signaling. It has been implicated as a lipid second messenger in a wide variety of cellular processes including activation of Ca^{2+}-independent PKC isoforms, regulation of growth and cytoskeleton in fibroblasts, and control of the respiratory burst in neutrophils (English 1996). Also when PA is on the outer leaflet of triggering cells it mediates stimulation of cells in direct contact. Under physiological conditions, PA can exist as a monoanion or dianion. For long-chain PA in a mostly phosphatidylcholine (PC) or phosphatidylethanoline (PE) bilayer, the pK_{a2} of the PA is ~7.8 (Swairjo et al 1994). This charge makes PA an effective metal chelator, and it can form complexes with Ca^{2+}; the interaction with Ca^{2+} can lead to partial lateral phase separation as well as fusion (as it will with most anionic phospholipids). If present in the extracellular medium, PA and lysoPA tend to be bound to serum albumins.

Lysophospholipids, products of PLA_2 cleavage, retain the same polar head group as the phospholipid but are considerably more hydrophilic than diacylphospholipids because of the removal of the sn-2 fatty acyl chain. In sufficient quantities (>20 mol%) they exhibit lytic properties because these phospholipids prefer to pack in micelles rather than bilayers. Hence cells usually want to dispose of these quickly to prevent cell damage and, potentially, death. One lysophospholipid, lysoPA, can be specifically produced by a PA-specific PLA_2. Like PA, the lysoPA has either -1 or -2 charge but is fairly water-soluble, forming micelles as opposed to bilayer vesicles at basic pH values. LysoPA is an extremely potent cell activator; it binds to specific receptors to trigger a series of responses.

Fatty acids, the other product of PLA_2 action, impart a negative charge to membranes, although the pK_a of the carboxylic acid is shifted up to 7–7.5 for fatty acids in a bilayer suggesting that a reasonable fraction of fatty acid is protonated in cell membranes and hence uncharged. Unlike the more polar phospholipids, fatty acids can undergo rapid transbilayer movement via the protonated form. They can rapidly distribute among different membranes. For this reason and because such behavior can make free fatty acids uncouplers of pH gradients, free fatty acids are in general minor components of cell membranes. The important fatty acid in signal transduction is arachidonic acid, a precursor of bioactive compounds involved in the inflammatory response. This fatty acid is primarily esterified to the *sn*-2 position in PC and PE.

Diacylglycerols, the hydrophobic PLC product of phospholipid cleavage, are also normally minor transient lipid components in membranes. They have unique roles as intermediates in lipid metabolism and as second messengers. The signal-activated formation of DAG is often biphasic: an initial rapid rise in DAG concentration that is transient and corresponds to the cleavage of phosphatidylinositol by PLC, followed by a slower release of DAG that is sustained over many minutes and derived by hydrolysis of PC (Nakamura 1996). The function of DAG in signaling is to activate PKC, which in turn plays a role in cellular regulation, tumor promotion, and perhaps oncogenesis (Nishizuka 1992). Rapid attenuation of DAG levels in cells requires diacylglycerol kinase, which converts DAG to PA. DAG is a weakly polar species and in moderate quantities can destabilize bilayers; these lipids also have a tendency to cluster and partially phase separate. DAG exhibits relatively rapid transbilayer flip-flop; it can also exchange between bilayers in model systems by a vesicle collision mechanism (Zhou and Roberts 1997). However, in intact cells, this exchange behavior translates to transport and mobilization of DAG between membranes only if protein carriers are available or if two membranes are juxtaposed.

GENERAL STRATEGIES FOR INTERFACIAL ENZYME KINETIC.   A variety of kinetic effects is common to these interfacially active enzymes: (1) "interfacial activation"—the preference of aggregated versus monomeric substrate (Verheij et al 1981), (2) inhibition due to "surface dilution" by nonsubstrate amphiphiles (Dennis 1973), and (3) processive catalysis—the binding of water-soluble enzymes to bilayer surfaces via nonactive site interactions, which allows for multiple catalytic cycles before the enzyme desorbs from the surface (Jain and Gelb 1991). These effects complicate kinetic analyses of the enzymes, as discussed below.

Lipases as well as phospholipases exhibit an enhanced V_{max}/K_m toward substrates in aggregated as opposed to monomeric forms (Verheij et al 1981). This activation is usually monitored using synthetic phospholipids with a critical micelle concentration in the mM range so that the comparison is for a monomeric versus a micellar substrate. The mechanism responsible for observed rate increases upon aggregation of substrate may not be unique to all phospholipases but may be quite different, depending on the individual phospholipase and its chemical specificity. The degree of activation observed toward phospholipid aggregates varies, depending on the source of the enzyme (i.e., pancreatic versus cobra venom PLA_2), specificity of the phospholipase (PLA_2 exhibits a 20–100 fold activation, while nonspecific and PI-specific bacterial PLC enzymes exhibit only a 2–3-fold and 5–6-fold activation, respectively, upon micellization), and type of aggregate employed (micelle versus bilayer) (De Haas 1971 Roberts 1991b). An example of substrate interfacial activation is shown in Figure 5-3A for PI-specific PLC from *B. thuringiensis* (Lewis et al 1993).

"Surface dilution" refers to the observation that phospholipase-specific activities decrease when a phospholipid aggregate (e.g., detergent mixed micelle) is diluted with detergents or other lipophilic molecules, keeping the total substrate phospholipid concentration constant. This can reflect competitive inhibition of the enzyme by the diluent binding to the active site or an inhibition because the surface concentration of the substrate has been reduced (Dennis 1973, Carman et al 1995). The same enzyme whose interfacial activation curve is displayed in Figure 5-3A exhibits surface dilution by Triton X-100, a nonionic detergent, as shown in Figure 5-3B. Analyses of surface dilution in a well-characterized mixed micelle system can provide K_m values in mole fraction (or mol%) of the substrate in the surface (Carman et al, 1995 and references therein).

The last of these generic kinetic features, processive catalysis (or "scooting mode" catalysis), has been well documented for pancreatic PLA_2 using small unilamellar vesicles of anionic phospholipids (Jain and Gelb 1991, Gelb et al 1995). Anchoring the phospholipase to the substrate surface via either a specific surface binding site or less specifically through electrostatic interactions, allows the enzyme to hydrolyze one substrate molecule after another before dissociating from the surface. Since the catalytic step occurs at the interface, the Michaelis-Menten complex of the substrate with enzyme is no longer a solitary monomeric species in the aqueous phase, but a complex in two dimensions. The overall catalysis now occurs in two discrete steps (Figure 5-4): (1) enzyme in the aqueous phase, E, binds to the substrate interface (usually promoting a conformational change, E*, that enhances catalysis); (2) E* binds a substrate molecule in the active site, hydrolyzes it, releases products, and is recycled as free E* still anchored at the interface. E* can bind another phospholipid in the active site and cleave the appropriate ester bond, or it can dissociate from the interface after the catalytic cycle. If E* hydrolyzes several phospholipid molecules before dissociating, it is working in a "scooting" mode, while if it dissociates from one interface and rebinds to another in between catalytic cycles, the enzyme is working in a "hopping" mode.

Understanding these different kinetic effects is important in documenting potent, specific inhibitors of phospholipases. Unlike the action of inhibitors on soluble enzymes, phospholipase inhibitors may function by preventing enzyme adsorption to the surface as well as by preventing substrate from binding in the active site. One can consider this a form of allosteric regulation.

IN VITRO ASSAY SYSTEMS.   While the natural substrates of phospholipases are normally phospholipids organized in a bilayer, the enzyme specific activities are often quite low. This may be critical for keeping phospholipases

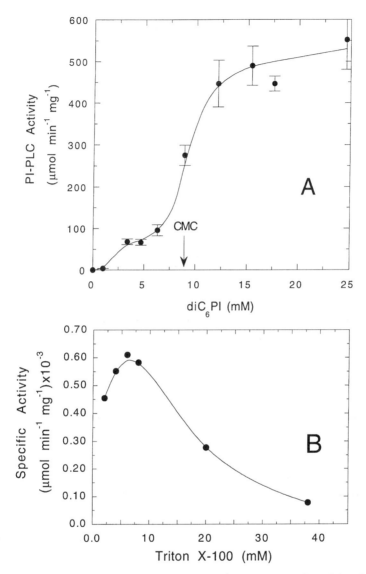

Figure 5-3. (A) Interfacial activation of sPI-PLC (*B. thuringiensis*) as shown by the dependence of enzyme activity on diC$_6$PI concentration; the CMC of the substrate in this buffer is indicated by the arrow (adapted from Lewis et al 1993). (B) Surface dilution kinetics of sPI-PLC hydrolyzing a fixed amount of bovine brain PI (4 mM) solubilized in varying concentrations of Triton X-100 (adapted from Zhou et al 1997b).

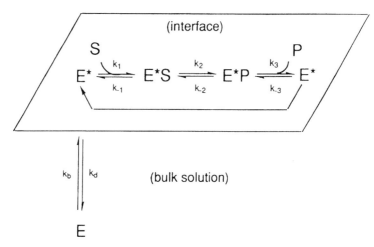

Figure 5-4. Minimum kinetic pathways to be considered for an interfacially active enzyme: S, phospholipid substrate; P, product; E*, surface-activated enzyme.

from randomly destroying cell membranes. The tighter packing of phospholipids in a model bilayer makes it difficult for these surface-active but not membrane-anchored enzymes to bind substrate productively. With this in mind, there is no universal assay system for phospholipases. Rather there are well-defined systems, each with advantages and disadvantages, that have been developed for specific purposes. The general types of phospholipid substrates (summarized in Figure 5-5) include: (1) phospholipid monolayers, (2) synthetic short-chain phospholipids that exist in solution as monomers and aggregate above a critical micelle concentration to form rod-shaped micelles, (3) detergent-mixed micelles, which present the phospholipid substrate in a micellar matrix surrounded by various amounts of detergents (e.g., Triton X-100, deoxycholate or cholate), (4) bilayer vesicles (further delineated by size: SUV, small unilamellar vesicle, and LUV, large unilamellar vesicle), composed either of a single phospholipid or of binary mixtures, and (5) natural membranes (e.g., autoclaved *E. coli*) providing accessible substrates for a number of phospholipases. For many phospholipases, chemical species that contain the appropriate ester bond (as well as a chromophore for ease in monitoring hydrolysis) but which do not aggregate can be used as well (a good example is a 7-hydroxycoumarin ester, which is a completely soluble substrate for cytosolic PLA_2, and p-nitrophenylphosphocholine, a substrate for some PC-specific PLC enzymes).

Monolayers are not useful for routine assays of phospholipases that would be needed in purifying such an enzyme. Aside from the surface film balance, which is not a routine piece of equipment, the substrate and enzyme must be pure and must not introduce any surface-active artifacts. However,

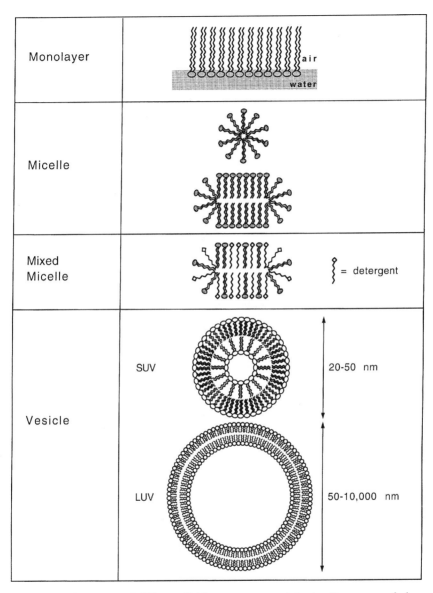

Figure 5-5. Summary of different lipid aggregates used for *in vitro* assays of phospholipases.

the technique is excellent for determining how products affect the interface packing and if the protein inserts into the phospholipid monolayer when interacting with substrate or substrate analogs (if a nonhydrolyzable analog is used, a change in surface pressure can be used to estimate how much of the protein could be inserted).

Synthetic short-chain phospholipids with fatty acyl chains composed of four or five carbons (diC_nPX where X is the base esterified to the phosphate, and n is the number of carbons in the fatty acyl chains) do not aggregate in aqueous solution at concentrations under 20 mM. For these phospholipids, the concentration for the monomer-to-micelle transition (critical micelle concentration or CMC) is extremely high, and the substrate is exclusively a monomer. While an unnatural substrate for phospholipases, diC_4PC or diC_5PC (or the other head group counterparts) can be used for mechanistic studies without the complication of an interface. Synthetic phospholipids with slightly longer chains provide the ideal kinetic system for determining if a given phospholipase displays "interfacial activation" (Roberts 1991b). The CMC values for diC_nPC, diC_nPA, diC_nPS (diacylphosphatidylserine), and diC_nPI compounds have been determined (Roberts 1991b, Lewis et al 1993, Garigapati et al 1995). They vary depending on acyl chain length (Bian and Roberts 1992). Above the CMC, these phospholipids form rod-shaped micelles whose size distribution is well understood (Lin et al 1987). DiC_6PC is an excellent candidate for interfacial activation studies since it has a moderately high CMC (14 mM) and is useful for assessing a K_m and V_{max} for the enzyme acting on a monomeric substrate (as long as the K_m is well below the CMC, e.g., for $K_m < 5$ mM). The anionic short-chain lipids diC_6PA and diC_6PS under low ionic strength conditions have similar CMCs and can be used to monitor the occurrence of interfacial activation of the enzymes towards negatively charged substrates. Enzyme activity with these substrates is usually measured by the pH-stat technique, which measures the amount of base needed to maintain a constant pH as fatty acid product is generated. ^{31}P NMR spectroscopy is also a useful technique for monitoring substrate hydrolysis, since the phosphorus resonances for all the products of different phospholipase actions have distinct chemical shifts from the substrate phospholipid (Roberts 1991a). This latter assay technique is appropriate for small unilamellar vesicles and mixed micelle substrates as well.

Detergent mixed micelles have been developed as easily prepared, excellent substrates for routine assays of phospholipases. For most phospholipases, zwitterionic phospholipids (PC or PE) packed in bilayer vesicles without other added cofactors are poor substrates. This is primarily due to poor adsorption of the enzyme on the vesicle surface (i.e., most of the enzyme exists in the E and not the E* state), and inaccessibility of the substrate (tighter packing of the headgroups and interchain interactions may make it difficult for a phospholipase to bind a substrate molecule productively). As detergent is added to a PC bilayer vesicle system, phospholipase activity initially increases as bilayers are solubilized in mixed micelles, then decreases as the surface concentration of substrate decreases (the "surface dilution" effect). Rapid exchange and mixing dynamics of phospholipids and detergent molecules in these micelles ensure that the enzyme is not interacting with a particle depleted in substrate (Soltys and Roberts 1994). While this system is excellent for monitoring phospholipases in purification schemes, it can also be used to extract detailed kinetic parameters (Carman et al 1995). When DAG is either a product, substrate, or cofactor for an enzyme reaction, solubilization and exchange behavior of this weakly polar lipid may be kinetically important, and detergent-mixed micelles are in general a better system than bilayers. One way of determining if DAG dynamics are relevant in a system is examining kinetics in different detergent mixed micelles. In general, Triton X-100 appears optimal for both DAG solubilization and exchange. Deoxycholate (as well as other bile salts such as taurocholate and CHAPS), a common detergent used in different assays, may be a poor choice because of its poor DAG capacity and slow exchange kinetics. However, if mixed micelles with this detergent exhibit comparable activity to other detergent assay systems, then micelle component exchange cannot have any kinetic significance (Zhou and Roberts 1997).

Bilayer vesicles have been problematic for kinetic analyses. PC vesicles are poor substrates for some phospholipases, in part because the enzyme has a lower affinity for a zwitterionic tightly packed surface. PLA_2 incubated with unilamellar vesicles of PC shows little activity initially, but after a lag phase, τ, rapid hydrolysis occurs (Biltonen et al 1991). The end of the lag phase correlates with product accumulation that increases the affinity of the enzyme for the surface as well as for the substrate (Sheffield et al 1995). There are other pathways to reduce τ, for example, the addition of anionic amphiphiles or acylation of the enzyme, or increasing vesicle curvature (SUVs are better substrates than LUVs). The nonspecific PLC from *Bacillus cereus* also exhibits a lag phase when large unilamellar vesicles are used as the substrate (Basanez et al 1996b). For PLA_2, lag phases are not observed when anionic phospholipids are incorporated into vesicles, and the activity observed is quite high. The higher apparent activity of the enzyme toward anionic vesicles has been attributed to the fact that the enzyme is effectively adsorbed onto the negatively charged vesicle surface and can hydrolyze many substrate molecules without dissociating; i.e., enzyme is now in the E* state and can work in a processive mode. Detailed methodologies have been developed to extract both surface and catalytic site binding parameters with this system (Gelb et al 1995). To compare substrate specificity or assess competitive inhibitors without the complications of affecting enzyme dissociation from the vesicle surface, this is probably the preferred assay system. Again, pH-stat techniques are usually used to monitor product release, although a number of fluorescently labeled substrates can be solubilized in nonsubstrate anionic vesicles for a more sensitive assay. This modified vesicle assay can be particularly useful for phospholipases that have very specific substrates (e.g., PI, PIP, or PIP_2 in PC vesicles) as long as the appropriate chromophore can be introduced into the substrate, and as long as the other components of the vesicle act as neutral diluents (i.e., have no affinity for the active site). More recently, mathematical analyses have been developed to extract kinetic parameters for systems in which the neutral diluent is imperfect and has some affinity for the active site (Yu et al 1997). This treatment may be applicable to a wide range of phospholipase systems.

Natural membranes prepared from intact cells (e.g., *E. coli* or erythrocytes) are less effective assay systems for extracting kinetic constants, since it is hard to control the surface concentration of a specific phospholipid, and the effects of nonphospholipid components may affect the packing of the bilayer. However, they can be extremely sensitive if the phospholipids are radiolabeled. Many investigators use these membranes because they feel that they best approximate what the enzymes are dealing with *in situ*. A variation on using natural membranes has been developed to monitor PLC and PLD activity in plasma membrane preparations (Lucas et al 1995). This method uses peroxidase-catalyzed chemiluminescent oxidation of luminol by the H_2O_2 derived from choline oxidation to monitor endogenous PLD activity. For PLC activity, the same preparation is treated with phosphatase to convert phosphocholine to choline. The H_2O_2 detected reflects the sum of PLC and PLD activities. This sensitive assay may prove very useful for these two activities since it does not resort to radiolabeling or extensive chromatographic separation steps.

STRUCTURE/FUNCTION STUDIES OF PHOSPHOLIPASES

A complete understanding of all of the kinetic hallmarks of phospholipases and the role of these enzymes in signal transduction can only be achieved with a molecular-level structural analysis of these enzymes and their diverse effectors. The following sections provide an overview of phospholipase structural information, catalytic mechanisms and known inhibitors, interaction/activation by other species (lipids, proteins, etc.), and a discussion of the role of these enzymes in signal transduction.

1. PHOSPHOLIPASE A_2

Three distinct types of fatty acylases have been found that cleave the fatty acid esterified to the *sn*-2 position of phospholipids: (1) secretory PLA_2 ($sPLA_2$), small disulfide cross-linked proteins secreted by cells that require mM Ca^{2+} for optimal activity (these are the best-studied PLA_2 and offer a paradigm for understanding interfacial activation on a molecular level), (2) cytoplasmic PLA_2 ($cPLA_2$), larger proteins with distinct Ca^{2+}-binding (requiring µM rather than mM Ca^{2+}) and catalytic domains (but not 3-D structure), and a mechanism which involves a covalent fatty acyl serine, and (3) intracellular Ca^{2+}-independent PLA_2 ($iPLA_2$), moderate-sized proteins with little known about their catalytic mechanism or structure. For a recent review of the superfamily of PLA_2 signal transduction enzymes, see Dennis (1997). A summary of the different types and characteristics of PLA_2 enzymes is presented in Table 5-1.

Venom and Pancreatic sPLA$_2$

STRUCTURE AND MECHANISM. Secreted forms of PLA_2 ($sPLA_2$) have been detected and isolated from a wide variety of sources, including snake, bee and wasp venoms, pancreas, kidney, bacteria, and virtually every mammalian tissue that has been studied. Typically, these enzymes have molecular weights between 13 and 15 kDa. There is substantial homology among the different types of secreted PLA_2 enzymes. Best studied are the enzymes from snake venoms and from pancreatic tissues. The physiological role of pancreatic PLA_2 is the initial digestion of phospholipid components in dietary fat, whereas the toxins in snake venoms serve to immobilize prey, promoting cell lysis and membrane disruption. The pancreatic $sPLA_2$ is produced as a proenzyme requiring cleavage of the terminus for activation. $sPLA_2$ enzymes have an absolute requirement for Ca^{2+}, typically needing concentrations of ~1 mM for maximum activity. This divalent metal ion is a key component of the active site. In some cases it may also promote binding of the enzyme to its substrate, although many $sPLA_2$ enzymes can bind to phospholipid aggregates without Ca^{2+} present.

The overall structure of all $sPLA_2$ enzymes examined is similar (Scott and Sigler 1995). The pancreatic enzymes are monomeric (Dijkstra et al 1983), while the venom enzymes tend to aggregate in dimer or trimer units (Brunie et al 1985). The oligomeric state for some of the venom enzymes appears to have no functional significance in the catalytic process. Rather, it may occur *in vivo* to maintain the protein in an inactive state until activity is required. These $sPLA_2$ proteins consist of three major and two minor α-helical segments, a double-stranded antiparallel β-sheet, a primary Ca^{2+}-binding loop (and in some structures a second Ca^{2+} site), and seven disulfide bonds (Figure 5-6A). Several of the enzymes have been crystallized with phospholipid analogs present, either an inhibitor (p-bromophenacyl bromide, covalently bound to the active site H48 at N^{δ} [Renetseder et al 1988]), a nonhydrolyzable substrate analog (Thunnissen et al 1990), or a transition state analog (L-1-O-octyl-2-heptylphosphonyl-*sn*-glycero-3-phosphoethanolamine), where a tetrahedral phosphonate replaces the *sn*-2 carbonyl group (White et al 1990). These x-ray structures all show a similar cleft lined with hydrophobic residues with H48 in the deepest part of the cleft, hydrogen bonded to a water molecule, and an aspartic acid residue in an orienta-

Table 5-1. Characteristics of phospholipase A_2-type enzymes

Type	Subunit MW (kDa)	Cofactor	Mechanism	Domain Structure	Regulation
$sPLA_2$ venom and pancreas	13–15	Ca^{2+} (mM)	Attack of H_2O to form a tetrahedral oxyanion transition state	Interfacial binding region distinct from catalytic site	Physical state of bilayer; affinity for anionic phospholipids
$hsPLA_2$	14	Ca^{2+} (mM)	Tetrahedral oxyanion transition state	Shallower catalytic site than venom enzymes	Ca^{2+}; binds to receptor
$sPS-PLA_2$	55	?	Lipase-like mechanism?	?	Specificity for PS and lyso-PS
$cPLA_2$	85	Ca^{2+} (μM)	Formation of fatty acyl intermediate via active site serine (lipase-like)	CaLB (or C2) distinct from active-site domain	Ca^{2+} binding to membrane; phosphorylation
$iPLA_2$	80	–	Lipase-like mechanism	Ankryin repeats	?
$iPLA_2$ (plasmalogen)	39	–	Lipase-like mechanism	?	Selectivity for plasmalogens
PAF-PLA_2					Interaction with ARK, Tec11a, dynamin, spectrin β-chain?
Ia	29, 30		Covalent acyl enzyme intermediate		
Ib	29, 30, 45	–			
II	40			WD-40 repeat	WD-40 sequence that may bind to PH domains

tion suggestive of what is observed in serine proteases. The active site Ca^{2+} is hepta-coordinated in a pentagonal bipyramidal cage formed by the backbone carbonyls of the calcium-binding loop, D49, and two oxygen atoms of the inhibitor. In the venom structure a secondary Ca^{2+} site is 6.6 Å from the catalytic Ca^{2+}, has five ligands instead of seven, and is more loosely bound.

Interactions of the substrate analogs with the protein shed light on the mechanism of the enzyme. With the exception of the phosphate group, there are few well-defined interactions of the lipid headgroup with the protein. The ethanolamine amino group of the transition state analog is hydrogen-bonded to N53; however, this interaction cannot occur with a choline headgroup when the substrate is PC. The lack of specific interactions is consistent with the lack of substrate specificity for these $sPLA_2$ enzymes. The sn-1 chain is more disordered than the sn-2 chain and has fewer interactions with the protein; methylene and methyl groups furthest from the glycerol are least well ordered. Both fatty acyl chains occupy a hydrophobic channel that extends 14 Å from the active site histidine. The left wall of the channel is formed from a potentially mobile segment that is held in place when substrate is bound. The presence of the hydrophobic channel provides an explanation for the interfacial activation of $sPLA_2$ enzymes. In order to reach H48 and the other catalytic machinery, a substrate must diffuse from the aggregate matrix into the hydrophobic channel (Scott et al 1990). Polar groups can easily diffuse into the channel, since the left wall of the channel is not completely formed until the substrate is bound and anchored to the Ca^{2+}. The "seal" provided by the binding of the enzyme to the interface allows the transfer of substrate into the active site without solvation of the hydrophobic alkyl substituents (Figure 5-6B). For a monomeric substrate, the lipid would be solvated, and this water would need to be removed. Experiments testing this concept using thiol cross-linkable phospholipids are consistent with this view (Soltys et al 1993).

None of the crystal structures can address the conformation of the enzyme in the presence of an interface directly, since only monomeric phospholipid analog could be bound (the two-dimensional order of micelles or bilayers makes it difficult to crystallize enzymes in their presence). However, solution NMR spectroscopy can provide conformational details for an enzyme this size. Verheij and coworkers have analyzed the solution structure of $sPLA_2$ for the protein free, bound to micelles, and in a micellar environment in the presence of a substrate analog (Van den Berg et al 1995a,b). The N-terminal portion of peptide occupies a slightly different and more disordered position in the solution structure than in the x-ray structure. When nonsubstrate micelles (an n-alkyl phosphorylcholine) are present, there are no changes in the protein structure. However, when the active-site ligand is present, the structure tightens up and bears a stronger resemblance to the x-ray structure. The N-terminal part of the protein moves inward to be part of an α-helix, and an unusual nuclear Overhauser effect (NOE) enhance-

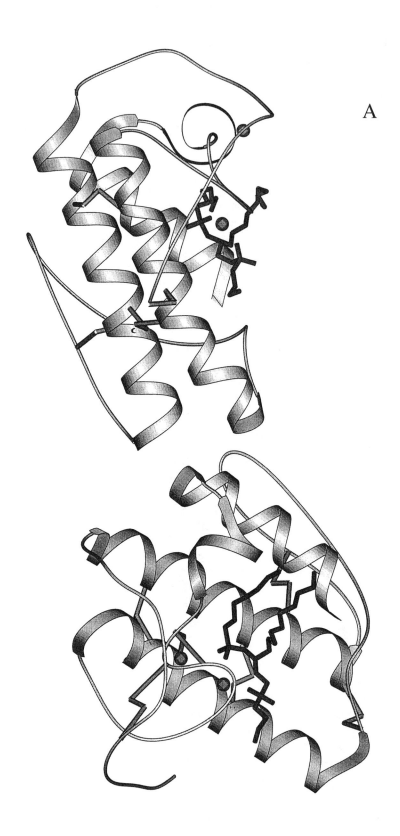

A

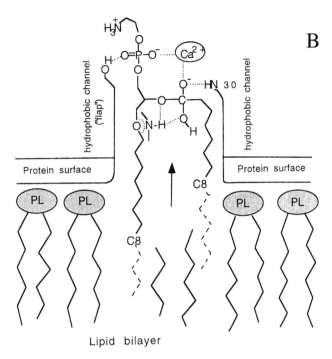

Figure 5-6. (A) A ribbon diagram of the human synovial fluid sPLA$_2$ complexed with a transition-state inhibitor and Ca^{2+} (round balls); the two protein monomers in the unit cell are shown (adapted from Scott et al 1991). (B) Model for interfacial catalysis derived from crystal structures of sPLA$_2$ enzymes (adapted from Scott et al 1990).

ment involving the A-1 amino group and a hydrogen bond for $^{15}N/^{1}H$ for H48-D99 is detected. N-H interactions of free amino groups are usually obscured by rapid exchange with water. If detected they indicate a relatively long-lived structure where exchange with solvent has been minimized. The net result of adding the ligand to the active site is that the sPLA$_2$ structure in solution is rigidified and shows the same interactions as in the x-ray structure. Thus, micelle binding not involving the active site appears to be a relatively nonspecific process that does not change the dynamics of the protein. Only when the active site is occupied is the optimal protein conformation achieved.

Mutagenesis studies of PLA$_2$ have concentrated on many of the catalytic residues, Ca^{2+} ligands and components of the hydrophobic channel (for a recent summary see Roberts 1996). The interfacial binding site of sPLA$_2$ has also been mutated. To separate kinetic effects at the active-site binding from those attributed to interfacial binding, a polymerized PC matrix has been developed by Cho and coworkers (Dua et al 1995). The fatty acyl chains of phospholipids are modified to contain a terminal lipoic acid residue, which can be oxidized to form a cross-linked vesicle matrix. The polymerized phospholipids are not hydrolyzed by PLA_2. A pyrene-PC (or other pyrene-labeled phospholipid) is inserted into the polymerized vesicle; this substrate is homogeneously distributed if it represents less than 10% of the total lipid. Bovine serum albumin is present to selectively bind the pyrene-labeled fatty acid product of PLA_2 action (extraction of fatty acid is not a rate-limiting step). Use of this system has shown that the $sPLA_2$ interfacial binding sites contain cationic residues. The spatial distribution of the charged residues differs with the $sPLA_2$ species. Hence, for interfacial binding, it is the surface electrostatic potential that is important for anchoring the enzyme to the interface, and a well-defined three-dimensional architecture ("lock/key"-type binding site) is less so (Han et al 1997).

PHYSICAL PROPERTIES OF THE BILAYER REGULATE $sPLA_2$. One of the more interesting kinetic hallmarks of $sPLA_2$ is the lag phase exhibited in the hydrolysis of PC vesicles. The end of the lag phase correlates with product accumulation that enhances binding of the enzyme to the PC vesicles (Sheffield et al 1995). Products reduce τ, but there is a threshold effect. For example with PLA_2 from *Agkistrodon piscivorus piscivorus* acting on large unilamellar vesicles of dimyristoyl-PC or dipalmitoyl-PC (DPPC), 0.07 mole fraction lyso-PC reduces τ from 1500 to 150–200 s; 0.08 mole fraction lyso-PC abolishes the lag. The PLA_2 products can enhance enzyme binding to the vesicle surface (characterized by an equilibrium binding constant K_B) or enhance the catalytic step (formation of E^*, a change in the enzyme bound to the vesicle that is characterized by an equilibrium constant K^*). Increasing K_B or K^* will increase enzyme binding and decrease τ. However, for the step characterized by K_B, as the concentration of substrate vesicles increases and approaches infinity, the observed specific activity should approach V_{max} if products alter K_B. Since PLA_2 activity before τ doesn't increase at high DPPC, binding to the surface can't be a major cause of the increased activity (Sheffield et al 1995). Fluorescence studies suggest that lyso-PC (and fatty acid synergistically) induces curvature in the bilayer and changes hydration around the phospholipid.

Mammalian $sPLA_2$

SECRETED ENZYMES AND THE INFLAMMATORY RESPONSE. Secreted PLA_2 activities with many characteristics similar to the venom and pancreatic enzymes also occur in mammalian tissues and have been well studied because of their potential role in the inflammatory response (Kudo et al 1993). An acid-stable, Ca^{2+}-dependent (requiring mM Ca^{2+} levels) $sPLA_2$ (14 kDa) is released from secretory vesicles when a variety of cells is stimulated in a dose-dependent fashion. This activity is often denoted $hsPLA_2$, if the enzyme is from human cells. $sPLA_2$ is overexpressed in patients with sepsis, pancreatitis, and rheumatoid arthritis, suggesting that it may have a role in those disease states (although it does not appear to be, the PLA_2 that generates arachidonic acid). $sPLA_2$ enzymes have extensive homology with the well-studied group II (snake venom) and group I (pancreatic) enzymes. Crystal structures exist for $sPLA_2$ from human rheumatoid arthritic synovial fluid (Wery et al 1991, Scott et al 1991). One noticeable difference between the mammalian and venom enzymes is that the hydrophobic channel is smaller in the synovial fluid enzyme. One synovial fluid PLA_2 structure (Wery et al 1991) has no bound Ca^{2+} and is quite similar to the structure of bovine pancreatic or snake venom enzymes. The active-site H48 is hydrogen-bonded to a water molecule, and the hydrophobic channel is quite distinct. The two structures from the Sigler group (Scott et al 1991) show the enzyme in a Ca^{2+}-bound form in the absence and presence of a transition-state analog. Although the structural features consistent with a catalytic mechanism similar to other extracellular PLA_2 enzymes are present, the shape of the hydrophobic channel appears to be uniquely modulated by substrate binding.

ROLE OF $sPLA_2$ IN SIGNAL TRANSDUCTION PATHWAYS.   The $hsPLA_2$ family of enzymes clearly plays a role in generating proinflammatory lipid mediators. Details of how the secretory or venom PLA_2 enzymes affect cellular activities are sparse, although receptors for extracellular PLA_2s (including type I, $proPLA_2$-I, and synovial PLA_2-II) have been identified (Ishizaki et al 1994, Lambeau et al 1994). A number of studies have shown that $sPLA_2$-type enzymes have receptor-mediated effects as well as phospholipid digestive functions. For example, the binding of $sPLA_2$-I to the receptor appears to induce cell proliferation and muscle contraction. Other effects of $sPLA_2$ include stimulating invasion of an artificial extracellular matrix by tumor cells that express a high-affinity PLA_2 receptor (Kundu and Mukherjee 1977). This latter effect provides a new physiological role for $sPLA_2$ enzymes. Another role for $sPLA_2$, the level of which is increased in the acute response phase, may be to promote phagocytosis of injured cells and tissue debris. This would enhance inflammation and tissue damage (Hack et al 1997).

Yet another way $hsPLA_2$ is involved in signaling pathways is in the generation of lyso-PA. Human $sPLA_2$-II has a high specificity for PA (Snitko et al 1997). The action of this enzyme on PA in the external leaflet of the membrane would produce lyso-PA, a potent lipid second messenger. $hsPLA_2$ could represent a major path to make this lipid.

NOVEL MAMMALIAN $sPLA_2$.   Recently, a new $sPLA_2$ has been identified that is quite different from all the characterized $sPLA_2$ enzymes. The enzyme, with a molecular weight of 55 kDa, is secreted by rat platelets and is phosphatidylserine (PS) specific (Sato et al 1997). The enzyme has the typical GXSXG sequence of lipases and is inhibited by DFP (diisopropyl fluorophosphate), suggesting that it works via a covalent fatty acyl intermediate (presumably serine).

Recombinant protein hydrolyzed the *sn*-1 fatty acid of lyso-PS and PS, but not PC, PE, PI, PA, or triglyceride. This is unique kinetic behavior compared to other members of the sPLA$_2$ family.

cPLA$_2$

STRUCTURE AND MECHANISM.   Intracellular Ca^{2+}-dependent PLA$_2$ has virtually no homology to sPLA$_2$ enzymes. These enzymes are much larger (typically ~85 kDa) than the sPLA$_2$ and appear to have discrete domains. Along with an active-site domain, cPLA$_2$ enzymes possess a Ca^{2+}-lipid binding domain (CaLB) that mutagenesis shows is responsible for binding the protein to the phospholipid interface in a Ca^{2+}-dependent manner (Clark et al 1991). The CaLB of cPLA$_2$ has homology with the CaLB or C$_2$ domain of other proteins involved in signal transduction, including protein kinase C, UNC-13, synaptotagmin, rabphillin-3A, GAP, and PLC-γ. The CaLB domain from synaptotagmin (residues 140–267) is characterized as a curved β-sandwich of two sheets (Sutton et al 1995). The most highly conserved stretch in other CaLB sequences is around the Ca^{2+}-binding ligands; in synaptotagmin, this region is at the top of the sandwich and is not well ordered. The fact that such a motif can subsequently interact with phospholipids in a membrane—either anionic ones or the phosphate moiety of PC, PE, or sphingomyelin—was used to model the cPLA$_2$ CaLB domain (Sutton et al 1995).

The cPLA$_2$ from platelets has a sequence reminiscent of the active site of lipases (Kramer et al 1991). Indeed, phospholipid hydrolysis proceeds by a different catalytic mechanism, one that generates a covalent fatty acyl intermediate (presumably at S228, since it is absolutely required for enzyme activity) (Sharp et al 1994). Unlike the sPLA$_2$, cPLA$_2$ enzymes are positionally sloppy and display lysophospholipase and transacylase activity (Kramer et al 1991, Reynolds et al 1993). The substituent in the *sn*-1 position of the substrate has little effect on enzyme specific activity. However, there is a marked preference for unsaturation at the *sn*-2 position. Arachidonic acid (20:4) is the optimal substituent in the *sn*-2 chain. Ca^{2+} ions activate the enzyme in a biphasic fashion (0.1–1 μM and 0.1–10 mM). The optimum activity occurs at pH 9–10, meaning that the enzyme is 50% active at physiological pH values.

MODES OF REGULATION.   cPLA$_2$ is regulated by intracellular Ca^{2+} and by phosphorylation (Clark et al 1991.) Both these effects involve changing the partitioning of the enzyme at the bilayer interface. In unstimulated cells

$cPLA_2$ is found in the cytosol, with some fraction associated with the membrane. Upon stimulation of cells, $cPLA_2$ associated with the membrane fraction is significantly increased. The Ca^{2+} functions by binding to the CaLB domain and the complex, then interacting with the bilayer. Consistent with this are the observations that (1) a mutant, ΔC_2, with the CaLB domain deleted still has catalytic activity but does not exhibit interfacial activation, and (2) a construct of CaLB fused to maltose-binding protein associates with membranes in a Ca^{2+}-dependent manner (Clark et al 1991).

While the CALB domain regulates $cPLA_2$ activity by anchoring protein to substrate membrane, phosphorylation also affects activity, but by a different mechanism. PKC and related kinases such as mitogen-activated kinase (MAP kinase) modulated by G-proteins phosphorylate $cPLA_2$ at S505. When phosphorylated by MAP kinase, $cPLA_2$ exhibits a 30% increase in the rate of hydrolysis of a sn-2 arachidonyl-PC (Bayburt and Gelb 1997). The molecular mechanism of the activation is unclear, since phosphorylation has no effect on the catalytic efficiency towards a soluble substrate. The phosphorylated enzyme also loses its Ca^{2+} sensitivity (Qiu et al 1993).

$iPLA_2$

Yet another distinct type of PLA_2 activity is found in cells. The defining characteristic of this class of PLA_2 is that it has no requirement for Ca^{2+} as a cofactor (hence the terminology $iPLA_2$ for Ca^{2+}-independent PLA_2). The most common $iPLA_2$ is a multimer of 270–350 kDa that is composed of 80–85 kDa subunits. Activity is specifically inhibited by bromoenol lactones (this inhibitor appears to be fairly diagnostic for $iPLA_2$). Originally isolated and cloned from CHO cells (Tang et al 1997), the enzyme has a lipase motif ($GXS_{465}XG$), suggesting a lipase-like mechanism involving the formation of a fatty-acyl-enzyme intermediate, and eight ankyrin repeats. Ankyrin repeat domains are known to bind tubulin and integral membrane proteins, specifically proteins involved in ion fluxes. Similar $iPLA_2$ proteins have also been found in P388D1 macrophages (Ackermann et al 1994, Balboa et al 1997), human platelets (Corssen 1996), and pancreatic islets (Ma et al 1997). The role of the $iPLA_2$ in P388D1 macrophages has been proposed to be regulation of basal phospholipid remodeling reactions. In general, $iPLA_2$ enzymes exhibit little specificity for the identity of the fatty acid esterified to the sn-2 position, although there may be a preference for specific headgroups. The enzyme from CHO cells has a marked preference for PA (generating lyso-PA); it also hydrolyzes PAF. CHO-derived $iPLA_2$ interacts with calmodulin and ATP, independent of regulatory proteins (Wolf and Gross 1996).

A smaller (~39 kDa) $iPLA_2$ enzyme is localized in the cytosol and is thought to play a role in releasing fatty acids during ischemic injury and reperfusion (Farooqui et al 1995). The enzyme exhibits specificity towards plasmalogens. The breakdown of plasmalogens in neural membranes during neurodegenerative diseases is receptor-mediated. The selective hydrolysis of ethanolamine plasmalogens by iPLA$_2$ could explain why ethanolamine plasmalogens are low in affected brain regions in Alzheimer's disease (Farooqui et al 1997). Such an iPLA$_2$ activity could be responsible for arachidonic acid release and accumulation of prostaglandins and lipid peroxides in such tissue. Glycosylaminoglycans are potent inhibitors of this iPLA$_2$, and these molecules may be the endogenous regulators of enzyme activity.

Another type of Ca^{2+}-independent PLA$_2$ has been identified in lung alveolar type II cells (Fisher et al 1992, 1994). This enzyme, termed yPLA$_2$ for lysosomal-type PLA$_2$), has an acidic pH optimum, is insensitive to pBPB (a good inhibitor of sPLA$_2$), and is inhibited *in vitro* by the transition state analog MJ33 (1-hexadecyl-3-trifluoroethylglycero-*sn*-2-phosphomethanol) and by surfactant protein A (SP-A). This PLA$_2$ does not have a preference for PCs with arachidonic acid esterified to the *sn*-2 position. Instead, yPLA$_2$ degrades dipalmitoyl-PC, a major lipid in lung cells. It has been suggested that it is this PLA$_2$-type activity that is responsible for degradation of internalized dipalmitoyl-PC. iPLA$_2$ has a role in disaturated PC (e.g., dipalmitoyl-PC) synthesis in granular pneumocytes (Fisher and Dodia 1997).

PAF-PLA$_2$

STRUCTURE AND MECHANISM.   Platelet activating factor (PAF) is a phospholipid that is a potent cell activator. A PLA$_2$-type activity specific for this lipid, labeled PAF acetylhydrolase (PAF-AH or PAF-PLA$_2$), modulates PAF levels in a wide variety of mammalian cells. The enzyme PAF-PLA$_2$ occurs as intra- and extracellular secreted isoforms (Hattori et al 1993 1994a, 1995): Ia, Ib, and II. Isoform Ia is a heterodimer of 29 kDa and 30 kDa subunits. Isoform Ib is a heterotrimeric enzyme with subunits of 29 (γ), 30 (β), and 45 (α) kDa; it is distinguished kinetically by its high specificity for PAF and its inability to cleave propionoyl or butyroyl chains at the *sn*-2 position. Isoform II is a multimer of a single 40-kDa polypeptide (termed α). It has a broader substrate specificity and hydrolyzes oxidized PC. Both serine and cysteine residues are critical for catalysis as judged by chemical modification. From the cDNA the likely active site S47 is found in a typical lipase sequence of GXSXG, with the region after the active site looking like the first transmembrane region of the PAF receptor.

REGULATION AND RELATIONSHIP TO SIGNAL TRANSDUCTION PATHWAYS. The subunit of isoform II contains a WD-40 repeat sequence similar to that found in many trimeric G-proteins. This motif appears to modulate protein-protein interactions. PAF-AH α interacts with ARK, spectrin β-chain, Tec 11a, and dynamin—all with distinct regulatory features in signal transduction pathways. This PAF-AH subunit matches the product of the causative gene for Miller-Dieker Syndrome (Hattori et al 1994b), a disease that causes smooth brain structures. Such an identification implies that PAF-AH has a key role in neuronal cell migration and brain development, even though it is found in a wide variety of tissues.

Specific Inhibitors of PLA$_2$ Enzymes

One of the developments that has allowed separation of the different PLA$_2$ activities in cells is the availability of relatively specific inhibitors for different classes. A summary of the specificity of available inhibitors is shown in Table 5-2. For sPLA$_2$, *p*-bromophenacyl bromide (pBPB) and related compounds are relatively potent inhibitors. They form covalent adducts at the active site and are not inhibitory for cPLA$_2$ enzymes; however, they have been shown to inhibit some iPLA$_2$ activities. MJ33, 1-hexadecyl-3-trifluoroethylglycero-*sn*-2-phosphomethanol, has a broad specificity for many types of PLA$_2$ enzymes, including most sPLA$_2$ forms and at least one type of iPLA$_2$ (from lung alveolar type II cells [Fisher et al 1992]).

For cPLA$_2$ several inhibitors have been identified, including N-ethyl-maleimide (potent but not very specific for cPLA$_2$) and arachidonyl trifluoromethyl ketone (AACOCF$_3$). Even though the enzyme forms an acyl serine as an intermediate (similar to the covalent intermediate of serine proteases), diisopropylfluorophosphate (DFP), a good inhibitor of serine proteases is not particularly effective towards cPLA$_2$. AACOCF$_3$ is a fluoroketone inhibitor that exists as a hydrate in solution and when it is initially bound to enzyme. Over time it forms an enzyme-bound hemiketal (Street et al 1993). Another good inhibitor of cPLA$_2$ (with unclear specificity, since it may be active towards other serine-based and arachidonic-acid-

Table 5-2. Specificity of different phospholipase A$_2$ inhibitors

Compound[a]	Inhibitor for			
	sPLA$_2$	cPLA$_2$	iPLA$_2$	PAF-PLA$_2$
AACOCF$_3$	−	+	+	+
MAFP	−	+	+	+
BEL	−	−	+	?
p-BPB	+	−	−/+[b]	−
MJ33	−	+	+	?
DFP	−	−	?	+

[a] Abbreviations: AACOCF$_3$, arachidonyl trifluoromethyl ketone; MAFP, methyl arachidonylfluorophosphate; BEL, bromoenol lactone; p-BPB, p-bromophenacyl bromide; MJ33, 1-hexadecyl-3-trifluoroethylglycerol-*sn*-2-phospho-methanol; DFP, diisopropylfluorophosphate

[b] iPLA$_2$ from lung aveolar type II cells is inhibited by this compound.

selective targets) is methyl arachidonylfluorophosphate (MAFP), which is an arachidonic-acid-site-directed phosphorylation reagent.

iPLA$_2$ activities are very specifically inactivated by bromoenol lactones (BEL), which act as suicide inhibitors. This type of compound is very specific for iPLA$_2$ and does not appear to inhibit either sPLA$_2$ or cPLA$_2$. When introduced into cells, it allows one to separate iPLA$_2$ effects on cell physiology from both sPLA$_2$ and cPLA$_2$ effects. However, several of the inhibitors that are potent for cPLA$_2$ (AACOCF$_3$ and MAFP) can also inhibit iPLA$_2$.

There are a number of inhibitors of PLA$_2$ activity that are often used in studying activation of these enzymes *in vivo*. Mepacrine and aristolochic acid as well as scalaradial inhibit arachidonic acid release, and hence are considered PLA$_2$ inhibitors (see Liu and Levy 1997). The exact point where and mechanism by which they inhibit PLA$_2$ activity are less clear. Rather than directly binding to the enzymes, they may act upstream in the signaling process.

Little is known about specific inhibitors of PAF hydrolase. Group II PAF-AH is Ca^{2+}-independent and resistant to *p*-bromophenacyl bromide, dithionitrobenzoate, and heat, but inhibited by serine protease inhibitors (DFP and PMSF). The observation that DFP inhibits PAF-PLA$_2$ can distinguish it from cPLA$_2$ and to some extent from iPLA$_2$, all of which involve a fatty acyl intermediate. Given the unusual structure of PAF, one could imagine a thio-PAF or amide-linked PAF analog that could be very specific for this PLA$_2$ activity. Aside from these substrate analogs, and once a structure is available for a PAF-PLA$_2$, other compounds may be developed that effectively fill the active site and block substrate binding.

2. PHOSPHOLIPASE A$_1$

Enzymes that cleave the fatty acid from lysophospholipids are designated as PLA$_1$ activities. However, such phospholipases usually hydrolyze phospholipids at the *sn*-2 chain as well, and the lack of specificity is related to the lipase-like mechanism by which hydrolysis proceeds. In this sense one could consider cPLA$_2$ or iPLA$_2$ as having PLA$_1$ activity. Is there a PLA$_1$ activity that is specific for the *sn*-1 chain? A soluble PLA$_1$ activity with a strong preference for PA has been isolated from mature bovine testis (Higgs and Glomset 1994). It is also found in brain but not liver, spleen, or heart. The specificity for PA suggests that this enzyme may regulate the level of PA and lyso-PA (both lipid second messengers that are potent cell activators) in spermatogenesis or in sperm function. Perhaps more intriguingly, this perceived specificity for the *sn*-1 chain may be the result of the conformation of PA in solution. PA is an unusual phospholipid in that it has a glycerol backbone orientation that is parallel to the bilayer plane rather than perpendicular to it, the more common orientation observed for PC, PE, and PS in bilayers (Pascher et al 1992, and references therein). In order for the fatty acyl chains to be packed together, one of the chains must have a kink near the carbonyl. For PC the kink occurs at C-2 of the *sn*-2 chain; for PA the kink occurs at the *sn*-1 C-2. Such a difference in packing of PA versus PC in a bilayer would place the *sn*-1 ester bond of PA at the interface and render it more accessible than the *sn*-2 ester, and could explain the "preferential" hydrolysis of PA and PLA_1 activity.

3. PHOSPHOLIPASE C

Cleavage of phospholipids by PLC enzymes generates the hydrophobic product diacylglycerol (DAG) as well as a water-soluble phosphate ester. Four general classes of this phospholipase have been identified on the basis of substrate specificity: (1) phosphatidylinositol-specific PLC (or PI-PLC), (2) nonspecific PLC (often referred to as PC-PLC), (3) PA-PLC (or PA phosphatase, PAP), and (4) glycosylphosphatidylinositol-specific PLC (GPI-PLC). A summary of the characteristics of PLC enzymes is presented in Table 5-3.

PLC enzymes with specificity for phosphoinositides are the critical components of phosphatidylinositol-mediated signaling pathways. The rapid production of DAG from PI activates PKC, which in turn phosphorylates a variety of proteins. As observed for PLA_2, the ubiquitous PI-PLC enzymes are found in secreted and cytosolic forms. Many microorganisms secrete a PI-PLC (sPI-PLC) that is important for infection by that bacterium (Ikezawa and Taguchi 1981, Low 1981), although the detailed mechanism is not known. These secreted enzymes are also able to cleave GPI-anchored proteins. Intracellular PI-PLC enzymes (cPI-PLC) are ubiquitous in mammalian cells. However, there are also nonspecific PLC enzymes (often characterized by their action towards PC molecules, and hence termed "PC-PLC" enzymes), secreted and probably intracellular as well. Other *sn*-3 phosphodiesterase activities have been characterized, notably a membrane-bound activity (PA phosphatase or PA-PLC) in yeast, which will specifically hydrolyze PA.

PI-PLC

EXTRACELLULAR ENZYMES (sPI-PLC).   Bacterial sPI-PLC enzymes from the human pathogens *Staphylococcus aureus* (Low 1981), *Listeria monocytogenes* (Camilli et al 1991, Menguad et al 1991, Leimeister-Wachter et al 1991), and *Clostridium novyi* (Taguchi and Ikezawa 1978), as well as nonpathogenic *Bacillus cereus, Bacillus thuringiensis* (Griffith et al 1991, Ikezawa and Taguchi 1981) and *Cytophaga* sp. (Artursson and Puu 1992, Jager et al 1991) are water-soluble and specific for nonphosphorylated PI. In *B. thuringiensis* transcription of the gene for PI-PLC is activated at the onset of stationary phase (Lereclus et al 1996), perhaps suggesting a role in organophosphate

Table 5-3. Classification of phospholipase C enzymes

Type	Subunit MW (kDa)	Cofactor	Mechanism	Domain Structure	Regulation
sPI-PLC *Bacillus* sp.	35	–	Two steps: (1) intramolecular phosphotransferase to form cIP; (2) cyclic phosphodiesterase to form IP	α/β barrel; deep active-site pocket	?
cPI-PLC α	57–70	Ca^{2+}	(May not really be a PLC)	?	?
cPI-PLC β	125–154	Ca^{2+}	Assumed similar to sPI-PLC (both cIP and IP products observed)	X, Y, CaLB (or C2), PH, SH3/SH3 domains	Ca^{2+} binding and allosteric binding of PIP_2; activation by Gq proteins
cPI-PLC γ	145	Ca^{2+}	Assumed similar to sPI-PLC (both cIP and IP products observed)	X, Y, PH domains; SH2 and SH3	Ca^{2+} binding; allosteric binding of PIP_2; tyrosine phosphorylation
cPI-PLC δ	85	Ca^{2+}	Similar to sPI-PLC with release of cIP slow compared to attack by H_2O	X, Y catalytic domains; PH and CaLB domains in N-terminal region	Ca^{2+} binding; allosteric binding of PIP_2 and possibly other membrane components
cPI-PLC ε	?	Ca^{2+}	?	?	?
GPI-PLC protozoan	39	–	Similar to sPI-PLC, but integral membrane protein		Covalent modification?
sPC-PLC *B. cereus*	28	Zn^{2+}-metalloenzyme	Attack of H_2O to form pentavalent phosphorus transition state	Nearly all α-helical compact structure; interfacial binding distinct from active-site binding	Physical state of bilayer
PA-PLC (PAP) NEM-sensitive (PAP-1)		Mg^{2+}	Active site thiol?		Oleic-acid-induced translocation from the cytosol to the ER
NEM-insensitive (PAP-2)	51–53 (decreased to 28 when deglycosylated)	–	?		Located in the plasma membrane; interacts with PLD

recycling. sPI-PLC enzymes also cleave GPI anchors (Ferguson et al 1985), often at a rate 10-fold higher than PI. This ability to cleave GPI anchors has made the use of these enzymes widespread in liberating proteins anchored to cell membranes in this fashion. sPI-PLC enzymes share significant sequence homology with one another, but little homology with mammalian intracellular cPI-PLC enzymes (Griffith et al 1991).

PI-PLC enzymes from *Bacillus* sp. catalyze the cleavage of the glycerophosphate linkage of PI in a stereospecific (Lin et al 1990) and Ca^{2+}-independent manner in two steps (Volwerk et al 1990, Griffith et al 1991): (1) an intramolecular phosphotransfer reaction to form inositol cyclic 1,2-monophosphate (cIP), and (2) hydrolysis of the cIP to produce inositol-1-phosphate. Use of substrate analogs has shown that the phosphotransferase step is absolutely required, since a phospholipid with 2-methoxyinositol, 2-fluoro- or 2-dihydroinositol moiety could not be hydrolyzed by the enzyme (Lewis et al 1993, Morris et al 1996). The second step is considerably less efficient than the phosphotransferase reaction, since the enzyme has a much higher K_m for cIP and a much lower V_{max} (Zhou et al 1997b). The enzyme exhibits a 5–6-fold kinetic preference "interfacial activation" for micellar PI compared to monomeric substrate (Lewis et al 1993). Crystal structures of PI-PLC from *B. cereus* have in the absence and presence of *myo*-inositol and a GPI moiety (Heinz et al 1995, 1996) provided mechanistic insights. The protein is an imperfect eight-stranded α/β barrel with no hydrogen bonds between β-5/β-6 strands. The N- and C-terminus of the protein are spatially close. The *myo*-inositol is in a deep active-site pocket. A structure of the enzyme with a fragment of a GPI-anchor, glucosaminyl(α1-6)-D-*myo*-inositol (Heinz et al 1996), is also available. The *myo*-inositol moiety is similar to that in the structure with inositol bound (well-defined contacts with the protein). The glucosaminyl group is less well defined and exposed to solvent at the entrance of the active site, and there are few specific contacts with the protein. An analysis of the structure of the bound glucosaminyl group predicts that the rest of a natural glycan does not contribute to interactions with PI-PLC. This explains the lack of specificity for the enzyme in cleaving GPI anchors. The active site bears a striking similarity to RNase with two histidines (H32 and H82) positioned to act as proton donors and acceptors in a mechanism similar to that of ribonuclease. Along with an aspartate, this catalytic triad is preserved in the PI-PLC from *Listeria monocytogenes*.

One of the more unusual features of this sPI-PLC is a novel type of interfacial activation of the enzyme towards cIP, a water-soluble substrate with no tendency to partition into interfaces. As shown in Figure 5-7, PC (or PE) micelles (as well as bilayers) activate the enzyme significantly by binding to the enzyme at a discrete site (Zhou et al 1997a, b) and allosterically affecting both K_m and V_{max}; K_m decreased (e.g., from 90 mM to 29 mM with diC$_7$PC micelles added), while V_{max} increased almost 7-fold in the presence of diC$_7$PC micelles. The enzyme efficiency (V_{max}/K_m) in the presence of

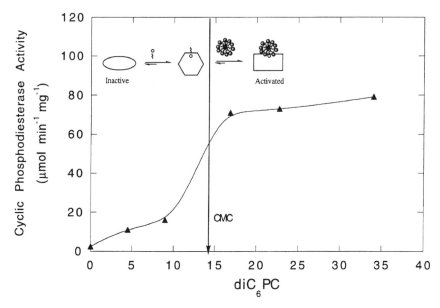

Figure 5-7. Change in cyclic phosphodiesterase activity of sPI-PLC toward 15-mM cIP as a function of added diC$_6$PC; the CMC of the PC is indicated by an arrow. Below the CMC, specific and tight binding of a PC monomer converts the enzyme to an intermediate form (indicated by a hexagon) with enhanced activity. Further anchoring of the PC-PI-PLC complex to a surface (micelle or bilayer) optimally activates the enzyme, indicated by the rectangle in the inset (adapted from Zhou et al 1997a).

diC$_7$PC increased more than 21-fold, but it was still 20-fold lower than initial phosphotransferase activity for monomeric diC$_6$PI. This type of phospholipid activation serves to anchor the enzyme to the aggregate surface in a more active form and allows processive catalysis to occur (shown schematically in Figure 5-7). PC binding also anchors the enzyme toward PI (actually a D-thio-DMPI) in a vesicle assay system (Hendrickson et al 1996) and enhances substrate turnover. Nonlipid molecules, notably the aminoglycoside G418, have also been identified to activate this enzyme (Morris et al 1996). G418 is thought to act on the E-S complex enhancing k$_{cat}$, but increasing K$_m$ 10-fold. Both PC and G418 contain a positively charged amino moiety, which appears to be the critical component in PI-PLC activation.

While the sPI-PLC from bacterial cells has received the most attention, secreted forms of the enzyme have also been detected in mammalian cells. An sPI-PLC appears to be secreted by intact Swiss 3T3 cell cultures. Kinetically the enzyme behaves like the well-characterized intracellular isozymes, as it is Ca^{2+}-dependent and has a pH optimum between 5 and 6 (Birrell et al 1995). Further characterization is needed to see if this is a new enzyme or a cytosolic activity that has been secreted.

CYTOSOLIC ACTIVITIES (CPI-PLC).   Mammalian PI-PLC enzymes also exhibit both intrinsic phosphotransferase and cyclic phosphodiesterase activities, with both products acting as second messengers. The characterization and regulation of PI-PLC isoforms have been extensively reviewed (Rhee and Choi 1992). These enzymes involved in PIP_2 metabolism regulate PI-mediated signaling cascades *in vivo* (Wu et al 1995). Early classifications based on separate gene products with sequence similarities were made for five immunologically distinct PI-PLC enzymes (Rhee et al 1989). These proteins, designated PI-PLC α through ε, have strikingly different molecular weights and regulatory behavior. The best-characterized PI-PLC isozymes, $-β$, $-γ$, and $-δ$, share three conserved regions (Figure 5-8): (1) an N-terminal pleckstrin homology (PH) domain, (2) the X-domain (~170 amino acids), and (3) the Y-domain (~260 amino acids). A brief summary of their characteristics is included in Table 5-3. PI-PLC isozymes appear to have different subcellular locations. This may be key to their roles in signaling. For example, in rat liver, PI-PLCβ1 is predominantly nuclear, the γ1 enzyme is largely detected in the cytoplasm with some association with the nucleus, and the δ1 enzyme is restricted to the cytoplasm (Bertagnolo et al 1995). The isozymes in the nucleus are associated with the nuclear matrix and lamina. This suggests that PI-PLCβ1 and γ1 have key roles in the nuclear phosphoinositide cycle.

The X- and Y-domains are necessary for catalysis, since in rat PI-PLCδ1 a construct containing only these two domains is catalytically active. The protein is monomeric, and Ca^{2+} binds to the protein in the absence of lipid (Grobler and Hurley 1996). Sequence alignments of the bacterial sPI-PLC show 27% similarity with mammalian PI-PLC region X (perhaps most strikingly, residues corresponding to H82, H32, and G83 are conserved). Since the histidines are proposed to be part of the catalytic mechanism in the bacterial enzyme, it is probable that eukaryotic PI-PLC enzymes may employ a similar mechanism. In PI-PLCδ, site-specific mutagenesis of resi-

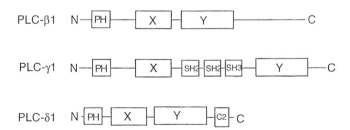

Figure 5-8. Breakdown (and alignment) of sequences for different cPI-PLC isozymes into domains.

dues in the X-domain located a histidine (311) and a tyrosine (314) critical for activity (Ellis and Katan 1995).

All isoforms contain a pleckstrin-homology (PH) domain at the C-terminus. PH domains have high affinity for the phosphoinositide PIP_2 and serve as modules for binding the protein to a bilayer. NMR structures of these domains are available. In the case of PI-PLCδ1, the interaction of the PH domain with a membrane has an allosteric effect on enzyme activity toward cIP (Wu et al 1997a). Other studies suggest a similar interaction. PH domains also bind βγ G-proteins very tightly (Harlan et al 1994). None of the PI-PLC isoforms contains a membrane-spanning sequence, and indeed most of the PI-PLC is purified from the cytosolic fraction. The purified membrane-associated PI-PLC has been shown to be the same as the cytosolic PI-PLC (Lee et al 1987). As with $cPLA_2$, a C_2 or CaLB domain is found in these phospholipases.

The PLCβ family, containing three members, has 400 amino acids in the C-terminal domain beyond the Y region. This extension interacts with G-proteins of the G_q class (Simon et al 1991, Smrcka et al 1991, Taylor et al 1991) and the βγ-complexes in particular (Carozzi et al 1993). Gα (15 and 16) proteins couple a wide variety of receptors to PI-PLCβ isoforms (Offermanns and Simon 1995). An example of how hormones stimulate PI-PLC through activation of a G-protein-coupled receptor is provided by histamine binding to the H1 receptor and stimulating PLC (a key part of this work is to isolate H1 and H2 receptors in cells, in order to separate adenylate cyclase signaling and PI signaling). $G\alpha_q$-type proteins are involved (Kuhn et al 1996).

PLCγ isozymes are abundant in many tissues and cell types. They contain a region between X and Y that appears related to sequences found in nonreceptor tyrosine kinases of the *src* family. Deletion of these regions, termed the SH2 and SH3 domains, produces an enzyme that still has catalytic activity but altered regulation (Emori et al 1989). SH2 and SH3 domains modulate protein-protein interactions in signal transduction pathways activated by protein tyrosine kinases (Schlessinger 1994). The PI-PLCγ SH2 domain binds to short phosphotyrosine-containing sequences in growth factor receptors. The SH3 domain targets proteins with proline and hydrophobic sequences. In PI-PLCγ these domains link receptor and cytoplasmic protein kinases to PI hydrolysis. PLCγ is the only PI-PLC isozyme that is activated by phosphorylation of specific tyrosine residues. The phosphorylation is carried out by activated tyrosine kinase growth factor receptors such as those for platelet-derived growth factor (PDGF) (Kim et al 1991), epidermal growth factor (EGF) (Margolis et al 1989), and fibroblast growth factor (Buergess et al 1990). Phosphorylation enhances the relocation of PLCγ from the cytosol to the plasma membrane, where presumably it is better able to interact with its phospholipid substrates (Kim et al 1991, Vega et al 1992). The catalytic activity of the phosphorylated form of PLCγ is increased compared with that of the unphosphorylated form, although this effect depends upon assay conditions and the physical state of the substrate (Kim et al 1991, Wahl et al 1992). Other parts of the protein may also be involved in regulating enzymatic activity. It has been proposed that ligation of the SH2 domain by growth factor receptors may mediate conformational changes in PI-PLC that alter enzyme activity (Koblan et al 1995). The SH3 domain has a different role, unrelated to catalysis: the SH3 domain must be present in PLCγ in order for the cell to initiate a mitogenic response (Smith et al 1989). Neither the PI-PLC catalytic activity nor the SH2 domain is required to initiate a mitogenic response (Huang et al 1995). Furthermore, the expression of PLCγ during phorbol-ester-induced differentiation may be down-regulated by posttranscriptional processing (Lee et al 1995). All these data imply that the activation of PLCγ might be one of the important pathways for growth factors to trigger mitogenic signals.

PI-PLCδ isozymes are expressed in brain, seminal vesicles, fibroblasts, and probably other tissues. Regulation of this class of PI-PLC does not involve G-proteins or phosphorylation; Ca^{2+}-binding to membrane phospholipids may regulate the activity of this enzyme. A crystal structure (Essen et al 1996) of the catalytic portion of rat PI-PLCδ1 (the PH domain, residues 1–132, have been removed) shows an active site that is part of a TIM-barrel-like domain and has significant structural similarity to the sPI-PLC structure from $B. cereus$ (Figure 5-9). A Ca^{2+} in the active site interacts with the C-2 of the inositol ring in crystal structures with IP_x products and cIP analogs bound (Essen et al 1997). This catalytic domain is connected at its C-terminus with a β-sandwich C_2-domain (two Ca^{2+} sites are seen in this region in the structure); the N-terminus connects with an EF-hand domain (Essen et al 1996). The mechanism proposed for the formation of IP_x by this enzyme is schematically illustrated in Figure 5-10; as with the bacterial enzyme, cIP is an intermediate. However, unlike the bacterial enzyme, PI-PLCδ action on PI generates both the cyclic inositol phosphodiester (cIP) and acyclic inositol phosphate (I-1-P). The mole fraction cIP appears constant over the time course of the hydrolysis. However, cIP by itself is a poor substrate for the enzyme (K_m ~20 mM, V_{max} ~1/10 that for PI hydrolysis [Wu et al 1997a]). A careful analysis of cIP and PI kinetics (both exhibiting a sigmoidal dependence on substrate concentration, although there is no evidence for oligomerization of the enzyme) indicates that PI hydrolysis follows a sequential mechanism with release of cIP being slow compared to the attack of a water molecule. cIP hydrolysis by PI-PLCδ is activated allosterically by the binding of enzyme to a PC surface (other phospholipid interfaces may also activate the enzyme). This interfacial activation (a twofold increase in V_{max}) has been localized to the PH domain of the enzyme, since hydrolysis of cIP by a PH-domain-truncated PI-PLCδ1 (Δ(1–132)PI-PLCδ1) from rat cannot be activated by phospholipids (Wu et al 1997a).

There are two other PI-PLC families thought to exist, PLCα and PLCε, whose properties and regulation appear to be different from the $-β, -γ$, and

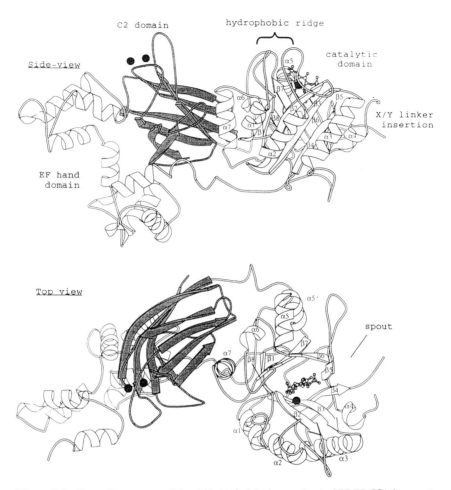

Figure 5-9. Overall structure of the $\Delta(1-132)$ deletion variant of PI-PLCδ1 from rat (reproduced with permission from Essen et al 1997). The shaded portion indicates the C2 domain; the dark spheres indicate Ca^{2+} ions. The active site is shown with 1,4,5-IP$_3$ bound. The "side" view shows the domain structure of the protein; a reorientation of 90° produces the "top" view, which corresponds to a view from the membrane surface.

$-\delta$ isozymes. PLCα enzymes have reported molecular weights of 57–70 kDa. Originally isolated from ram seminal vesicles, these PI-PLC enzymes are the most abundant form found in liver and many fibroblast cell lines. A sequence of PLCα has been published (Bennett et al 1988). There is no obvious sequence homology of this protein to the $-\beta$, $-\gamma$, or $-\delta$ isoforms. More recent work expressing the protein encoded by the cDNA previously

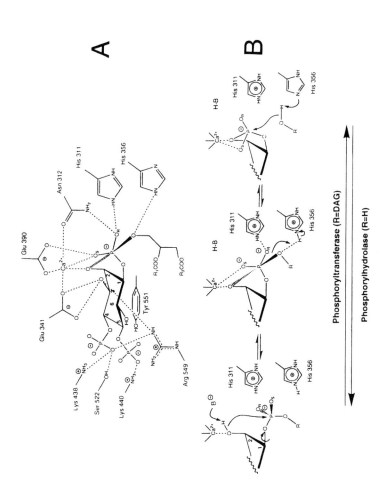

Figure 5-10. Proposed reaction mechanism for PI-PLCδ1 (reproduced with permission from Essen et al 1997). In (A) is shown a model indicating protein contacts with the proposed transition state. In (B) the reaction scheme for general acid/base catalysis along with the proposed catalytic residues involved is shown.

identified as PLCα showed that the candidate (62–68-kDa) protein) was a thiol:protein-disulfide oxidoreductase and exhibited no PLC activity (Srivastava et al 1991). Whether the other PLCα activities in this class are truly distinct from $-\beta$, $-\gamma$, and $-\delta$ isozymes remains to be seen.

Kinetic specificities vary among the cPI-PLC forms with $-\beta$ and $-\delta$ isozymes exhibiting a 40- to 60-fold preference for PIP_2 over PI and 1 mM Ca^{2+}, while the $-\gamma$ enzyme shows only a two- to threefold preference for PIP_2. Since PIP_2 has different surface behavior than PI, such a kinetic preference is hard to interpret unambiguously. Soluble glycerol inositol phosphates provide a simpler system for assessing the importance of substrate phosphorylation on kinetic parameters of the enzyme (Wu et al 1997a). For PI-PLCδ1, GPIP is a much better (at least 100 time better) substrate than GPI, indicating the importance of 4′-phosphate in the binding and catalysis by PI-PLCδ1. In comparison with GPIP, $GPIP_2$ shows fourfold increase in V_{max} with little change of K_m, indicating that the 5′-phosphate is important for catalysis, but not important for substrate binding. Similar analyses with soluble GPI_x for the other PI-PLC isozymes could also determine whether phosphates on the inositol ring are critical to substrate binding, catalysis, or both.

A NOVEL REGULATORY MECHANISM FOR cPI-PLC.   An intriguing path for regulating intracellular phospholipases is the activation by specific cytosolic proteins. Using permeabilized cells (prepared with streptolysin O) to study the coupling of G-proteins with PI-PLC activity, Cockcroft and coworkers (Cockcroft 1992, Cockcroft and Thomas 1992) have shown that the addition of GTPγS activates PLC activity (as measured by IP_3 production). If the cell system is depleted of cytosol in such a way that cells still have cellular architecture but not soluble proteins, the addition of GTPγS does not lead to PLC activation. However, activation of PI-PLC can occur by inclusion of a PI-transfer protein (PI-TP). PI-TP has PC or PI bound (the affinity for PI is 16 times higher than PC) and moves lipids around *in vitro*. The loss of responsiveness to GTPγS occurs as cells are depleted of intracellular PI-TP. One interpretation of these results is that PIP_2 synthesis occurs with PI bound to the transfer protein (Fensome et al 1996). If substrate (PIP_2) in the membrane is not accessible to PLC, then perhaps the preferred substrate is PI-TP-bound PIP_2. PI-TP also stimulates PI-4-kinase activity. Modulation of PI-PLC by this protein is relatively novel. It explains the lack of activity of the enzyme toward the large amount of PI over PIP_2 in membranes by making protein-bound PIP_2 accessible and optimally oriented for hydrolysis by the enzyme.

GPI-PLC

A separate class of PI-PLC isozymes is membrane-bound with a marked specificity for GPI-anchors. Many glycoproteins are anchored to the plasma membrane via glucosylaminyl-phosphatidylinositol (GPI) linkages. Parasites such as *Trypanosoma brucei* rapidly remodel these surface glycoproteins as a way to outwit the immune system of the host. A membrane-bound GPI-specific PLC localized on the cytoplasmic leaflet of the intracellular membranes cleaves intermediates of GPI biosynthesis (Mensa-Wilmot and Englund 1992). This activity must be regulated to ensure survival of the parasite (degradation of GPI anchors must be controlled; otherwise, the cell surface would be stripped of glycoproteins). The protozoan enzyme requires a GlcN(1α6) linkage with an unmodified amino group for optimum liberation of DAG (Morris et al 1995). As with the other PLC enzymes that hydrolyze PI and derivatives, the inositol 2-hydroxyl group is not required for substrate recognition, but is required for catalysis (Morris et al 1996).

Nonspecific PLC

While it is well established that the initial breakdown of PIP_2 by a cPI-PLC rapidly generates the DAG to activate protein kinase C, a sustained activation of PKC by DAG could in some cases involve hydrolysis of PC by a general nonspecific PLC (or by PLD and subsequent hydrolysis of PA by PA phosphatase [Exton 1990, Billah and Anthes 1990, Pelech and Vance 1989]). Since no IP_3 is generated by these activities, there is no Ca^{2+} mobilization. The evidence for a PC-PLC is weak, but it has been suggested that bacterial sPLC may be a useful model for this mammalian activity, since it can mimic mammalian PLC activity in enhancing prostaglandin synthesis (Levine et al 1988). The bacterial enzyme may also be antigenically similar to mammalian PLC (Clark et al 1986). sPLC also induces the aggregation and subsequent diffusion of liposomes (Nieva et al 1993); this property may be useful for transport of molecules in drug therapies. Nonspecific PLC in bacteria also plays a role quite distinct from signaling. In several pathogenic bacteria, PC-PLC along with the PI-PLC is needed for pathogenesis. The N-terminal domain of the α-toxin from *Clostridium perfringens* is homologous to the PC-PLC from *B. cereus*. This PLC activity is essential for lethality *in vivo* and for mediating platelet aggregation *in vitro* (Guillouard et al 1996).

sPLC from *B. cereus* is the prototypical nonspecific PLC enzyme. The enzyme is translated as a 283-residue precursor with a 24-residue signal peptide and a 14-residue propeptide (Johansen et al 1988). The secreted form contains 245 amino acids and is a monomeric Zn^{2+} metalloenzyme. This PLC is extremely stable as long as Zn^{2+} is bound; removal or substitution of the Zn^{2+} ions causes inactivation of the enzyme (Little 1981). Unfolding of the PLC in guanidinium hydrochloride or urea can be reversed by gradual removal of the denaturant and subsequent addition of Zn^{2+}. This strategy has been used recently in cloning the enzyme in *E. coli* by renaturing protein packed in inclusion bodies (Tan et al 1997). Although PC-PLC is often described as nonspecific, phospholipids with a headgroup containing either a nucleophile or a positive charge (PS, PC, PE, PG) are good substrates, while phospholipids such as phosphatidic acid or phosphatidylmethanol are poor substrates when presented in a mixed micelle assay system (C. Tan and M.F. Roberts, unpublished results). This bacterial enzyme is not very effective against large unilamellar vesicles composed of PC. In fact, long-chain PCs solubilized in detergent mixed micelles are a 50–100-fold better substrate than the same PC packed in bilayer vesicles. The explanation for this kinetic phenomenon lies with the decreased area of the PC headgroup packed in bilayer versus micellar systems, as well as with accessibility of the phosphodiester bond to be cleaved. In terms of interfacial kinetics, conversion of the enzyme from a E to E* state appears hindered with PC packed in vesicles. There is no lag in substrate hydrolysis with micelle or SUVs. However, a variable lag period is observed for hydrolysis of PC packed in LUVs. LUVs have little curvature, and phospholipids are well packed, without defects. Often membrane perturbants (e.g., small amounts of detergents, DAG) can activate the B. cereus sPLC towards PC bilayers. The lag period ends when vesicle aggregation begins (Basanez et al 1996b). Gangliosides inhibit PLC (Daniele et al 1996) by altering k_{cat} of the adsorbed enzyme as well as the availability of substrate in a suitable organization. These membrane components stabilize the bilayer lamellar phase as well as interact with PLC (Basanez et al 1996a). These observations suggest that PC-PLC needs defects or DAG-rich patches in order to hydrolyze phospholipids in a bilayer. Thus, under normal circumstances, a cytoplasmic enzyme similar to the PLC from B. cereus would be inactive in cells.

The crystal structure of B. cereus sPLC (Hough et al 1989) reveals 10 α-helical regions folded into a tightly packed single domain. The helices contain 66% of the residues, the remainder forming loops between them. The molecule surface is smooth except for a shallow (8 Å deep by 5 Å wide) cleft that includes the active site. The metal ion cluster in the active site is similar to that observed in alkaline phosphatase from E. coli. Both hydrophobic (F, I, N, L, Y) and hydrophilic (S, T, D, E) residues surround the active site. A comparison of the structures for the enzyme complexed with phosphate (Hansen et al 1992) and a nonhydrolyzable PC substrate analog (Hansen et al 1993) with predictions based on computer modeling leads to an interesting dilemma. In the crystal structure of the complex, all three Zn^{2+} ions interact with the phosphonate moiety, which is in a strained and unusual conformation, with the choline moiety folded back and nearly parallel to the fatty acyl chains. Were such an orientation to occur with substrate, the strain in the phosphate linkage could provide the energy for hydrolysis. In this strained complex, a water molecule, located apical to the DAG leaving group, would function as the attacking nucleophile. The modeling studies (Sundell et al 1994) suggested that the phosphonate (which has a K_i roughly 10-fold higher than the K_m for monomeric substrate) may not be bound in the conformation adopted by the actual substrate. Rather, substrate is bound in a conformation consistent with the low energy state of normal phospholipids. Cleavage is initiated by an in-line attack of an activated water molecule on the phosphodiester linkage, to generate a pentacoordinated phosphorus in the transition state. In the substrate analog structure, the strain induced in the phosphodiester would aid in bond cleavage; in the modeling, the phosphate group is thought to be reoriented after cleavage to stabilize the developing negative charge on the alkoxy leaving group. After bond cleavage, the phosphocholine (phosphomonoester) moiety diffuses out of the active site, leaving the DAG alkoxide behind for protonation and eventual release. This ordered release of products is constant with earlier kinetic observations.

Recent advances in the cloning of the nonspecific bacterial PLC have further qualified these mechanisms. On the basis of the crystal structure with the phosphonate bound, the side chain of E146 was suggested to stabilize the water molecule for an attack on the phosphate. Martin and coworkers (Martin et al 1996) have shown that the E146Q mutant still has activity (1.6% WT) but altered protein stability. More interestingly, if this mutant was assayed against PC in Tris buffer, a change in hydrolytic specificity was observed (C. Tan and M.F. Roberts, unpublished results). Tris is known to bind to the active site of PLC in the vicinity of $Zn(2)$; when it binds to the mutant but not to the WT enzyme, the sn-2 fatty acyl chain is cleaved from PC to yield lyso-PC. This change of specificity suggests that the active site of PLC can be adapted for several types of hydrolytic chemistry.

ROLE OF PC-PLC IN SIGNAL TRANSDUCTION. Are there examples of a PC-PLC involved in a signaling pathway? Activated macrophages produce NO, a mediator of a variety of biological functions. The inducible NO-synthase involves a PKC-dependent pathway. It has been suggested (on the basis of PC metabolism and DAG synthesis) that in this system the DAG needed to activate PKC is provided by an unusual PC-specific PLC, and not by PI-PLC or PLD (Sands et al 1994).

A PC-PLC also appears to generate DAG in rat-1 fibroblasts upon PDGF (platelet-derived growth factor) activation of cells (van Dijk et al 1997). One observes PA, choline, DAG, and phosphocholine, products of PLD and PLC, upon stimulation. Phorbol esters block protein kinase C and completely block PLD activity (but not MAPK activation). However, DAG and phosphocholine are still produced. If PC-PLC (*B. cereus*) is added to the cells, the net response mimics PDGF activation. If exogenous PLD is added, a different response is observed. A reasonable conclusion is that PC-PLC, rather than PLD, plays an important role in PDGF-activation of the MAPK pathway.

Other experiments suggest that chronic elevation of intracellular DAG generated by PLC hydrolysis of PC is oncogenic. Johansen et al (1994) formed transfectants of NIH 3T3 cells expressing the *B. cereus* PLC gene.

These modified cells show behavior characteristic of transformation: independent growth in agar, loss of contact inhibition, etc.

PA-PLC

Several PLC-type enzymes with hydrolytic activity specifically toward phosphatidic acid (i.e., phosphatidate phosphohydrolase or PAP activities) have also been purified. The enzyme from yeast, which is an integral membrane protein and Mg^{2+}-dependent, appears to play a role in phospholipid remodeling (Carman et al 1995). It is inhibited by sphingosine, which might suggest that it has a role in signal transduction in yeast. In mammalian cells, two types of PA-PLC have been studied: (1) an N-ethylmaleimide (NEM) sensitive, Mg^{2+}-dependent enzyme (also referred to as PAP-1) that is regulated by oleate-induced translocation to the membrane (in this case the endoplasmic reticulum), and (2) an NEM-insensitive, Mg^{2+}-independent membrane-associated glycoprotein in the plasma membrane activity (often referred to as PAP-2). PAP-2 activity from rat liver has been purified and characterized (Waggoner et al 1995, Brindley and Waggoner 1996). In treating cells with EGF, there is a reduction in PA-PLC that coprecipitates with the EGF-receptor. An increase in PA-PLC activity is associated with a particular isoform of PKC (ε). Such an observation ties PA-PLC to this signaling component as a pathway to provide DAG from PA rather than PIP_2 (Jiang et al 1996).

PA-PLC can also dephosphorylate PA, lyso-PA, sphingosine-1-P, and ceramide-1-phosphate. It converts these signaling lipids to DAG, sphingosine, and ceramide. Sphingosine has been shown to inhibit PA-PLC and increase PA concentrations. This behavior is consistent with PA-PLC functioning to mitigate mitotic signals as well as to generate other lipophilic second messengers such as DAG, ceramide, or sphingosine. This enzyme could supply a path for cross-talk between the two lipid signaling pathways.

Inhibitors of PLC Enzymes

One of the major advances in understanding the roles of the different PLA_2 enzymes in signaling pathways has been the availability of selective inhibitors. Unfortunately for PLC enzymes, no such isozyme-specific inhibitors exist yet. The inhibitors developed for PLC enzymes are mostly substrate analogs with mechanism-based modifications. Lipophilic PC analogs with a methylene group instead of the anhydride oxygen that is the site of cleavage exhibit a K_i that is 10-fold higher than the substrate K_m. Other types of substitutions (S for O, NH for O) produce lipids with K_i values similar to K_m for phospholipid substrates, but these are not particularly specific for PLC. For PI-PLC from B. *cereus*, dioctyl methylphosphate has a $K_i = 12$ μM (Ryan et al 1996). For this enzyme the best inhibitors were the most hydrophobic compounds; molecules without the lipid moieties were 1000-fold poorer inhibitors. However, these compounds will bind to many other proteins that interact with phospholipids. Selective PI analogs for PI-PLC were also examined but are impractical for use with cells, because they will affect any proteins that interact with phosphoinositides. Water-soluble inhibitors in general are not effective at inhibiting these activities. For PI-PLC, glucosaminyl (α1-6)-D-myo-inositol was more potent than myo-inositol or inositol-1-phosphate, but these polar molecules are poor inhibitors compared to lipophilic analogs.

Vanadate is rare among water-soluble inhibitors of phospholipases in that it has a K_i roughly the same as the K_m for PC substrate (Tan and Roberts 1996). Vanadate forms a pentacoordinate complex with the enzyme that mimics the high-energy transition state of phosphate ester hydrolysis. By using vanadate as a water-soluble inhibitor of a surface-active enzyme, one also removes the concerns about the inhibitor partitioning between bulk solution and the interface. One of the results of a careful analysis of vanadate inhibition of the enzyme is that there must be discrete interfacial binding that is distinct from a substrate binding at the active site. PLC binds to the micelle and works in a "scooting" mode, catalyzing the hydrolysis of several PCs before dissociating in solution. This is true in the case of monomeric substrate where there is no interface present. However, in the presence of an interface (micelle), vanadate might still bind to the enzyme active site while the PLC interacted with the interface at a binding site distinct from the catalytic site. This behavior is the first evidence that a discrete surface binding step must occur with the nonspecific PLC. However, vanadate cannot be used as a specific inhibitor of PC-PLC (or PI-PLC) in cells, since 0.1 mM vanadate will also inhibit protein tyrosine phosphatases.

Often cations, either polyvalent inorganic or organic ions (La^{3+}, Al^{3+}, neomycin, polyamines), are used to inhibit PI-PLC. These ions, rather than interact specifically with the enzymes, form complexes with the phosphoinositide substrates and inhibit PI-PLC activity by altering the accessibility/susceptibility of substrate (McDonald and Mamrack 1995). Since these complexes can alter the distribution of phosphoinositides in the bilayer and potentially effect partial phase separation, they may interfere with a wide variety of other membrane-localized participants in signaling pathways.

There is clearly a need to develop selective inhibitors of PI-PLC enzymes for use in cells. Perhaps the use of combinatorial chemistry will uncover new compounds with tight and selective binding to these enzymes.

4. PLD

Phospholipase-D-type activities cleave the distal phosphodiester bond of phospholipids, generating PA and a free base; PC is usually the preferred substrate. As shown in Table 5-4, PLD activities have been studied in plant,

Table 5-4. Characteristics of phospholipase-D-type enzymes

Type	Subunit MW (kDa)	Specificity	Cofactor	Mechanism	Regulation
sPLD (GPI-PLD)	100–110	GPI anchors	Ca^{2+} (not catalytic); Zn^{2+} (catalytic)	Unclear; specific for GPI	Physical state of bilayer; affinity for anionic phospholipids
hPLD PLD1	120	PC	Ca^{2+}, PIP_2	Ping-pong mechanisms with a covalent phosphatidyl-enzyme intermediate likely	Activated by G-proteins, PKC, PI-TP; GDI
PLD2					Inhibition of constitutive activity; mechanism unclear

Abbreviations: hPLD, cloned human PLD; PI-TP, phosphatidylinositol transfer protein; PKC, protein kinase C; GDI, guanine nucleotide exchange protein

bacterial, and mammalian tissues (Heller 1978, Wang 1993). A hallmark of these enzymes is that they also catalyze a transphosphatidylation reaction in the presence of moderate to high concentrations of primary alcohols, a property often used to prepare phospholipids with the same acyl chain distribution and different headgroups. Transphosphatidylation implies the existence of a phosphoryl-enzyme intermediate and serves as a sensitive and selective way to monitor the presence of PLD in a variety of cells. PLD enzymes from plant and bacterial sources have been fairly well studied. The bacterial enzymes may act as virulence determinants (McNamara et al 1995). In yeast, activation of PLD is an early event in sporulation (Ella et al 1995). Bacterial and plant PLDs show significant sequence similarities to each other and to two other classes of phospholipid-specific enzymes: bacterial cardiolipin synthases, and eukaryotic and bacterial phosphatidylserine synthases (Ponting and Kerr 1996).

The few kinetic and mechanistic studies of PLD have concentrated on plant enzymes. For example, PLD from cabbage exhibits interfacial activation toward monomeric short-chain PC substrate. However, this is abolished at high Ca^{2+} (Allgyer and Wells 1979). This latter effect is complex since the high Ca^{2+} will lower the CMC of product PA and complicate the analysis. An unusual kinetic feature of the enzyme partially purified from cabbage and a bacterial PLD from *Streptomyces chromofuscus* that may have relevance in signal transduction is that cyclic lyso-PA is formed as an intermediate upon PLD hydrolysis of lyso-PC (Friedman et al 1995). Cyclic lyso-PA appears to be an obligatory part of the mechanism and is eventually hydrolyzed by the enzyme to form lyso-PA. Kinetic studies using vanadate as an inhibitor strongly suggest that the phospholipase and phosphodiesterase activities may be distinct in this enzyme.

PLD IN SIGNAL TRANSDUCTION.   Mammalian PLD enzymes have been implicated in membrane trafficking, and the regulation of mitosis as well as signal transduction (for a recent review see Exton, 1997). A mammalian PLD activity was first observed in rat brain tissue in 1975. This membrane-bound activity was independent of Ca^{2+} and activated by free fatty acids *in vitro* (Kanfer 1980). Methods to solubilize it from rat membranes have been outlined, and a variety of sensitive assays have been developed on the basis of the transphosphatidylation reaction (Danin et al 1993). The assays in particular have allowed studies of the enzyme *in situ* and its production of second messenger PA. The basal activity of PLD is quite low in many cell types, although it can be very rapidly activated. While this low basal activity has hindered purification of mammalian PLD enzymes, it has spawned a large number of studies examining the different factors that activate the enzyme.

Clearly, the production of PA has an effect on a variety of cell functions (as well as serving as the substrate for a PLA_2 to generate lyso-PA). The Ca^{2+}-dependent enzyme occurs in multiple forms (as has been seen for the other phospholipases as well) and has a kinetic preference for PC. In many instances it is difficult to separate PA activation from that of lyso-PA, a major intracellular second messenger, since the latter can be generated by PLA_2 action. An example where this has been successful is the generation of superoxide in neutrophils, where the PA from PLD action induces the interaction of G-protein with membrane-bound reduced NADH oxidase. *In vitro*, G-proteins of the ARF (ADP-ribosylation factor) and Rho families activate PLD; these G-proteins may also activate PLD *in vivo*. ARFs are a family of 21-kDa GTP-binding proteins that have been implicated as ubiquitous regulators of multiple steps in both exocytic and endocytic membrane traffic in mammals and yeast. ARFs bind PIP_2 specifically; only when bound to lipid can they exchange nucleotide (Kahn et al 1996). There are two potential phospholipid binding sites on ARF, each of which is coupled to the nucleotide binding site. There is also evidence for control of PLD and Rho proteins by soluble tyrosine kinases and unusual serine/threonine kinases.

A PLD activity is important in the formation of Golgi coated vesicles. This vesiculation requires the activation of ARF to initiate coat assembly. Interestingly, cytosolic ARF is not necessary for formation of vesicles if membranes have high constitutive PLD (Ktistakis et al 1996). If exogenous PLD is added and PA formed, the coatomer is bound to the Golgi membrane. In model systems the coatomer binds selectively to vesicles with PA and PIP_2. Thus, the role of PLD may be to produce lipid anchors for coat protein.

Neutrophils and HL60 cells secrete lysosomal enzymes from granules. PLD appears to have a role in this secretion. Neutrophil activation occurs via initial receptor-linked activation of PLC to generate DAG and IP_3. The Ca^{2+} flux along with DAG activates PKC ($\beta1$ isoform), which in turn phosphorylates a plasma membrane component that activates PLD (Olson and Lambeth 1996). GTP proteins RhoA and ARF are involved along with a 50-kDa cytosolic regulatory factor. In HL60 cells, PLD activity can be detected in both cytosolic and membrane fractions. Membrane-localized PLD has significant basal activity and is inhibited by Ca^{2+}; cytosolic PLD is activated by Ca^{2+} and shows little activity in the absence of GTP. Significant PLD activity is detected upon stimulation of cells with GTPγS. The activity in the cytoplasm co-chromatographs with small GTP-binding proteins (Siddiqi et al 1995). A separate membrane PLD exists whose endogenous regulator is RhoA. In both neutrophils and HL60 cells, PIP_2 appears to be required for maximum activation of PLD. PIP_2 stimulates PLD in brain and in permeabilized U937 cells *in vitro*. Decreasing the PIP_2 in cells (by inhibiting PI-4-kinase with antibodies, or by treating the cells with neomycin, a high-affinity ligand of PIP_2) blocks PLD activation (Pertile et al 1995). Thus, PIP_2 is likely to be a cofactor (perhaps this points to the existence of a PH domain in PLD?). Membrane-localized PLD in different cells probably prefers micellar- to bilayer-packed PCs, since it was found that $diC_{10}PC$ was a much better substrate than dipalmitoyl-PC (or $diC_{16}PC$) (Vinggaard et al 1996). More interestingly, when added to membrane preparations containing the enzyme, it obviated the need for PIP_2. This suggests that the role of PIP_2 is to tether the enzyme to the membrane.

HUMAN MPLD.   The first report of a cloned human-membrane-localized PLD appeared in 1995 (Hammond et al 1995). The protein (PLD1) is membrane-associated (but not an integral membrane protein) and more specifically localized to per-nuclear regions: the ER, Golgi apparatus, and late endosomes. The sequence has a predicted molecular weight of 124 kDa, consistent with the molecular weight of the expressed recombinant material (~120 kDa). Analysis of the sequence indicates that it is devoid of recognizable domain structures. PLD1 is selective for PC, stimulated by PIP_2, activated by ARF-1, and inhibited by oleate. The interaction of PLD1 with ARF appears to be direct, although no details are known. PLD1 is probably the PLD activity that has been the subject of most studies. A second PLD (PLD2) with different properties and regulation was cloned recently (Colley et al 1997). PLD2 is constitutively active and may be regulated by inhibition. PLD2 is localized at the plasma membrane. PLD2 undergoes redistribution in serum-stimulated cells, suggesting a role in cytoskeletal regulation or in endocytosis.

PLD AND CROSS-TALK WITH PHOSPHOLIPASES AND OTHER SIGNALING PATHWAYS. In intact cells, PLD activity is often linked to other phospholipases. For example, in human neutrophils, PLD-mediated PA and DAG induce $cPLA_2$ as measured by enhanced arachidonic acid release (Bauldry and Wooten 1997). The exact mechanism of the $cPLA_2$ activation is unknown in this system. Work by Dennis and coworkers (Balsinde et al 1997) has shown that in macrophages activated by PMA, PLD activation involves PKC, and the effect of PAF on PLA_2 is potentiated. A key to separating out the roles of the two enzymes is the use of specific phospholipase inhibitors.

Phospholipase products include a number of glycerolipid-derived second messengers that stimulate cell division. In some cases these enzymes can be regulated by sphingolipid-derived lipid second messengers that inhibit cell division and induce apoptosis (Brindley et al 1996). This intermingling of the phospholipases with apoptosis activators represents a novel way for the cell to control vesicle movement, cell division, and death. DAG stimulates PKC; PA and lyso-PA stimulate tyrosine kinases and activate the Ras-Raf-mitogen-activated protein kinase pathway. Ceramide, the analog of DAG, inhibits PLD by decreasing its interactions with the G-proteins (ARF and Rho) necessary for activation. Hydrolysis to sphingosine and sphingosine-1-phosphate stimulate PLD (hence these lipids are mitogenic). Ceramides also stimulate phosphatase action on the PA, lyso-PA, ceramide phosphate, and sphingosine phosphate, while sphingosine inhibits this activity (for a recent review see Spiegel et al 1996). Other components of signaling systems can interact with PLD. Ceramide interferes with PKC-mediated activation of PLD in cells (Venable et al 1996). The mechanism appears to be inhibition of the translocation of G-proteins and protein kinase C isoforms required for PLD activity to membranes (Aboulsalham et al 1997). Other sphingoid bases have the opposite effect. Sphingosine and sphingosine-1-phosphate stimulate PLD, leading to increased PA and cell growth (Spiegel and Milstien 1996). Again, this is an example of cross-talk between sphingolipid turnover and DAG cycle.

GPI-PLD. A phospholipase D enzyme specific for GPI anchors has been identified at high levels in mammalian serum and plasma (Davitz et al 1987). The enzyme is a monomer of 110 kDa. It is inhibited by thiol reagents (e.g., p-chloromercuriphenylsulfonic acid) and Zn^{2+} chelators (1,10-phenanthroline). The physical state of the substrate appears to modulate enzyme activity. If purified enzyme is added to HeLa cells, no release of GPI-linked alkaline phosphatase is detected unless 0.1% Nonidet is added (Low and Huang 1991). The requirement for loosening or micellization of the bilayer for PLD to access GPI-linked proteins may account for the observation that endothelial and blood cells retain GPI-anchored proteins on their surfaces in the presence of high concentrations of GPI-PLD in the serum. This specialized PLD is also strongly inhibited by product PA or lyso-PA (Low and Huang 1993).

INHIBITORS OF PLD ACTIVATION.   While mechanism-based inhibitors of PLD have not been developed, a number of compounds have been shown to inhibit PLD activation *in vivo*. Often they provide clues as to the role of the phospholipase in the particular cell process. Fodrin, the nonerythroid spectrin, has been suggested to be an endogenous PLD inhibitor (Lukowski et al 1996). This protein (a dimer of molecular weight 1000 kDa) is effective at 1 nM and has no effect on adenylate cyclase. It does, however, inhibit GTPγS-stimulated PLA$_2$ and PLC activities, although at slightly higher inhibitor concentrations. Its extreme potency for inhibiting PLD activation suggests that the cytoskeleton has a role in secretion and cell proliferation. A number of natural products, notably wortmannin and quercetin (Corssen 1996), also inhibit PLD activation in cells. Again, the details of how they inhibit PLD activation are unclear, but they are extremely useful is deconvoluting the role of different phospholipases in cells.

An example of how different phospholipase inhibitors can be used to dissect which phospholipase is the primary activator leading to a particular cellular event is provided by examining the effect of recombinant basic fibroblast growth factor on arachidonic acid release from rat pancreatic acini (Hou et al 1996). Using a battery of different phospholipase inhibitors including mepacrine and aristolochic acid (PLA$_2$ inhibitors), U73122 (a PLC inhibitor), and wortmannin (an inhibitor of growth-factor-stimulated PLD), it was shown that arachidonic acid release in these cells depended on sequential action of tyrosine kinase, PLC, PKC, and DAG lipase, but not on PLA$_2$ or PLD activation (Corssen 1996).

SUMMARY

Recent advances in phospholipase biochemistry and physiology have shown that these enzymes have intertwining roles in signal transduction in cells. They are responsible for generation and dissipation of most of the lipid-derived second messengers in cells, as well as for remodeling membranes in some cases. An attempt to summarize all the interconnections of phospholipases and products is shown in Figure 5-11. For each ester cleavage site there are discrete forms of the appropriate phospholipase localized in the cytosol, membrane-associated, and in some cases secreted. Often the product of one phospholipase is an activator or inhibitor of one with different specificity. Regulation of phospholipases may also be achieved by sphingoid lipids, products of hydrolytic chemistry involved in apoptosis. While our understanding of structures for some of the secreted forms of the enzymes may have relevance for mechanisms of cytosolic activities, the molecular details of most of the cytosolic enzymes (cPLA$_2$, different cPI-

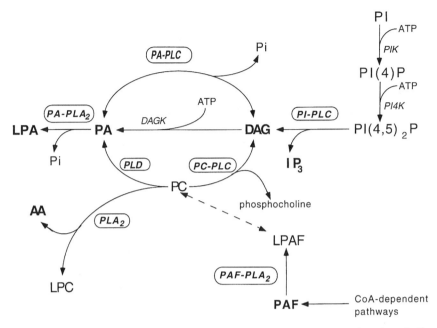

Figure 5-11. Generation and dissipation of lipid second messengers by phospholipases.

PLC isozymes, and a cPLD-type enzyme) remain a challenge. Once these are known, the next step in fully understanding the role of these proteins in signal transduction is to map out regulatory protein/phospholipase interactions on a molecular level. This information should allow for the development of highly selective inhibitors that affect discrete points along a signal transduction pathway.

REFERENCES

Aboulsalham A, Liossis C, O'Brien L, and Brindley DN. 1997. Cell-permeable ceramides prevent the activation of phospholipase D by ADP-ribosylation factor and RhoA. J Biol Chem 272:1069–1075.

Ackermann EJ, Kempner ES, and Dennis EA. 1994. Ca^{2+}-independent cytosolic phospholipase A_2 from macrophage-like P388D1 cells. Isolation and characterization. J Biol Chem 269:9227–9233.

Allgyer TT and Wells MA. 1979. Phospholipase D from savoy cabbage: purification and preliminary kinetic characterization. Biochemistry 18:5348–5353.

Artursson E and Puu G. 1992. A phosphatidylinositol-specific phospholipase C from *Cytophaga*: Production, purification, and properties. Can J Microbiol 38:334–1337.

Balboa MA, Balsinde J, Jones SS, and Dennis EA. 1997. Identity between the Ca^{2+}-independent phospholipase A_2 enzymes from P388D1 macrophages and Chinese hamster ovary cells. J Biol Chem 272:8576–8580.

Balsinde J, Balboa MA, Insel PA, and Dennis EA. 1997. Differential regulation of phospholipase D and phospholipase A_2 by protein kinase C in P388D1 macrophages. Biochem J 321:805–809.

Basanez G, Fidelio GD, Goni FM, Maggio B, and Alonso A. 1996a. Dual inhibitory effect of gangliosides on phospholipase C-promoted fusion of lipidic vesicles. Biochemistry 35:7506–7513.

Basanez G, Nieva JL, Goni FM, and Alonso A. 1996b. Origin of the lag period in the phospholipase C cleavage of phospholipids in membranes. Concomitant vesicle aggregation and enzyme activation. Biochemistry 35:15183–15187.

Bauldry SA and Wooten RE. 1997. Induction of cytosolic phospholipase A_2 by phosphatidic acid and diglycerides in permeabilized human neutrophils: interrelationship between phospholipase D and A_2. Biochem J 322:353–363.

Bayburt T and Gelb MH. 1997. Interfacial catalysis by human 85 kDa cytosolic phospholipase A_2 on anionic vesicles in the scooting mode. Biochemistry 36:3216–3231.

Bennet CF, Balcarek JM, Varrichio A, and Crooke ST. 1988. Molecular cloning and complete amino acid sequence of form-1 phosphoinositide-specific phospholipase C. Nature 334:268–270.

Bertagnolo V, Mazzoni M, Ricci D, Carini C, Neri LM, Previati M, and Capitani S. 1995. Identification of PI-PLC β1, γ1, and δ1 in rat liver: subcellular distribution and relationship to inositol lipid nuclear signaling. Cell Signal 7:669–678.

Bian J and Roberts MF. 1992. Comparison of surface properties and thermodynamic behavior of lyso- and diacylphosphatidylcholines. J Coll Int Sci 153:420–428.

Billah MM and Anthes JC. 1990. The regulation and cellular functions of PC hydrolysis. Biochem J 269:281–291.

Biltonen RL, Heimburg TR, Lathrop BK, and Bell JD. 1991. Molecular aspects of phospholipase A_2 activation. In: Mukherjee AB, ed. Biochemistry, molecular biology, and physiology of phospholipase A_2 and its regulatory factors. New York: Plenum Publishing Corp. 86–103.

Birrell GB, Hedberg KK, and Griffith OH. 1995. An extracellular inositol phospholipid-specific phospholipase C is released by cultured Swiss 3T3 cells. Biochem Biophys Res Commun 211:318–324.

Brindley DN, Abousalham A, Kikuchi Y, Wang CN, and Waggoner DW. 1996.[66] Cross talk between the bioactive glycerolipids and sphingolipids in signal transduction. Biochem Cell Biol 74:469–476.

Brindley DN and Waggoner DW. 1996. Phosphatidate phosphohydrolase in signal transduction. Chem Phys Lipids 80:45–57.

Brunie S, Bolin J, Gewirth D, and Sigler PB. 1985. The refined crystal structure of dimeric phospholipase A_2 at 2.5Å. Access to a shielded catalytic center. J Biol Chem 260:9742–9749.

Buergess WH, Dionne CA, Kaplow J, Mudd R, Friesel R, Zilberstein A, Schlessinger J, and Jaye M. 1990. Characterization and cDNA cloning of phospholipase C-γ, a major substrate for heparin-binding growth factor 1 (acidic fibroblast growth factor)-activated tyrosine kinase. Mol Cell Biol 10:4770–4777.

Camilli A, Goldfine H, and Portnoy DA. 1991. *Listeria monocytogenes* mutants lacking phosphatidylinositol-specific phospholipase C are avirulent. J Exp Med 173:751–754.

Carman G, Deems RA, and Dennis EA. 1995. Lipid signaling enzymes and surface dilution kinetics. J Biol Chem 270:18711–18714.

Carozzi A, Camps M, Gierschik P, and Parker P. 1993. Activation of phosphatidylinositol lipid-specific phospholipase C-β_3 by G-protein $\beta\gamma$ subunits. FEBS Lett 315:340–342.

Clark JD, Lin L, Kriz RW, Ramesha CS, Sultzman LA, Lin AY, Milona N, and Knopf JL. 1991. A novel arachidonic acid-selective cytosolic PLA_2 contains a Ca^{2+}-dependent translocation domain with homology to PKC and GAP. Cell 65:1043–1051.

Clark MA Shorr RGL and Bomalski JS. 1986. Antibodies prepared to *Bacillus cereus* phospholipase C crossreact with a phosphatidylcholine preferring phospholipase C in mammalian cells. Biochem Biophys Res Commun 140:14–119.

Cockcroft S. 1992. G-protein regulated phospholipases C, D, and A_2-mediated signaling in permeabilized neutrophils. Biochim Biophys Acta 1113:135–160.

Cockcroft S and Thomas GMH. 1992. Inositol-lipid-specific phospholipase C isoenzymes and their differential regulation by receptors. Biochem J 288:1–14.

Colley WC, Sung TC, Roll R, Jenco J, Hammond SM, Altshuller Y, Bar-Sagi D, Morris AJ, and Frohman MA. 1997. Phospholipase D2, a distinct phospholipase D isoform with novel regulatory properties that provokes cytoskeletal reorganization. Curr Biol 7:191–201.

Corssen J. 1996. Phospholipase activation and secretion: evidence that PLA_2, PLC, and PLD are not essential to exocytosis. Am J Physiol 270:C1153–C1163.

Daniele JJ, Maggio B, Bianco ID, Goni FM, Alonso A, and Fidelio GD. 1996. Inhibition by gangliosides of *Bacillus cereus* phospholipase C activity against monolayers, micelle, and bilayer vesicles. Eur J Biochem 239:105–110.

Danin M, Chalifa V, Mohn H, Schmidt U-S, and Liscovitch M. 1993. Rat brain membrane-bound phospholipase D. In: Fain JN, ed. Lipid metabolism in signaling systems. New York: Academic Press. 14–24.

Davitz MA, Hereld D, Shak S, Krakow J, Englund PT, and Nussenzweig V. 1987. A glycanphosphatidylinositol-specific phospholipase D in human serum. Science 238:81–84.

DeHaas GH, Bonsen PPM, Pieterson WA, and Van Deenen LLM. 1971. Studies on phospholipase A_2 and its zymogen from porcine pancreas. Biochim Biophys Acta 239:252–266.

Dennis EA. 1997. The growing phospholipase A_2 superfamily of signal transduction enzymes. Trends Biochem Sci 22:1–2.

Dennis EA. 1973. Phospholipase A_2 activity towards phosphatidylcholine in mixed micelles: surface dilution kinetics and the effects of thermotropic phase transitions. Arch Biochem Biophys 158:485–493.

Dijkstra BW, Kalk KH, Hol WGJ, and Drenth J. 1981. Structure of bovine pancreatic phospholipase A$_2$ at 1.7 Å. J Mol Biol 147:97–123.

Dijkstra BW, Renetseder R, Kalk KH, Hol WGJ, and Drenth J. 1983. Structure of porcine pancreatic phospholipase A$_2$ at 2.6Å resolution and comparison with bovine phospholipase A$_2$. J Mol Biol 168:163–179.

Dua R, Wu S-K, and Cho W. 1995. A structure-function study of bovine pancreatic phospholipase A$_2$ using polymerized mixed liposomes. J Biol Chem 270:263–268.

Ella KM, Dolan JW, and Meier KE. 1995. Characterization of a regulated form of phospholipase D in the yeast *Saccharomyces cerevisiae*. Biochem J 307:799–805.

Ellis MV and Katan M. 1995. Mutations within a highly conserved sequence present in the X region of phosphoinositide-specific phospholipase C δ1. Biochem J 307:69–75.

Emori Y, Homma Y, Sorimachi H, Kawasaki H, Nakanishi O, Suzuki K, and Takenawa T. 1989. A second type of rat phosphoinositide-specific phospholipase C containing a src-related sequence not essential for phosphoinositide hydrolyzing activity. J Biol Chem 264:21885–21890.

English D. 1996. Phosphatidic acid: a lipid messenger involved in intracellular and extracellular signaling. Cell Signal 8:341–347.

Essen L-O, Perisic O, Cheung R, Katan M, and Williams R. 1996. Crystal structure of a mammalian phosphoinositide-specific phospholipase Cδ. Nature 380:595–602.

Essen L-O, Perisic O, Katan M, Wu Y, Roberts MF, and Williams R. 1997. Structural mapping of the catalytic mechanism for a mammalian phosphoinositide-specific phospholipase C. Biochemistry 36:1704–1718.

Exton J. 1990. Signaling through PC breakdown. J Biol Chem 265:-4.

Exton JH. 1997. Phospholipase D: Enzymology, mechanisms of regulation, and function. Physiol Rev 77:303–320.

Farooqui AA, Rapoport SI, and Horrocks LA. 1997. Membrane phospholipid alterations in Alzheimer's disease: deficiency of ethanolamine plasmalogens. Neurochem Res 22:523–527.

Farooqui AA, Yang HC, and Horrocks LA. 1995. Plasmalogens, phospholipases A$_2$ and signal transduction. Brain Res Rev 21:152–161.

Fensome A, Cunningham E, Prosser S, Tan SK, Swigart P, Thomas G, Hsuan J, and Cockcroft S. 1996. ARF and PITP restore GTPγS-stimulated protein secretion from cytosol-depleted HL60 cells by promoting PIP$_2$ synthesis. Curr Biol 6:730–738.

Ferguson MA Low MG and Cross GAM. 1985. Glycosyl-*sn*-1,2-dimyristoyl-phosphatidylinositol is covalently linked to *Trypanosoma brucei* variant surface glycoprotein. J Biol Chem 260:14547–14555.

Fisher A, Dodia C, and Chander A. 1994. Inhibition of lung calcium-independent phospholipase A$_2$ by surfactant protein A. Am J Physiol 267:L335–L341.

Fisher A, Dodia C, Chander A, and Jain M. 1992. A competitive inhibitor of phospholipase A_2 decreases surfactant phosphatidylcholine degradation by the rat lung. Biochem J 288:407–411.

Fisher AB and Dodia C. 1997. Role of acidic Ca^{2+}-independent phospholipase A_2 in synthesis of lung dipalmitoyl phosphatidylcholine. Am J Physiol 272:L238–L243.

Friedman P, Markman O, Haimovitz R, Roberts MF, and Shinitzky M. 1995. Conversion of lysophospholipids to cyclic lysophosphatidic acid by phospholipase D. J Biol Chem 271:953–957.

Garigapati V, Bian J, and Roberts MF. 1995. Synthesis and characterization of short-chain diacyl phosphatidic acids. J Coll & Int Sc. 169:486–492.

Gelb MH, Jain MK, Hanel AM, and Berg OG. 1995. Interfacial enzymology of glycerolipid hydrolases: Lessons from secreted phospholipases A_2. Annu Rev Biochem 64:653–688.

Griffith OH, Volwerk JJ, and Kuppe A. 1991. Phosphatidylinositol-specific phospholipase C from *Bacillus cereus* and *Bacillus thuringiensis*. Methods Enzymol 197:493–502.

Grobler JA and Hurley JH. 1996. Expression, characterization, and crystallization of the catalytic core of rat phosphatidylinositide-specific phospholipase C δ1. Protein Sci 5:680–686.

Guillouard I, Garnier T, and Cole ST. 1996. Use of site-directed mutagenesis to probe structure-function relationships of α-toxin from *Clostridium perfringens*. Infect Immunol 64:2440–2444.

Hack CE, Wolbink GJ, Schalkwijk C, Speijer H, Hermens WT, and van den Bosch H. 1997. A role for the secretory phospholipase A_2 and C-reactive protein in the removal of injured cells. Immunol Today 18:111–115.

Hammond SM, Altshuller YM, Sung TC, Rudge SA, Rose K, Engebrecht J, Morris AJ, and Frohman MA. 1995. Human ADP-ribosylation factor-activated phosphatidylcholine-specific phospholipase D defines a new and highly conserved gene family. J Biol Chem 270:29640–29643.

Han SK, Yoon ET, Scott DL, Sigler PB, and Cho W. 1997. Structural aspects of interfacial adsorption. A crystallographic and site-directed mutagenesis study of the phospholipase A_2 from the venom of *Agkistrodon piscivorus piscivorus*. J Biol Chem 272:3573–3582.

Hansen S, Hansen LK, and Hough E. 1992. Crystal structures of phosphate, iodide, and iodate-inhibited phospholipase C from *Bacillus cereus* and structural investigations of the binding of reaction products and a substrate analogue. J Mol Biol 225:543–549.

Hansen S, Hough E, Svensson LA, Wong Y-L, and Martin SF. 1993. Crystal structure of phospholipase C from *Bacillus cereus* complexed with a substrate analog. J Mol Biol 234:79–187.

Harlan JE, Hajduk PJ, Yoon HS, and Fesik SW. 1994. Pleckstrin homology domains bind to phosphatidylinositol-4,5-bisphosphate. Nature 371:168–170.

Hattori K, Hattori M, Adachi H, Tsujimoto M, Artai H, and Inoue K. 1995. Purification and characterization of platelet-activating factor acetylhydrolase II from bovine liver cytosol. J Biol Chem 270:22308–22313.

Hattori M, Adachi H, Tsujimoto M, Arai H, and Inoue K. 1994. The catalytic subunit of bovine brain platelet-activating factor acetylhydrolase is a novel type of serine esterase. J Biol Chem 269:23150–23155.

Hattori M, Adachi H, Tsujimoto M, Arai H, and Inoue K. 1994. Miller-Dieker Iissencephaly gene encodes a subunit of brain platelet-activating factor. Nature 370:216–218.

Hattori M, Arai H, and Inoue K. 1993. Purification and characterization of bovine brain platelet-activating factor acetylhydrolase. J Biol Chem 268: 8748–18753.

Heinz DW, Ryan M, Bullock T, and Griffith OH. 1995. Crystal structure of the phosphatidylinositol-specific phospholipase C from *Bacillus cereus* in complex with *myo*-inositol. EMBO J 14:3855–3863.

Heinz DW, Ryan M, Smith MP, Weaver LH, Keana JF, and Griffith OH. 1996. Crystal structure of phosphatidylinositol-specific phospholipase C *from Bacillus cereus* in complex with glucosaminyl(α1-6)-D-*myo*-inositol, an essential fragment of GPI anchors. Biochemistry 35:9496–9504.

Heller M. 1978. Phospholipase D. Adv Lipid Res 16:267–326.

Hendrickson HS, Banovetz C, Kirsch MJ, and Hendrickson EK. 1996. Kinetics of phosphatidylinositol-specific phospholipase C with vesicles of a thiophosphate analogue of phosphatidylinositol. Chem Phys Lipids 84:87–92.

Higgs HN and Glomset J A. 1994. Identification of a phosphatidic acid-preferring phospholipase A_1 from bovine brain and testis. Proc Natl Acad Sci USA 91:9574–9578.

Homma Y, Emori Y, Shibasaki F, Suzuki K, and Takenawa T. 1990. Isolation and characterization of a γ-type phosphoinositide specific phospholipase C (PLC-γ2). Biochem J 269:13–18.

Hou W, Arita Y, and Morisset J. 1996. Basic fibroblast growth factor-stimulated arachidonic acid release in rat pancreatic acini: sequential action of tyrosine kinase, phospholipase C, protein kinase C, and diacylglycerol lipase. Cell Signal 8:487–496.

Hough E, Hansen LK, Birknes B, Jynge K, Hansen S, Hordvik A, Little C, Dodson E, and Derewenda Z. 1989. High-resolution (1.5Å) crystal structure of phospholipase C from *Bacillus cereus*. Nature 338:57–360.

Huang PS, Davis L, Huber H, Goodhart PJ, Wegrzyn RE, Oliff A, and Heimbrook DC. 1995. An SH3 domain is required for the mitogenic activity of microinjected phospholipase C-γ1. FEBS Lett 358:287–292.

Ikezawa H and Taguchi T. 1981. Phosphatidylinositol-specific phospholipase C from *Staphylococcus aureus*. Methods Enzymol 71:731–741.

Ishizaki J, Hanasaki K, Higashino K, Kishino J, Kibuchi N, Ohara O, and Arita H. 1994. Molecular cloning of pancreatic group I phospholipase A_2 receptor. J Biol Chem 269:5897–5904.

Jager K, Stieger S, Brodbeck U. 1991. Cholinesterase Solubilizing factor from cytophaga sp. is a phosphatidylinositol-specific phospholipase C. Biochim. Biophys. Acta, 1074:45–51

Jain MK and Gelb MH. 1991. Phospholipase A_2 catalyzed hydrolysis of vesicles: uses of interfacial catalysis in the scooting mode. Methods in Enzymol 197:112–125.

Jiang Y, Lu Z, Zang Q, and Foster DA. 1996. Regulation of phosphatidic acid phosphohydrolase by epidermal growth factor. Reduced association with the EGF receptor followed by increased association with protein kinase C ε. J Biol Chem 271:29529–29532.

Johansen T, Bjorkoy G, Overvatn A, Diaz-Meco MT, Traavik T, and Moscat J. 1994. NIH 3T3 cells stably transfected with the gene encoding phosphatidylcholine-hydrolyzing phospholipase C from *Bacillus cereus* acquire a transformed phenotype. Mol Cell Biol 14:646–654.

Johansen T, Holm T, Guddal PH, Sletten K, Haugli FB, and Little C. 1988. Cloning and sequencing of the gene encoding the phosphatidylcholine-preferring phospholipase C of *Bacillus cereus*. Gene 65:293–304.

Kahn RA, Terui T, and Randazzo PA. 1996. Effects of acid phospholipids on ARF activities: potential roles in membrane traffic. J Lipid Mediat Cell Signal 14:209–214.

Kanfer J. 1980. The base exchange enzymes and phospholipase D of mammalian tissue. Can J Biochem 58:1370–1380.

Kim HK, Kim JW, Zilberstein A, Margolis B, Kim JG, Schlessinger J, and Rhee SG. 1991. PDGF Stimulation of inositol prospholipid hydrolysis requires PLC-gamma 1 phosphorylation on tyrosine residues 783 and 1254. Cell, 65:435–441.

Kim JW, Ryu SH, and Rhee SG. 1989. Cyclic and noncyclic inositol phosphates are formed at different ratios by phospholipase C isozymes. Biochem Biophys Res Commun 163:177–182.

Koblan KS, Schaber MD, Edwards G, Gibbs JB, and Pompliano DL. 1995. Src-homology 2 (SH2) domain ligation as an allosteric regulator: Modulation of phosphoinositide-specific phospholipase Cγ1 structure and activity. Biochem J 305:745–751.

Kramer RM, Roberts EF, Manetta J, and Putnam JE. 1991. The Ca^{2+}-sensitive cytosolic phospholipase A_2 is a 100 kDA protein in human monoblast U937. J Biol Chem 266:5268–5272.

Ktistakis NT, Brown HA, Waters MG, Sternweis PC, and Roth MG. 1996. Evidence that phospholipase D mediates ADP ribosylation factor-dependent formation of Golgi coated vesicles. J Cell Biol 134:295–306.

Kudo I, Murakami M, Hara S, and Inoue K. 1993. Mammalian non-pancreatic phospholipase A_2. Biochim Biophys Acta 1170:217–231.

Kuhn B, Schmid A, Harteneck C, Gudermann T, and Schultz G. 1996. G-proteins of the G_q family couple the H2 histamine receptor to phospholipase C. Mol Endocrinol 10:1697–1707.

Kundu GC and Mukherjee AB. 1997. Evidence that porcine pancreatic phospholipase A_2 via its high affinity receptor stimulates extracellular matrix invasion by normal and cancer cells. J Biol Chem 272:2346–2353.

Lambeau G, Ancian P, Barhanin J, and Lazdunski M. 1994. Cloning and expression of a membrane receptor for secretory phospholipases A_2. J Biol Chem 269:1757–1578.

Lee KY, Ryu SH, Suh PG, Choi WC, and Rhee SG. 1987. Phospholipase C associated with particulate fractions of bovine brain. Proc Natl Acad Sci USA 84:5540–5544.

Lee YH, Lee HJ, Lee S-J, Min DS, Baek SH, Kim YS, Ryu SH, and Suh P-G. 1995. Down-regulation of phospholipase C-γ1 during the differentiation of U937 cells. FEBS Lett 358:105–108.

Leimeister-Wachter M, Domann E, and Chakraborty T. 1991. Detection of a gene encoding a phosphatidylinositol-specific phospholipase C that is coordinately expressed with listeriolysin in *Listeria monocytogenes*. Mol Microbiol 5:361–366.

Lereclus D, Agaisse H, Gominet M, Salamitou S, and Sanchis V. 1996. Identification of a *Bacillus thuringiensis* gene that positively regulates transcription of the phosphatidylinositol-specific phospholipase C gene at the onset of the stationary phase. J Bacteriol 178:2749–2756.

Levine L, Xiao DM, and Little C. 1988. Increased arachidonic acid metabolites from cells in culture after treatment with the phosphatidylcholine-hydrolyzing phospholipase C from *Bacillus cereus*. Prostaglandins 34:633–642.

Lewis K, Garigapati V, Zhou C, and Roberts MF. 1993. Substrate requirements of bacterial phosphatidylinositol-specific phospholipase C. Biochemistry 32:8836–8841.

Lin G, Bennett CF, and Tsai MD. 1990. Phospholipids chiral at phosphorus. Stereochemical mechanism of reactions catalyzed by phosphatidylinositide-specific phospholipase C from *Bacillus cereus* and guinea pig uterus. Biochemistry 29:2747–2757.

Lin T-L, Chen S-H, and Roberts MF. 1987. Thermodynamic analyses of the growth and structure of asymmetric linear short-chain lecithin micelles based on small angle neutron scattering data. J Amer Chem Soc 109:2321–2328.

Little C. 1981. Effect of some divalent metal cations on phospholipase C from *Bacillus cereus*. Acta Chem Scand B35:39–44.

Liu Y and Levy R. 1997. Phospholipase A_2 has a role in proliferation but not in differentiation of HL-60 cells. Biochim Biophys Acta 1335:270–280.

Low MG. 1981. Phosphatidylinositol-specific phospholipase C from *Bacillus thuringiensis*. Methods Enzymol 71:741–746.

Low MG and Huang K-S. 1993. Phosphatidic acid, lysophosphatidic acid and lipid A are inhibitors of glycosylphosphatidylinositol-specific phospholipase D: Specific inhibition of a phospholipase by product analogues? J Biol Chem 268:8480–8490.

Low MG and Huang KS. 1991. Factors affecting the ability of glycosylphosphatidylinositol-specific phospholipase D to degrade the membrane anchors of cell surfaces. Biochem J 279:483–493.

Lucas M, Sanchez-Margalet V, Pedrera C, and Bellido ML. 1995. A chemiluminescence method to analyze phosphatidylcholine-phospholipase activity in plasma membrane preparations and in intact cells. Anal Biochem 231:277–281.

Lukowski S, Lecomte MC, Mira JP, Marin P, Gautero H, Russo-Marie F, and Geny B. 1996. Inhibition of phospholipase D activity by fodrin. An active role for the cytoskeleton. J Biol Chem 271:24164–24171.

Ma Z, Ramanadham S, Kempe K, Chi XS, Ladenson J, and Turk J. 1997. Pancreatic islets express a Ca^{2+}-independent phospholipase A_2 that contains a repeated structural motif homologous to the integral membrane protein binding domain of ankyrin. J Biol Chem 272:11118–11127.

Margolis B, Rhee SG, Felder S, Mervic M, Lyall R, Levitzki A, Ullrich A, Zilberstein A, and Schlessinger J. 1989. EGF induces tyrosine phosphorylation of phospholipase C-II: A potential mechanism for EGF receptor signaling. Cell 57:1101–1107.

Martin SF, Spaller MR, and Hegenrother PJ. 1996. Expression and site-directed mutagenesis of the phosphatidylcholine-preferring phospholipase C of *Bacillus cereus*: probing the role of the active site Glu146. Biochemistry 35: 12970–12977.

McDonald LJ and Mamrack MD. 1995. Phosphoinositide hydrolysis by phospholipase C modulated by multivalent cations La^{3+}, Al^{3+}, neomycin, polyamines, and melittin. J Lipid Mediat Cell Signal 11:81–91.

McNamara PJ, Cuevas WA, and Songer JG. 1995. Toxic phospholipases D of *Corynebacterium pseudotuberculosis, C. ulcerans* and *Arcanobacterium haemolyticum*: cloning and sequence homology. Gene 156:113–118.

Mengaud J, Braun-Breton C, and Cossart P. 1991. Identification of phosphatidylinositol-specific phospholipase C activity in *Listeria monocytogenes*: A novel type of virulence factor? Mol Microbiol 5:367–372.

Mensa-Wilmot K and Englund PT. 1992. Glycosyl phosphatidylinositol-specific phospholipase C of *Trypanosoma brucei*: expression in *Escherichia coli*. Mol Biochem Parasitol 56:311–322.

Morris JC, Lei P-S, Shen TY, and Kojo MW. 1995. Glycan requirements of glycosylphosphatidylinositol phospholipase C from *Trypanosoma brucei*. J Biol Chem 270:2517–2524.

Morris JC, Ping-Sheng L, Zhai HX, Shen TY, and Mensa-Wilmot K. 1996. Phosphatidylinositol phospholipase C is activated allosterically by the aminoglycoside G418. 2-deoxy-2-fluoro-*scyllo*-inositol-1-O-dodecylphosphonate and its analogs inhibit glycosylphosphatidylinositol phospholipase C. J Biol Chem 271:15468–15477.

Nakamura S. 1996. Phosphatidylcholine hydrolysis and protein kinase C activation for intracellular signaling network. J Lipid Mediat Cell Signal 14:197–202.

Nieva JL, Goni FM, and Alonso A. 1993. Phospholipase C-promoted membrane fusion: Retroinhibition by the end product diacylglycerol. Biochemistry 32:1054–1058.

Nishizuka Y. 1992. Intracellular signaling by hydrolysis of phospholipids and activation of protein kinase C. Science 258:607–614.

Offermanns S and Simon MI. 1995. Gα15 and Gα16 couple a wide variety of receptors to phospholipase C. J Biol Chem 270:15175–15180.

Olson SC and Lambeth JD. 1996. Biochemistry and cell biology of phospholipase D in human neutrophils. Chem Phys Lipids 80:3–19.

Pascher I, Lundmark M, Nyholm P-G, and Sundell S. 1992. Crystal structures of phospholipids. Biochim Biophys Acta 1113:339–373.

Pelech SL and Vance DE. 1989. Signal transduction via phosphatidylcholine cycles. Trends Biochem Sci 14:28–30.

Pertile P, Liscovitch M, Chalifa V, and Cantley LC. 1995. Phosphatidylinositol 4,5-bisphosphate synthesis is required for activation of phospholipase D in U937 cells. J Biol Chem 270:5130–5135.

Ponting CP and Kerr ID. 1996. A novel family of phospholipase D homologues that includes phospholipid synthases and putative endonucleases: identification of duplicated repeats and potential active site residues. Protein Sci 5:914–922.

Qiu Z-H, de Carvalho MS, and Leslie CC. 1993. Regulation of phospholipase A_2 activation by phosphorylation in mouse peritoneal macrophages. J Biol Chem 268:24506–24513.

Renetseder R, Dijkstra BW, Huizing K, Kalk KH, and Drenth J. 1988. Crystal structure of bovine pancreatic phospholipase A_2 covalently inhibited by p-bromophenacylbromide. J Mol Biol 200:181–188.

Reynolds LJ, Hughes LL, Louis AI, Kramer RM, and Dennis EA. 1993. Metal ion and salt effects on the phospholipase A_2, lysophospholipase, and transacylase activities of human cytosolic phospholipase A_2. Biochim Biophys Acta 1167:272–280.

Rhee SG and Choi KD. 1992. Regulation of inositol phospholipid-specific phospholipase C isozymes. J Biol Chem 267:12393–12396.

Rhee SG, Suh PG, Ryu S-H, and Lee KY. 1989. Studies of inositol phospholipid-specific phospholipase C. Science 244:546–550.

Roberts MF. 1991a. Using NMR spectroscopy to assay phospholipases. Methods Enzymol 197:31–48.

Roberts MF. 1991b. Using short-chain phospholipids to assay phospholipases. Methods Enzymol 197:95–112.

Roberts MF. 1996. Phospholipases: structural and functional motifs for working at an interface. FASEB J 10:1159–1172.

Ryan M, Smith MP, Vinod TK, Lau WL, Keana JF, and Griffith OH. 1996. Synthesis, structure-activity relationships, and the effect of polyethylene glycol on inhibitors of phosphatidylinositol-specific phospholipase C from *Bacillus cereus*. J Med Chem 39:4366–4376.

Sands WA, Clark JS, and Liew FY. 1994. The role of a phosphatidylcholine-specific phospholipase C in the production of diacylglycerol for nitric oxide synthesis in macrophages activated by IFN-γ and LPS. Biochem Biophys Res Commun 199:461–466.

Sato T, Aoki J, Nagai Y, Dohmae N, Takio K, Doi T, and Inoue K. 1997. Serine phospholipid-specific phospholipase A that is secreted from activated platelets. A new member of the lipase family. J Biol Chem 272:2192–2198.

Schlessinger J. 1994. SH2/SH3 signaling proteins. Curr Opin Genet Dev 4:25–30.

Scott DL, White SP, Otwinowski Z, Yuan W, Gelb MH, and Sigler PB. 1990. Interfacial catalysis: The mechanism of phospholipase A_2. Science 250:1541–1546.

Scott DL and Sigler PB. 1995. Structure and catalytic mechanism of secretory phospholipase A_2. Adv Protein Chem 45:53–88.

Scott DL, White SP, Browning JL, Rosa JJ, Gelb MH, and Sigler PB. 1991. Structures of free and inhibited human secretory phospholipase. A_2 from inflammatory exudate. Science 254:1007–1010.

Sharp JD, Pickard RT, Chiou XG, Manetta JV, Kovacevic S, Miller JR, Varshavsky AD, Roberts EF, Strifler BA, Brems DN, and Kramer RM. 1994. Serine 228 is essential for catalytic activities of 85-kDa cytosolic phospholipase A_2. J Biol Chem 269:23250–23254.

Sheffield MJ, Baker BL, Owen NL, Baker ML, and Bell JD. 1995. Enhancement of *Agkistrodon piscivorus piscivorus* venom phospholipase A_2 activity toward phosphatidylcholine vesicles by lysolecithin and palmitic acid: Studies with fluorescent probes of membrane structure. Biochemistry 34:7796–7806.

Siddiqi AR, Smith JL, Ross AH, Qiu RG, Symons M, and Exton JH. 1995. Regulation of phospholipase D in HL60 cells. Evidence for a cytosolic phospholipase D. J Biol Chem 270:8466–8473.

Simon MI, Strathman MP, and Gautam N. 1991. Diversity of G-proteins in signal transduction. Science 252:802–808.

Smith MR, Ryu S-H, Suh PG, Rhee SG, and Kung H-F. 1989. S-phase induction and transformation of quiescent NIH 3T3 cells by microinjection of phospholipase C. Proc Natl Acad Sci USA 86:3659–3663.

Smrcka AV, Hepler JR, Brown KO, and Sternweis PO. 1991. Regulation of polyphosphoinositide-specific phospholipase C activity by purified Gq. Science 250:804–807.

Snitko Y, Yoon ET, and Cho W. 1997. High specificity of human secretory class II phospholipase A_2 for phosphatidic acid. Biochem J 321:737–741.

Soltys CE and Roberts MF. 1994. Fluorescence studies of phosphatidylcholine micelle mixing: Relevance to phospholipase kinetics. Biochemistry 33:11608–11617.

Soltys CE, Bian J, and Roberts MF. 1993. Polymerizable phosphatidylcholines: Importance of phospholipid motions for optimum phospholipase A_2 and C activity. Biochemistry 32:9545–9552.

Spiegel S and Milstien S. 1996. Sphingoid bases and phospholipase D activation. Chem Phys Lipids 80:27–36.

Spiegel S, Foster D, and Kolesnick R. 1996. Signal transduction through lipid second messengers. Curr Opin Cell Biol 8:159–167.

Srivastava SP, Chen NQ, Liu YX, and Holtzman JL. 1991. Purification and characterization of a new isozyme of thiol:protein-disulfide oxidoreductase from rat hepatic microsomes. J Biol Chem 266:20337–20344.

Street IP, Lin H-K, Laliberte F, Ghomashchi F, Wang Z, Perrier H, Tremblay NM, Huang Z, Weech PK, and Gelb MH. 1993. Slow- and tight-binding inhibitors of the 85-kDa human phospholipase A$_2$. Biochemistry 32:5935–5940.

Sundell S, Hansen S, and Hough E. 1994. A proposal for the catalytic mechanism in phospholipase C based on interaction energy and distance geometry calculations. Prot Engineer 7:571–577.

Sutton RB, Davtetov BA, Berhuis AM, Sudhof TC, and Sprang SR. 1995. Structure of the first C-2 domain of synaptotagmin. A novel Ca^{2+}/phospholipid binding mode. Cell 80:929–938.

Swairjo M, Seaton BA, and Roberts MF. 1994. Effect of vesicle composition and curvature on the dissociation of phosphatidic acid in small unilamellar vesicles—a ^{31}P NMR study. Biochim Biophys Acta 1191:354–361.

Taguchi R and Ikezawa H. 1978. Phosphatidylinositol-specific phospholipase C from *Clostridium novyi* type A. Arch Biochem Biophys 186:196–201.

Tan T, Hehir MJ, and Roberts MF. 1997. Cloning, overexpression, refolding, and purification of the non-specific phospholipase C from *Bacillus cereus*. Prot Expr & Purif 10:365–372.

Tan T and Roberts MF. 1996. Vanadate is a potent competitive inhibitor of phospholipase C from *Bacillus cereus*. Biochim Biophys Acta 1298:58–68.

Tang J, Kriz R, Wolfman N, Shaffer M, Seehra J, and Jones SS. 1997. A novel cytosolic calcium-independent phospholipase A$_2$ contains eight ankyrin motifs. J Biol Chem 272:8567–8575.

Taylor SJ, Chae HZ, Rhee SG, and Exton JH. 1991. Activation of the β1 isozyme of phospholipase C by α subunits of the Gq class of G-proteins. Nature 350:516–518.

Thunnissen MMGM, Ab E, Kalk KH, Drenth J, Dijkstra BW, Kuipers OP, Dijkman R, de Haas GH, and Verheij HM. 1990. X-ray structure of phospholipase A$_2$ complexed with a substrate derived inhibitor. Nature 347:689–691.

Van den Berg B, Tessari M, Boelens R, Dijkman R, Kaptein R, de Haas GH, and Verheij HM. 1995a. Solution structure of porcine pancreatic phospholipase A$_2$ complexed with micelles and a competitive inhibitor. J Biomol NMR 5:110–121.

Van den Berg B, Tessari M, de Haas GH, Verheij HM, Boelens R, and Kaptein R. 1995b. Solution structure of porcine pancreatic phospholipase A$_2$. EMBO J 14:4123–4131.

van Dijk MCM, Muriana FJG, de Widt J, Hilkmann H, and van Blitterswijk WJ. 1997. Involvement of phosphatidylcholine-specific phospholipase C in platelet-derived growth factor-induced activation of the mitogen-activated protein kinase pathway in rat-1 fibroblasts. J Biol Chem 272:11011–11016.

Vega QC, Cochet C, Filhol O, Chang CP, Rhee SG, and Gill GN. 1992. A site of tyrosine phosphorylation in the C terminus of the epidermal growth factor receptor is required to activate phospholipase C. Mol Cell Biol 12:128–135.

Venable ME, Bielawska A, and Obeid LM. 1996. Ceramide inhibits phospholipase D in a cell-free system. J Biol Chem 271:24800–24805.

Verheij HM, Slotboom AJ, and De Haas GH. 1981. Phospholipase A$_2$: a model for membrane-bound enzymes. Rev Physiol Pharmacol 91:91–203.

Vinggaard AM, Jensen T, Morgan CP, Cockcroft S, and Hansen HS. 1996. Didecanoyl phospatidylcholine is a superior substrate for assaying mammalian phospholipase D. Biochem J 319:861–864.

Volwerk JJ, Shashidhar MS, and Kuppe A. 1990. PI-specific PLC from *Bacillus cereus* combines intrinsic phosphotransferase and cyclic phosphodiesterase activities: a ^{31}P NMR study. Biochemistry 29:8056–8062.

Waggoner DW, Martin A, Dewald J, Gomez-Munoz A, and Brindley DN. 1995. Purification and characterization of a novel plasma membrane phosphatidate phosphohydrolase from rat liver. J Biol Chem 270:19422–19429.

Wahl MI, Jones GA, Nishibe S, Rhee SG, and Carpenter G. 1992. Growth factor stimulation of phospholipase C-γ1 activity. J Biol Chem 267:10447–10456.

Wang X. 1993. Phospholipases. In: Moore TS, ed. Lipid metabolism in plants. Boca Raton: CRC Press. 505–512.

Wery J-P, Schevitz RW, Clawson D, Bobbitt JL, Dow ER, Gamboa G, Goodson T, Hermann RB, Kramer RM, McClure DB, Mihelich ED, Putnam JE, Sharp J, Stark DH, Teater C, Warrick MW, and Jones ND. 1991. Structure of recombinant human rheumatoid arthritic synovial fluid phospholipase A$_2$ at 2.2 Å resolution. Nature 352:79–82.

White SP, Scott DL, Otwinowski Z, Gelb MH, and Sigler PB. 1990. Crystal structure of cobra-venom phospholipase A$_2$ in a complex with a transition-state analogue. Science 250:1560–1563.

Wolf M and Gross RW. 1996. Expression, purification, and kinetic characterization of a recombinant 80-kDa intracellular calcium-dependent phospholipase A$_2$. J Biol Chem 271:30879–30885.

Wu L, Niemeyer B, Colly N, Socolich M, and Zuker CS. 1995. Regulation of PLC-mediated signaling *in vivo* by CDP-diacylglycerol synthase. Nature 373:216–222.

Wu WI, Lin YP, Wang E, Merrill AHJ, and Carman GM. 1993. Regulation of phosphatidate phosphatase activity from the yeast saccharomyces cerevisiae by sphingoid bases. J Biol Chem 268:13830–13837.

Wu Y, Williams RL, Katan M, and Roberts MF. 1997a. Phosphatidylinositol-specific phospholipase C δ1 activity towards micellar substrates, inositol 1,2-cyclic phosphate and other water soluble substrates: A sequential mechanism and allosteric activation. Biochemistry 36:11223–11233.

Wu Y, Zhou C, and Roberts MF. 1997b. Stereocontrolled syntheses of water soluble inhibitors of phosphatidylinositol-specific phospholipase C: Inhibition enhanced by an interface. Biochemistry 36:356–363.

Yu BZ, Ghomashchi F, Cajal Y, Annand RR, Berg OG, Gelb MH, and Jain MK. 1997. Use of an imperfect neutral diluent and outer vesicle layer scooting mode hydrolysis to analyze the interfacial kinetics, inhibition, and substrate preferences of bee venom phospholipase A$_2$. Biochemistry 36:3870–3881.

Zhou C and Roberts MF. 1997. Diacylglycerol partitioning and mixing in detergent micelles: Relevance to enzyme kinetics. Biochim Biophys Acta 1348:273–286.

Zhou C, Qian X, and Roberts MF. 1997a. Allosteric activation of phosphatidylinositol-specific phospholipase C: Phospholipid binding anchors the enzyme to the interface. Biochemistry 36:10089–10097.

Zhou C, Wu Y, and Roberts MF. 1997b. Activation of phosphatidylinositol-specific phospholipase C towards inositol 1,2-(cyclic)-phosphate. Biochemistry 36:347–355.

6

Cyclic Nucleotide Phosphodiesterases: Structural and Functional Aspects

Dolores J. Takemoto

INTRODUCTION

As progress was being made on the mechanism of G-protein activation of adenylate cyclase, similar studies were being conducted to determine the mechanism of degradation of cyclic AMP. The term "cyclic nucleotide phosphodiesterase" includes a variety of enzymes that catalyze the reaction: $3',5'$-cyclic nucleotide monophosphate + $H_2O \rightarrow 5'$-nucleotide monophosphate. These enzymes are specific for the cyclic diester and will not hydrolyze the $2'3'$-cyclic nucleotides that occur as intermediates of nucleic acids.

Since the discovery of cyclic AMP phosphodiesterase activity (Butcher and Sutherland 1962) it has been reported that the enzymatic activity displayed anomalous kinetic behavior that was suggestive of multiple enzymes (Appleman and Kemp 1966, Blecher et al 1968, Brooker et al 1968, Senft et al 1968). Since that time the field has progressed from the first separation of multiple forms of phosphodiesterase (Thompson and Appleman 1971) to current studies using cloned and sequenced phosphodiesterases (see Yan et al 1996 and Bloom and Beavo 1996, to list a few).

This chapter has been divided into two sections based on the developments in the field. Section One will give a historical perspective of the field of phosphodiesterase research up to the time when enough evidence had accumulated to suggest that the multiple forms of the enzymes were due to distinct proteins. Section Two will extend this to current knowledge of the structure of these enzymes, including recent data on the domain organization of each

Introduction to Cellular Signal Transduction
A. Sitaramayya, Editor
©1999 Birkhäuser Boston

form and the current classification scheme. This chapter is not meant to be all-inclusive, and the reader is referred to several excellent reviews on this topic throughout this chapter. In addition, there are now two Web sites for current information on these enzymes (*http://www.ibls.gla.ac.uk/IBLS/Staff/m-houslay/houslaylab.html* and *http//www.weber.u.washington.edu~pde/*).

MULTIPLE FORMS

The Assay

In any scientific field, nothing can be done until a method of measurement is devised. This was true for the discovery of cyclic AMP and for subsequent work on the phosphodiesterases. In the late 1950s Sutherland and Rall (1958) discovered a molecule, cyclic AMP, that was found to mediate the actions of certain hormones such as epinephrine (Sutherland and Rall, 1958). This discovery led to the development of the second messenger hypothesis for cyclic AMP, which means that the primary messenger, say, a hormone, binds to the cell surface receptor and causes the production of a second messenger such as cyclic AMP. It is now well established that the production of cyclic AMP is effected by adenylate cyclase that is coupled to a G-protein. This subject is covered in another chapter within this text.

At about the same time that cyclic AMP phosphodiesterase was first measured in beef heart (Butcher and Sutherland 1962), it was also measured in rabbit tissue (Drummond and Perrott-Yee, 1961), dog heart (Nair 1966), and rat brain (Cheung, 1967). This work suggested that most mammalian tissues contain a soluble phosphodiesterase activity with a K_m of around 10^{-4} M for cyclic AMP and that this activity is inhibited by methylxanthines. The actual K_m turned out to be much more in tune with the cellular levels of cyclic AMP, once the K_m was measured using a more sensitive isotopic phosphodiesterase assay (Brooker et al 1968). This assay, developed during the same time as the isotope dilution assay was set up to measure cyclic nucleotides, was used to suggest that there is more than a single form of phosphodiesterase present in a single tissue. It was found, using very low substrate concentrations, that rat brain homogenates contain enzymatic activities with two apparent K_ms of 100 μM and 1 μM (Brooker et al 1968).

Prior to that time phosphodiesterase products were directly measured by using radioactive substrate and chromatographic separation of 5'AMP (Cheung 1967a, 1969, Mandel and Kuehl 1967). In the assay of Brooker et al (1968) the 5'AMP was further reacted with excess 5'nucleotidase to yield the nucleoside and inorganic phosphate. Since the nucleoside was labeled, it could be separated from the phosphate-containing other products by some type of ion exchange resin. The choice of a source of 5'nucleotidase was determined by analyzing many different samples of snake venom. It was finally decided that king cobra (*Ophiophagus hana*) provided the best results, although many labs use rattlesnake venom (*Crotalus atrox*). Both are now commercially available. Separation of products was accomplished by anion exchange columns (Hynie et al 1966, DeLange et al 1968, Beavo et al 1970), paper chromatography (Cheung et al 1967a, Rosen 1970, Gulyassy and Oken 1971, Schroder and Plageman 1972), and by precipitation (Sobel et al 1968, Poch 1971, Schonhofer et al 1972). Although other methods were suggested, such as high-speed liquid chromatography (Pennington 1971), the use of highly specific radioactive cyclic nucleotides and snake venom nucleotidase was generally adopted, because it was sensitive, specific, and convenient. The separation of product is now generally accomplished by anion exchange. At the time of the isotopic assay performed by Brooker et al, this was only just beginning to be available. At one time during the assay setup, batches of resin showed extremely high backgrounds. It was later discovered that the supplier had to change the commercial synthesis method in order to conform to new OSHA specifications. This new method produced batches of resin that were not usable until it was discovered that including ethanol in the resin cut the background back to normal levels. At a much later time, when a very rapid assay was required for the retinal phosphodiesterase, a proton release assay became a method of choice for the retinal enzyme (Cheung 1969, Liebman and Evanczuk 1982).

When I entered the lab of Dr. Michael Appleman, in 1974, the assay was already well established and routine. Nevertheless, as an entering graduate student I was told that, come the next day, I would have to milk the rattlesnakes for their 5'nucleotidase. After a rather sleepless night I was informed that this substance was already commercially available.

At that time the optimal conditions for measuring phosphodiesterase in tissues were not fully established. Thus, multiple assays were run daily, without the now commonly used throwaway tubes. Some of the parameters being worked out were: pH (7.5–8.5), divalent metal requirement (Mg^{2+} or Mn^{2+} at around 1 mM), and substrate concentration (anywhere from 1 mM to 0.1 µM).

Histological assays for cyclic nucleotide phosphodiesterases were also established at that time (Goren et al 1971), using lead to detect the inorganic phosphate released from the 5'AMP by alkaline phosphatase, and using a coupled enzyme system of myokinase (Christiansen and Monn 1971). Because of the high substrate requirements, these assays did not always detect high-affinity phosphodiesterases.

In 1963 cyclic GMP was discovered in rat urine (Ashman et al 1963) and subsequently in other tissues (Robison et al 1971, Goldberg et al 1973, Goldberg and Haddox 1977). Although a cyclic GMP phosphodiesterase was first measured in 1970 (Beavo et al 1970), it was not until 1973 (Russell et al 1973) that a separate enzymatic activity having cyclic GMP as a preferred substrate was discovered. This assay used the same procedure, but with isotopically-labeled cyclic GMP.

By the early 1970s evidence for multiple forms of phosphodiesterase in a single tissue had accumulated. For example, rat brain preparations showed kinetic behavior that was suggestive of more than a single type of phosphodiesterase (Brooker et al 1968). Extrapolation of the data yielded two K_ms of 100 μM and 1 μM for cyclic AMP. The rat brain preparation also hydrolyzed cyclic GMP, but this cyclic nucleotide did not interfere with cyclic AMP hydrolysis at low concentrations, further suggesting the presence of more than a single form of phosphodiesterase.

PRESENCE OF MULTIPLE FORMS OF PHOSPHODIESTERASE

In 1970, Rosen (1970) showed that chromatographically separable forms of phosphodiesterase were present in frog erythrocytes. Before this was achieved the kinetic data was only suggestive of multiple forms. Multiple forms of phosphodiesterase were first found in rat brain cortex using agarose gel filtration (Thompson and Appleman, 1971a). This was also found to be true of rat adipose tissue, heart, skeletal muscle, and kidney (Thompson and Appleman, 1971b). This early work also suggested that the forms were either multisubunit species or aggregates. The higher molecular weight form, 400,000, had a K_m of 100 μM for cyclic AMP, and this was separable from a higher affinity form with a K_m of around 5 μM for cyclic AMP. Only the higher molecular weight form could hydrolyze cyclic GMP, and with a slightly higher affinity than for cyclic AMP. The lower molecular weight form was quite specific for cylic AMP and showed negative cooperativity (Russell et al 1972). The presence of multiple forms of phosphodiesterase with differing substrate specificities was confirmed by other groups shortly thereafter (Klotz et al 1972; Kakiuchi et al 1971). The presence of a single lower molecular weight species of 200 kDa, which exhibited a high affinity for cyclic AMP and negative cooperativity, was of particular interest to these early studies. It is well known that this type of kinetic behavior could mean either that there was a single enzyme species with true negative cooperativity or that more than a single species was present in the peak obtained from the agarose gel filtration column. However, further purification on gel filtration (Sephadex G-200) and DEAE-cellulose ion exchange columns suggested that this was, indeed, a single enzyme form (Russell et al 1972). On the basis of computer modeling, Russell et al. (1972) proposed a negatively cooperative phosphodiesterase with allosteric regulation. This hypothesis suggested that binding of cyclic AMP to the first high-affinity site would induce a conformational change that could lead to lower affinity at another site. Thus, a hormone would activate adenylate cyclase, raising cyclic AMP levels. First, the phosphodiesterase would act in concert with the cyclase to establish the new steady state level of cyclic AMP, then, the lower affinity site would more slowly return the cyclic AMP levels to the basal resting state. It was felt that a negatively cooperative enzyme would allow cyclic AMP levels to remain elevated long enough to maintain a hormone response; thus, the concept of "kinetic buffering" for a phosphodiesterase was established. These early studies suggested that phosphodiesterases could be hormonally regulated, as was later shown to occur with insulin (see Makino et al 1992, for a review of this subject). These findings were very exciting as it was formerly believed that the phosphodiesterases did not have special regulatory mechanisms, but were just "housekeeping" enzymes.

From all of these studies it became apparent that there were at least three types of phosphodiesterases that were found in most tissues. The most common way to separate these was by use of DEAE-cellulose and elution with a sodium acetate gradient from 0 to 1 M in 0.05 M Tris-acetate (pH 6.0). A generic DEAE-cellulose profile that became the basis of phosphodiesterase classification is shown in Figure 6-1. Fraction I was found to hydrolyze cyclic GMP preferentially at 1 μM. Fraction II hydrolyzed both cyclic AMP and cyclic GMP with relatively low affinities and with positive cooperativity (Beavo et al 1970 and Russell et al 1973). The cyclic AMP hydrolytic activity was activated by low concentrations of cyclic GMP (Russell et al 1973). This behavior had been previously reported for preparations from

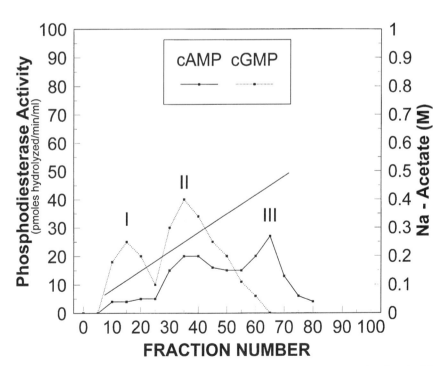

Figure 6-1. DEAE-profile of a hypothetical cellular fraction assayed using 1 μM cGMP or 1 μM cAMP as substrate.

liver (Beavo et al 1971), for thymic lymphocytes (Franks and MacManus 1971), and for adipose tissue (Klotz and Stock 1972). The original model proposed a polymeric enzyme composed of cyclic AMP and cyclic GMP phosphodiesterase subunits. As we shall see, this was not the case. Fraction III was the cyclic AMP phosphodiesterase that exhibited negative cooperativity. These fractions were often also referred to as DI, DII, and DIII phosphodiesterases, and their regulation and tissue distribution were found to vary among different tissues. The greatest progress in this area was in brain tissue.

Cheung (1970b, 1971) measured total phosphodiesterase activity and found that the activity lost during purification could be restored by a protein factor present in the extract. At the same time, it was found that the low K_m enzymes were greatly stimulated by Ca^{2+} at around 10 μM (Kakiuchi and Yamazaki 1970 and Kakiuchi et al 1971). Sensitivity to calcium was under the control of a heat-stable nondialyzable factor that was present in the brain extracts. A similar nondialyzable activating factor was also found in bovine heart (Goren and Rosen 1971, Teo et al 1973). This activator was purified and characterized as a heat-stable and pH-stable glycoprotein of 18 kDa with an isoelectric point of 4.0 (Stevens et al 1976, Watterson et al 1976). Subsequent studies showed that this factor, which we now know is calmodulin, bound two molecules of calcium, then complexed with the phosphodiesterase, lowering the K_m for cyclic AMP and increasing the V_{max} for cyclic GMP (Teo et al 1973, Wang et al 1975). The DI enzyme, which was activated by calmodulin, was partially purified (Watterson and Vanaman 1976, Ho et al 1977, Klee et al 1978) and found to have a molecular weight of 230 kDa that contained three separable bands on denaturing gels, 61 kDa, 59 kDa, and 18 kDa. The first two were later shown to be the catalytic subunits.

At the same time that the properties of the separable peaks of phosphodiesterase activities were being measured, the subcellular distribution and the effects of various activators and inhibitors were being studied. It was generally found that the low K_m cyclic AMP phosphodiesterase (DIII) was membrane-associated. This enzyme was observed in nerve endings and postsynaptic regions of brain (De Robertis et al 1967, Weiss and Costa 1968, Florendo et al 1971). The low-affinity cyclic AMP-cyclic GMP phosphodiesterase (DII) and the cyclic GMP phosphodiesterase (DI) were found to be soluble in rat liver (Russell et al 1973, Thompson et al 1973).

The pharmacological agents that were first used to inhibit phosphodiesterases in adenylate cyclase assays were the methylxanthines such as caffeine, theophylline, theobromine, and aminophylline (Sutherland and Rall 1958, Butcher and Sutherland 1962). Theophylline, for example, is a competitive inhibitor of phosphodiesterases (Butcher and Sutherland 1962, Cheung 1967a, Huang and Kemp 1971). The antibiotic puromycin was shown to be a competitive inhibitor of cyclic AMP phosphodiesterase and was useful in early studies to suggest a separate regulatory site on that enzyme (Appleman and Kemp 1966). From these early studies a very productive area of pharmacological drug design has now emerged.

At this time the question of whether the different peaks of phosphodiesterases were separate proteins had not been resolved. Lymphocytes appeared to have only DIII (Thompson et al 1976), and brain had been reported to have six enzymes, which were separable on isoelectric focusing gels (Pledger et al 1974 and Uzunov and Weiss 1972). The possibility of interconversion of subunits to create multiple forms was suggested by the finding of aggregate formation after ultracentrifugation (Pichard and Cheung 1976). In addition it became apparent that certain tissues had very distinct types of phosphodiesterases, which were not commonly present in other tissues. In 1972, Pannbacker et al (Pannbacker et al 1972) found high cyclic nucleotide phosphodiesterase activity in mammalian photoreceptors. This enzyme was found to be a light-activated enzyme that hydrolyzed cyclic GMP and required GTP and a GTPase for activity (Miki et al 1975, Wheeler and Bitensky 1977). This phosphodiesterase, which is unique to rod outer segments, was purified and characterized by Baehr et al in 1979 (Baehr et al 1979).

As of 1982, some of the multiple forms of phosphodiesterases had been purified to homogeneity on SDS gels. These initial results suggested that the phosphodiesterases were separate enzymes and not interconvertible forms of a single enzyme. At the Symposium on Cyclic Nucleotide Phosphodiesterases, the subcommittee on nomenclature (Drs. Appleman, J. Beavo, J. Hardman, C.B. Klee, and W.J. Thompson) suggested the following classification scheme:

Type I: Calmodulin-sensitive cyclic nucleotide phosphodiesterase
Type II: cGMP-sensitive cyclic nucleotide phosphodiesterase
Type III: Rhodopsin-sensitive cyclic GMP phosphodiesterase
Type IV: Cyclic AMP phosphodiesterase
(from Greengard and Robison, 1984)

Of these enzymes the least well characterized was the Type IV, which appeared to be membrane-bound and to have insulin sensitivity. The Type I (calmodulin-dependent) phoshodiesterase had been extensively documented for about 10 years by that time, and the enzyme's affinity for calmodulin had been used as a method for purification of the enzyme from brain and heart (Klee et al 1979, Klee and Krinks 1978, Watterson and Vanaman 1976). The enzyme initially copurified with another calmodulin-dependent protein, calcineurin, a phosphatase (Stewart et al 1982). However, this phosphatase could be separated from the phosphodiesterase by chromatofocusing, yielding a band of 58 kDa from brain, and one of around 57 kDa from heart (Krinks et al 1984). However, tryptic peptide maps of these enzymes suggested a large degree of homology (Krinks et al 1984).

The calmodulin-sensitive, Type I, enzyme could be treated by limited prote-olysis to yield an active, calmodulin-insensitive domain, the catalytic do-main, and a calmodulin binding domain. A model was proposed in which binding of calmodulin altered the conformation of the phosphodiesterase, thus increasing catalytic activity. This was the first report of a structural analysis of the domains of the Type I enzyme (Krinks et al 1984). The cyclic GMP-stimulated phosphodiesterase, Type II, had also been purified at that time from bovine adrenal and heart (Martins et al 1982). This enzyme apparently existed as a 240 kDa homodimer of two large subunits of around 105 kDa. Binding studies using several cyclic nucleotide analogs suggested that first, the cyclic GMP, bound to separate allosteric site(s), then, an induced conformational change caused a shift in kinetic behavior and acti-vation of cyclic AMP hydrolytic activity (Erneux et al 1984).

The Type III enzyme, which has been subsequently reclassified into several forms, could be separated on DEAE-cellulose as a broad peak with a high affinity for cyclic AMP (Thompson and Appleman, 1971). In most tissues it was membrane-bound, and in some instances insulin sensitivty could be measured (Marchmont et al 1981). This enzyme was purified from rat liver and was found to be phosphorylated on the catalytic 52 kDa subunit in response to insulin (Marchmont et al 1981).

Finally, at this time the retinal light-activated phosphodiesterase was purified (Baehr et al 1979), and this enzyme contained two catalytic subunits of 84 kDa and 88 kDa, respectively, and an inhibitory 9 kDa subunit.

The results of these studies suggested that the multiple phosphodi-esterases, respectively, first isolated on the gel filtration columns were, in-deed, different enzymes. Finally, two pieces of evidence further suggested that the multiple phosphodiesterases were different proteins. Peptide maps of the calmodulin-sensitive, insulin-sensitive, and retinal enzymes were not identical (Figure 6-2; Takemoto et al 1982, Takemoto et al 1984), and mono-clonal antibodies raised against the cyclic GMP-stimulated, the calmodulin-activated, and the light-activated phosphodiesterases did not cross-react (Figure 6-3; Hurwitz et al 1984).

Thus, by 1984, it was already well established that the multiple phos-phodiesterases were, in most cases, separate enzymes with different tissue distribution and different regulatory mechanisms. This was subsequently proven using molecular cloning techniques.

PHOSPHODIESTERASE STRUCTURES

In this section we will summarize the current knowledge of the known structures of the multiple phosphodiesterases (PDEs). As of this writing, very limited information is available on other than the primary structures of the phosphodiesterases. The NMR structure of the membrane-binding re-gion of PDE4A1 (a splice variant of a cAMP-specific PDE) has been com-

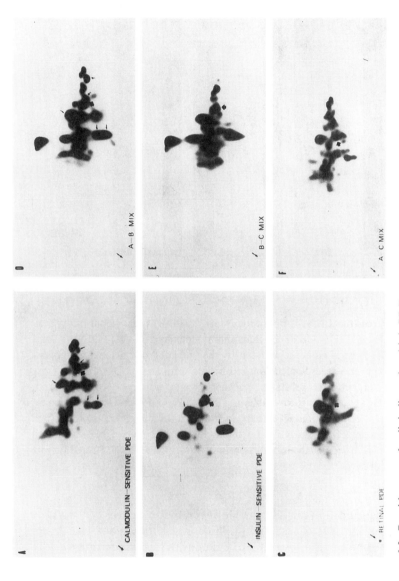

Figure 6-2. Peptide maps of radioiodinated multiple PDE. Lower-left arrow is sample origin. Light arrows indicate spots common to mixtures of either the calmodulin-sensitive PDE (A), the insulin-sensitive PDE (B), or the retinal rod outer segment PDE α and β subunit combined (C). Heavy arrows indicate spots that were common to all PDEs (from Takemoto et al, 1984, with permission).

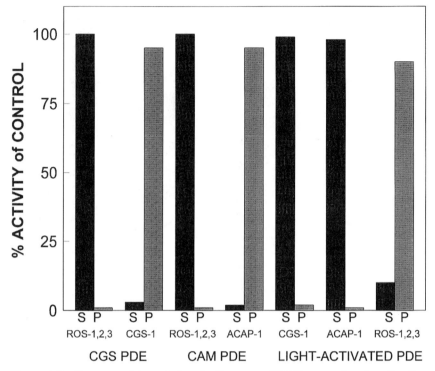

Figure 6-3. Cross-reactivity of phosphodiesterase (PDE) monoclonal antibodies. cGMP-stimulated PDE (CGS PDE), calmodulin-dependent (CAM) PDE, and light-activated PDE were immunoprecipitated by the indicated monoclonal antibodies. The supernatant and resuspended, washed pellets were assayed for PDE activity using 40 μM [³H] cGMP as substrate. Appropriate antisera were used to immuno-precipitate PDEs that were recoverable in a pellet (P), and not in the supernatant (S). ROS-1,2,3 = combination of three monoclonal antisera to light-activated retinal rod outer segment (ROS) PDE. CGS-1 = monoclonal antibody to cGMP-stimulated PDE. ACAP-1 = monoclonal antibody to calmodulin-activated PDE (from Hurwitz et al, 1984, with permission).

pleted and is included in this section. However, the sequences of the PDEs have not been included. These are available in genebanks and can be ac-cessed via the home pages given at the beginning of this chapter. As with the first section, we have attempted to summarize the current understanding of the multiple forms and their domain structures. I will use as the of this summary basis the DI, DII, and DIII enzymes, then expand this information to include the seven current PDE classifications. A current classification scheme is shown in Table 6-1. It is apparent from this that additional forms and splice variants will be isolated and sequenced in the future.

Calmodulin-dependent PDE (DI or PDE1)

The calmodulin-stimulated phosphodiesterase is one of the most extensively studied calmodulin-dependent proteins (Wu et al 1992). Calmodulin-stimulated phosphodiesterases, which had been purified from bovine brain, heart, lung, and testes, exhibited distinct molecular and catalytic properties as well as different sensitivities to calmodulin and to phosphorylation (Hansen and Beavo 1986, Mutus et al 1985, Sharma and Wang 1985, Sharma and Wang 1986). Earlier work with brain indicated that there were distinct isozymes of calmodulin-sensitive phosphodiesterases that had molecular weights of 63 kDa, 63 kDa plus 60 kDa, and 60 kDa (Sharma et al 1984). That these were not identical proteins was established on the basis of their differing reactivity with monoclonal antibodies and their peptide maps (Sharma et al 1984).

As shown in Table 6-1, at least three separate gene products and a number of splice variants with distinct primary structures have been identified. The first sequences and a proposed domain structure based upon these sequences were done using the 61-kDa and 63-kDa species from bovine brain and the 59-kDa species from bovine heart (Charbonneau et al 1991, Novack et al 1991). The 61-kDa brain species has 529 residues, and, by limited proteolysis, a fully active and calmodulin-independent species of 36 kDa was isolated and found to comprise residues 136–450. (Charbonneau et al 1991). This catalytic domain was found to be conserved in other phosphodiesterases from other species. With the use of limited proteolysis and synthetic peptides, the calmodulin binding sites were identified within residues 23–41. Sequence comparisons of the 63-kDa and 59-kDa species suggested that both different genes (the 63-KDa and 61-kDa species) and splice variants (the 61-KDa and 59-kDa species) existed (Bentley et al 1992, Novack et al 1991). Thus, the 63-kDa and the 61-kDa species were distinct gene products, while the 59-kDa species was a splice variant of the 61-kDa species. The catalytic domains were almost identical. The domain structures of these enzymes are shown in Figure 6-4.

Table 6-1: Cyclic nucleotide phosphodiesterase gene families

Short Name	PDE Isozyme Gene Families	Number of Gene Products Identified	Number of Splice Products Identified
PDE1	CaM-dependent PDEs	3	9+
PDE2	cGMP-stimulated PDEs	1	2
PDE3	cGMP-inhibited PDEs	2	2+
PDE4	cAMP-specific PDEs	4	15+
PDE5	cGMP-specific PDEs	2	2
PDE6	Photoreceptor PDEs	3	2
PDE7	High-affinity; cAMP-specific	1	1

There are currently at least six different CAM-PDE (calmodulin-sensitive phosphodiesterase) isozymes, including, 59-, 61-, 63-, 68-, and 75-kDa forms as well as olfactory-enriched forms. Three different genes have been identified, PDE1A, PDE1B, and PDE1C. The two splice variants of PDE1A, PDE1A1, and PDE1A2 encode the heart 59-kDa and bovine brain 61-kDa isozymes, and differ only in their N-termini (Novack et al 1991). The PDE1B1, which encodes the 63-kDa bovine brain isozyme, has only a single mRNA (Yan et al 1996). The PDE 1C gene appears to be very complex in expression. Thus far, at least five splice variants have been identified, PDE1C1–PDE1C5 (Yan et al 1996). A form of PDE1C (PDE1C2) is the olfactory form (Yan et al 1996).

Recently, a systematic determination of the kinetic parameters, calcium sensitivity, inhibition by pharmacological agents, and tissue distribution has revealed that each splice variant differs (Yan et al 1996). The K_m for cAMP varies from 1 to 110 μM, the EC50 for calcium from 0.27 to 2.5 μM, and the inhibition of phosphodiesterase activity by vinpocetine exhibits a K_i of from 10 to 40 μM. Of special interest is the fact that there is differing expression of the splice variants in different tissues. For example, olfactory epithelium has mostly PDE1C2, while testes has PDE1C5 (Yan et al 1996). Differing localization is observed in brain as well, using *in situ* hybridization (Yan et al 1996). These studies suggest that the different PDE1 enzymes may serve different regulatory functions.

In the future, "knockouts," "knockdowns," and overexpression systems may help to elucidate the roles of specific variants of PDE1 in specific cell functions.

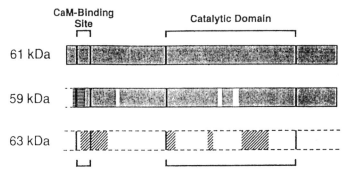

Figure 6-4. Schematic diagram illustrating the structural relationships among the 59-, 61-, and 63-kDa CaM-PDEs (calmodulin-activated PDEs). Segments with identical sequences are shaded with the same pattern. Open areas represent unknown sequences (now known), whereas shaded areas represent regions of known sequences. The positions of the conserved catalytic domain and a segment with calmodulin-binding properties are indicated (from Novack et al, 1991, with permission).

Cyclic GMP-Stimulated PDE and Cyclic GMP-Binding PDEs (DII or PDE2, PDE5, and PDE6)

The cyclic GMP-binding phosphodiesterases are a heterogeneous group of isozymes that exhibit allosteric cyclic GMP binding sites that are distinct from the catalytic sites. This group consists of at least three classes: the cGMP-stimulated PDEs (DII, PDE2), the photoreceptor cGMP PDEs of rods and cones (PDE6), and the cGMP-binding, cGMP-specific PDEs (PDE5). The domain structures of these PDEs are shown in Figure 6-5.

By 1990, most of the sequences of the cGMP-binding PDEs had been completed (see Stroop and Beavo 1992 for a review). The primary structures indicated that all had similar catalytic domains and cGMP-binding domains (sequences can be found elsewhere [Genebank]).

The cGMP-stimulated PDE had been purified to homogeneity in 1982 (Martins et al 1982 and see Beavo 1988 and Manganiello et al 1990 for reviews) and was found to exist as a homodimer of 210 kDa (Yamamoto et al 1983, Martins et al 1982). The PDE hydrolyzed both cAMP and cGMP, but micromolar cGMP stimulated cAMP hydrolysis about 10-fold (Beavo et al 1971). The positive cooperatively of this enzyme is thought to play a role in modulating the effects of cyclic nucleotides in heart and adrenal tissues (Hartzell and Fischmeister 1986). Structural studies on the domains of this type of PDE come from isolation of the catalytic fragments by limited proteolysis and from identification of the cGMP binding sites by use of photolabeling with [^{32}P]cGMP (Stroop et al 1989). The conserved catalytic fragment is isolated as a 36-kDa chymotryptic fragment of around 259 amino acids, at the C-terminus. The catalytic fragment is not sensitive to cGMP. Analyses of the amino acid sequences of catalytic domains of the PDEs do not suggest a catalytic mechanism; however, there is a region of 20 amino acids, which is invariant, and there are a number of invariant histidines (Charbonneau 1990, Charbonneau et al 1986).

A 60-kDa fragment can also be isolated from the chymotryptic treatment of the cGMP-stimulated PDEs (see Stroop and Beavo 1992 for a review). This was further broken down to a 39-kDa and a 28-kDa fragment, of which the latter was found to be labeled with cGMP and to be contained within the N-terminal part of the protein (see Figure 6-5C). The 36-kDa catalytic fragment no longer formed a dimer, and this suggested that the dimerization site was also within the N-terminal region of this protein. A similar domain structure has been suggested for the cGMP-binding PDEs (Stroop and Beavo 1992).

The retinal rod outer segment and cone PDEs have very different primary sequences and subunit compositions (Baehr et al 1979). The primary sequences are known, and both rod and cone forms consist of heterodimeric catalytic subunits. In the rods, they are designated as alpha (α) and beta (β subunits, and, although their sequences are similar, they are

cGMP PDE Monomer

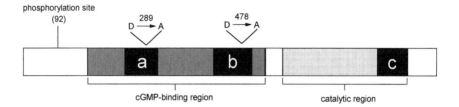

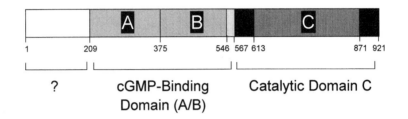

MODEL

PDA α DOMAINS

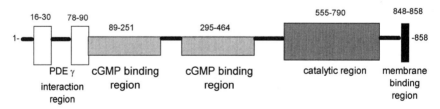

Figure 6-5. Domain structures of cGMP-binding PDEs. *Top.* Domain organization of the cGMP-binding PDE monomer. The cGMP-binding region (residues 142–526) is shown as a box with diagonal lines. Internally homologous repeats, a(228–311) and b(410–500) within the cGMP-binding region are shown as black boxes. C, also shown as a black box, denotes the putative cGMP-binding region of the catalytic domain (from McAllister-Lucas et al, 1995, with permission). *Middle.* Domain organization of the cGMP-stimulated PDE subunit (from Trong et al, 1990, with permission). *Bottom.* Model of PDE α domains of the retinal rod PDE (from Oppert et al, 1991, with permission).

encoded by different genes (Ovchinnikov et al 1986, Ovchinnikov et al 1987, Lipkin et al 1990, Li et al 1990, Pittler et al 1990, Baehr et al 1991). The rod and cone PDEs are extremely complex, and, to date, have not been expressed in fully active recombinant form. The catalytic domain of the alpha subunit of the retinal PDE was found to also be contained within a C-terminal 34-kDa fragment that was fully active (Oppert et al 1991). Both the rod and cone PDEs are activated in a light-dependent fashion when rhodopsin activates the retinal G-protein transducin (Yamazaki 1992). The active GTP-transducin alpha then interacts with, and removes, the inhibitory constraint of the small inhibitory protein, designated PDEγ. This subunit has been found to bind to the N-terminus of the catalytic subunits (Oppert et al 1991). The domain structure of the rod PDE alpha is shown in Figure 6-5C. Both the catalytic and PDEγ binding have been identified by limited proteolysis, but the cGMP binding sites have only been surmised from primary sequence comparisons with the other PDE forms. Finally, because this enzyme is found associated with the membrane, it also has a C-terminal membrane binding site (Oppert et al 1991).

Some discussion of the inhibitory subunit should be included herein, because the rod and cone types of PDEs appear to be unique in possessing this form of control. The inhibitory PDEγ subunit has been sequenced and found to contain 87 amino acids (Ovchinnikov et al 1986). Use of synthetic peptides identified that the binding region of PDEγ for the catalytic subunits resides within residues 24–45 of PDEγ (Morrison et al 1987; Morrison et al 1989, Takemoto et al 1992). Both partial proteolysis and deletion mutation revealed that the inhibitory region was in the C-terminal 10 amino acids (Lipkin et al 1988, Brown and Stryer 1989). The protein appears to be altered by phosphorylation, and this increases the inhibitory activity of the PDEγ subunit (Udovichenko et al 1994). Initial NMR structural studies indicated that the PDEγ subunit existed as a random coil (Takemoto and Cunnick 1992). This was also suggested by resonance energy transfer studies, which also indicated that the protein had a random structure in solution and did not attain a more ordered structure upon binding to transducin alpha (Berger et al 1997).

The role of the cGMP binding at noncatalytic sites is not known. In the rod outer segment the affinity for cGMP at these sites is $K_d = 100$ nM and 1–6 μM (Yamazaki et al 1996). These sites bind more than 90% of the cellular cGMP. Upon light activation of the protoreceptor cell, the activated PDE hydrolyzes cGMP to 5′GMP. The recovery of the cell to the dark state involves a rod-specific guanylate cyclase and calcium-binding proteins referred to as GCAP1 and GCAP2 (Dizhoor et al 1995). However, it has recently been suggested that the restoration of the cGMP concentration to the level in the dark-adapted cell could also occur when the cGMP, which is bound to the noncatalytic sites on PDE, is released. Thus, in the rod cell, the noncatalytic cGMP sites could play a role in recovery (Yamazaki et al 1996).

The structure of the cGMP binding site has not been determined other than the primary sequence. The amino acid sequences of all known cGMP-binding PDEs contain internally homologous repeats that are 80–90 residues in length and are arranged in tandem repeats within the hypothesized binding domains. In the bovine lung PDE5A, these repeats span residues 228–311 and 410–500 (McAllister-Lucas et al 1995). Within these regions there are invariant aspartic acid residues at positions 289 and 478. When mutations are examined at those positions, they are found to be critical for cGMP binding, and this suggests that there are two classes of binding sites, a fast and a slow binding site (McAllister-Lucas et al 1995). Further mutational analyses found that Asn276, Lys277, and Asp289 were critical for cGMP binding to a putative NKX$_n$D motif. No differences were found in catalytic activity of the wild-type or of these mutants, suggesting that cGMP binding does not alter the catalytic properties of the PDE5A isozymes (Turko et al 1996). A suggested model for binding of cGMP to the motif is shown in Figure 6-6.

cAMP-specific PDEs (PDE 3 and 4 or DIII)

There are two types of cAMP-specific PDEs, PDE3, which has a very high affinity for cAMP but shows competitive inhibition by cGMP, and PDE4, which also has a high affinity for cAMP but is not affected by cGMP. Members of the PDE3 family are activated through the action of agents that increase cAMP, such as glucagon. The diversity of these families is due to both different genes and splice variants.

The PDE3 family is important for the antilipolytic activity of insulin in fat cells (see Manganiello et al 1992 for a review). Isoforms of this enzyme family contain the conserved catalytic domain contained by all PDEs and a large hydrophobic domain at the N-terminus with multiple predicted membrane spanning domains (Leroy et al 1996, Meacci et al 1992). The membrane binding domain and catalytic domain are separated by putative regulatory domains that have several cAMP-dependent protein kinase consensus sequences for phosphorylation.

In intact adipocytes, lipolytic hormones, such as glucagon, increase the activity of a particulate PDE, PDE3, with little effect on other soluble PDEs. This occurs through activation of cAMP-dependent protein kinase, which results in the phosphorylation of Ser302 (Rahn et al 1996). This cAMP-induced activation of PDE3 may represent a type of feedback mechanism for regulation of cAMP turnover, steady-state concentrations of cAMP, and the activation state of the kinase. The different forms of PDE3, from the different gene products and splice variants, could produce an even more finely tuned tuned control system.

The PDE4 gene family is found to contain four distinct genes that code for PDEs of 68 to 77 kDa. This family of PDEs is known for sensitivity to rolipram, a potent antidepressant (Wachtel et al 1983). Houslay's group has

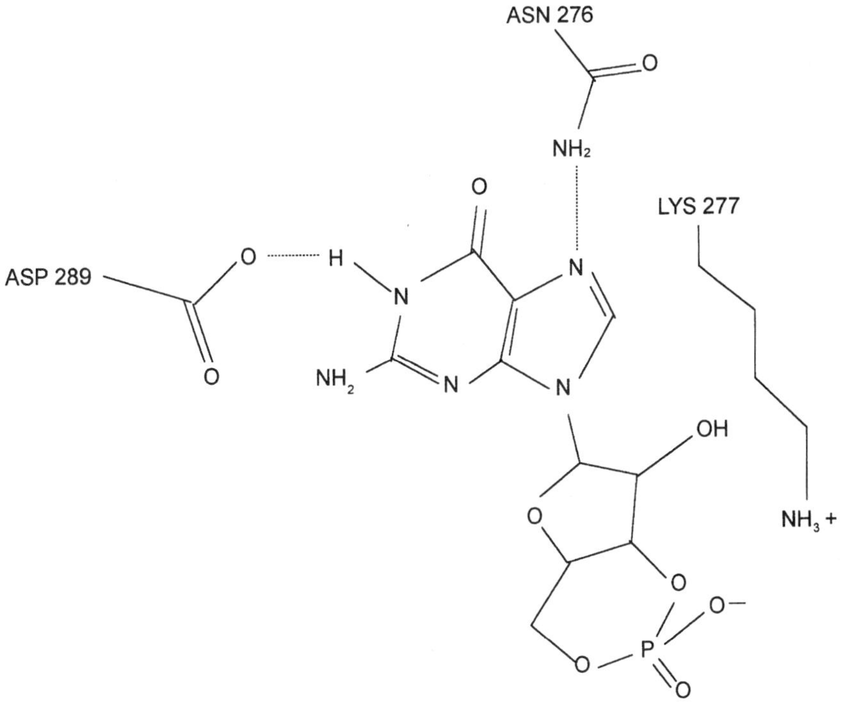

Figure 6-6. Possible interactions of cGMP with a putative NKXnD motif based on comparison with interactions of GTP in GTP-binding proteins (from Turko et al, 1996, with permission).

studied the structure and domain swapping of PDE4A. The core enzyme is a highly active and soluble enzyme to which are spliced variable N-terminal regions. One of these enzymes, RD1, has an N-terminal 25 amino acid segment spliced onto the "core" PDE (Smith et al 1996). This confers membrane localization to the enzyme, using a COS cell transfection system. This N-terminal splice region forms a distinct structure consisting of two independently folded helical regions separated by a hinge region (not shown in the picture) (Figure 6-7) (Smith et al 1996). The membrane binding domain is a compact and very hydrophobic domain encompassing residues Pro14–Trp20 of the PDE. It is hypothesized that the domain does not insert into the membrane but recognizes a membrane anchor protein. However, the mechanism of membrane binding is not known at present. This structure is the first three-dimensional information available for this class of enzymes.

PDE4 (PDE4D3) is activated in rat thyroid cells by thyroid-stimulating hormone through a cAMP-dependent phosphorylation (Sette and Conti 1996). This is a short-term activation that may be involved in the termination

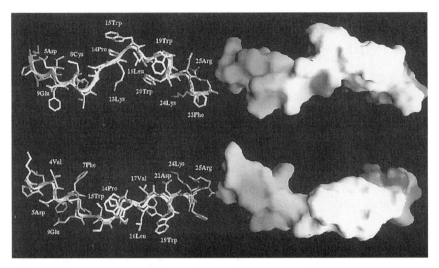

Figure 6-7. The mean calculated structure for peptide 1–25 RD1. The figure shows two orientations of the average calculated structure for peptide 1–25 RD1. On the left-hand side, both backbone and side-chain atoms are shown (no hydrogen atoms) for all residues. The helical path of the backbone is emphasized by the white ribbon. On the right-hand side, molecular surface maps are shown in the corresponding orientations. The surface is colored according to charge (red, negative; blue, positive). It is clear that the residues responsible for biological activity (Pro[14]–Trp[20]) form a hydrophobic cluster (white), flanked on either side by charged residues (from Smith et al, 1996, with permission).

of hormonal stimulation or in desensitization. Phosphorylation alters the V_{max} without altering the K_m for cAMP (Sette and Conti 1996). The inhibition by rolipram is also altered. The residue that was phosphorylated by cAMP-dependent protein kinase is Ser54. Thus, as with all other PDEs, the regulatory domain appears to reside within the N-terminal region of the protein.

The PDE7 family is the most recently identified PDE family. This group consists of high-affinity cAMP-PDEs that are not sensitive to PDE4 inhibitors such as rolipram. The protein has a molecular weight of 52 kDa. Recently, splice variants of this PDE have been isolated in skeletal muscle (Bloom and Beavo 1996).

SUMMARY

In the future, it is certain that genes for PDEs and more splice variants of each class will be identified. As more knowledge is attained about PDEs it is becoming apparent that these are not simple housekeeping enzymes.

Rather, they are a carefully controlled, regulated, and expressed group of proteins that undoubtedly have unique functions within individual cells. In the future, two areas will bloom in this field. The first is knowledge of the tertiary and quarternary structure of the PDEs. To date very little is known about this, and, given that many are targets for pharmacologically important drugs, this area is of crucial importance. Second, with the use of knockouts, knockdowns, and overexpression systems, it may be possible to determine the functions of unique forms of PDEs in cellular homeostasis.

ACKNOWLEDGMENTS

The author wishes to thank Drs. Joe Beavo, Michael Appleman, Joe Thompson, and Miles Houslay for their contributions, discussions, and suggestions in the writing of this chapter.

REFERENCES

Appleman M and Kemp G. 1966. Puromycin: A potent metabolic effect independent of protein synthesis. Biochem Biophys Res Comm 24:564–568.

Ashman, Lipton R, Melicow M, and Price T. 1963. Isolation of adenosine 3'5'-monophosphate and guanosine 3'5' - monophosphate from rat urine. Biochem Biophys Res Comm 11:330–334.

Baehr W, Champagne M, Lee A, and Pittler S. 1991. Complete cDNA sequences of mouse rod photoreceptor cGMP phosphodiesterase alpha-and-beta-subunit isozyme produced by alternative splicing of the beta-subunit gene. FEBS Lett 278:107–114.

Baehr W, Devlin M, and Applebury M. 1979. Isolation and characterization of cGMP phosphodiesterase from bovine rod outer segments. J Biol Chem 254:11669–11677.

Beavo J. 1988. Multiple isozymes of cyclic nucleotide phosphodiesterase. Advances in Second Messenger and Phosphoprotein Res 22:1–28.

Beavo J, Hardman J, and Sutherland E. 1970. Hydrolysis of cyclic guanosine and adenosine 3',5'-nucleotide phosphodiesterase isolated from frog erythrocytes. Arch Biochem Biophys 137:435–441.

Beavo J, Hardman J, and Sutherland E. 1971. Stimulation of adenosine 3'5'-monophosphate hydrolysis by guanosine 3'5'-monophosphate. J Biol Chem 246:3841–3846.

Bentley J, Kadlecek A, Sherbert C, Seger D, Sonnenburg W, Charbonneau H, Novack J and Beavo J. 1992. Molecular cloning of cDNA encoding a "63"-kDa calmodulin-stimulated phosphodiesterase from bovine brain. J Biol Chem 267:18676–18682.

Berger A, Cerione R, and Erickson J. 1997. Real time conformational changes in the retinal phosphodiesterase δ subunit monitored by resonance energy transfer. J Biol Chem. 272:2714–2721.

Blecher M, Merlino N, and Ro'Ane J. 1968. Control of the metabolism and lipolytic effects of cyclic 3'5' adenosine monophosphate in adipose tissue by insulin, methyl xanthines, and nicotinic acid. J Biol Chem 243:3973–3982.

Bloom T and Beavo J. 1996. Identification and tissue-specific expression of PDE 7 phosphodiesterase splice variants. Proc Natl Acad Sci 93:14188–14192.

Brooker G, Thompson W, and Appleman M. 1968. The assay of adenosine 3'5'-cyclic monophosphate and guanosine 3'5'-cyclic monophosphate in biochemical materials by enzymatic radioisotopic displacement. Biochem 7:4177–4181.

Brown R and Stryer L. 1989. Expression in bacteria of functional inhibitory subunit of retinal rod cGMP phosphodiesterase. Proc Natl Acad Sci 86:4922–4926.

Butcher R and Sutherland E. 1962. Adenosine 3'5'-monophosphate in biological materials. J Biol Chem 237:1244–1250.

Charbonneau H. 1990. Structure-function relationships among cyclic nucleotide phosphodiesterases. In: Beavo J and Houslay M, eds. Isozymes of cyclic nucleotide phosphodiesterases. Sussex, England: John Wiley and Sons. 267–291.

Charbonneau H, Beier N, Walsh K, and Beavo J. 1986. Identification of a conserved domain among cyclic nucleotide phosphodiesterase from diverse species. Proc Natl Acad Sci 83:9308–9312.

Charbonneau H, Kumar S, Novack J, Blumenthal D, Griffin P, Shabanowitz J, Hunt D, Beavo J, and Walsh K. 1991. Evidence for domain organization within the 61-kDa calmodulin-dependent cyclic nucleotide phosphodiesterase from bovine brain. Biochem 30:7931–7940.

Cheung W. 1967. Properties of cyclic 3'5'-nucleotide phosphodiesterase from rat brain. Biochem 6:1079–1087.

Cheung W. 1967a. Cyclic 3'5'-nucleotide phosphodiesterase: Pronounced stimulation by snake venom. Biochem Biophys Res Comm 29:478–482.

Cheung W. 1969. Cyclic 3'5'-nucleotide phosphodiesterase: Preparation of a partially inactive enzyme and its subsequent stimulation by snake venom. Biochem Biophys Acta 191:303–315.

Cheung W. 1970. Cyclic 3'5'-nucleotide phosphodiesterase. Biochem Biophys Res Comm 38:533–538.

Cheung W. 1971. Cyclic 3'5'-nucleotide phosphodiesterase: Evidence for and properties of a protein activator. J Biol Chem 246:2859–2869.

Christiansen R and Mann E. 1971. Adenosine 3'5'-monophosphate phosphodiesterase: Multiple molecular forms. Science 173:540–541.

DeLange R, Kemp R, Riley W, Cooper R, and Krebs E. 1968. Activation of skeletal muscle phosphodiesterase kinase by adenosine triphosphate and adenosine 3'5'-monophosphate. J Biol Chem 243:2200–2208.

De Robertis E, DeLores-Arnaiz G, and Alberici M. 1967. Subcellular distribution of adenyl cyclase and cyclic phosphodiesterase in rat brain cortex. J Biol Chem 242:3487–3493.

Dizhoor A, Olshevskaya E, Henzel W, Wong S, Stults J, Ankoudinova I, and Hurley J. 1995. Cloning, sequencing, and expression of a 24-kDa Ca^{2+}-binding protein activating photoreceptor guanylate cyclase. J Biol Chem 270:25200–25206.

Drummond G and Perrott-Yee S. 1961. Enzymatic hydrolysis of adenosine 3'5'-phosphoric acid. J Biol Chem 236:1126–1129.

Erneux C, Couchie D, Dumont J, and Justorf B. 1984. Cyclic nucleotide derivatives as probes of phosphodiesterase catalytic and regulatory sites. In: Strada S and Thompson W, eds. Advances in cyclic nucleotide and protein phosphorylation research. Vol 16. New York: Raven Press. 107–118.

Florendo N, Barnett R, and Greengard P. 1971. Cyclic 3'5'-nucleotide phosphodiesterase: cytochemical localization in cerebral cortex. Science 173:745–747.

Franks D and MacManus J. 1971. Cyclic GMP stimulation and inhibition of cyclic AMP phosphodiesterase from thymic lymphocytes. Biochem Biophys Res Comm 42:844–849.

Goldberg N and Haddox M. 1977. Cyclic GMP metabolism and involvement in biological regulation. Ann Rev Biochem 46:823–896.

Goldberg N, O'Dea R, and Haddox M. 1973. Cyclic GMP. In: Dummond G, Greengard P, and Robison G, eds. Advances in cyclic nucleotide research. Vol 3. New York: Raven Press. 155–223.

Goren E and Rosen O. 1971. The effect of nucleotides and a non-dialyzable factor on the hydrolysis of cyclic AMP by a cyclic nucleotide phosphodiesterase from beef heart. Arch Biochem Biophys 142:720–723.

Gulyassy P and Oken R. 1971. Assay of cyclic 3'5'-nucleotide phosphodiesterase in tissue homogenates. Proc Soc Exp Biol Med 137:361–365.

Hansen R and Beavo J. 1986. Differential recognition of calmodulin-enzyme complexes by a conformation specific anticalmodulin monoclonal antibody. J Biol Chem 261:14636–14645.

Hartzell H and Fischmeister R. 1986. Opposite effects of cyclic GMP and cyclic AMP on Ca^{2+} current in single heart cells. Nature 323:273–275.

Ho H, Wuch E, Stevens F, and Wang J. 1977. Purification of a Ca^{2+}-activatable cyclic nucleotide phosphodiesterase from bovine heart by specific interaction with its Ca^{2+} dependent modulator protein. J Biol Chem 252:43–50.

Huang Y and Kemp R. 1971. Properties of a phosphodiesterase with high affinity for adenosine 3'5'-cyclic phosphate. Biochem 10:2278–2283.

Hurwitz R, Hansen S, Harrison S, Martins T, Mumby M, and Beavo J. 1984. Immunological approaches to the study of cyclic nucleotide phosphodiesterases. In: Strada S and Thompson W, eds. Advances in cyclic nucleotide and protein phosphorylation research. Vol 16. New York: Raven Press. 89–106.

Hynie S, Krishna G, and Brodie B. 1966. Theopylline as a tool in studies of the role of cyclic adenosine 3'5'-monophosphate in hormone-induced lipolysis. J Pharmacol & Exp Therap 153:90–96.

Kakiuchi S and Yamazaki R. 1970. Calcium dependent phosphodiesterase activity and its activating factor (PAF) from brain. Biochem Biophys Res Comm 41:1104–1110.

Kakiuchi S, Yamazaki R, and Teshima Y. 1971. Cyclic 3'5'-nucleotide phosphodiesterase IV. Biochem Biophys Res Comm 42:968–974.

Klee C, Crouch T, and Krinks M. 1978. Subunit structure and catalytic properties of bovine brain cyclic nucleotide PDE. Fed Proc Abst. 188:1302.

Klee C, Crouch T, and Krinks M. 1979. Subunit structure and catalytic properties of bovine brain Ca^{2+}-dependent cyclic nucleotide phosphodiesterase. Biochem 18:722–729.

Klee C and Krinks M. 1978. Purification of a cyclic 3'5'-nucleotide phosphodiesterase inhibitory protein by affinity chromatography on activator protein coupled to Sepharose. Biochem 17:120–126.

Klotz U, Berndt S, and Stock K. 1972. Characterization of multiple cyclic nucleotide phosphodiesterase activities of rat adipose tissue. Life Sciences 11:7–17.

Klotz U and Stock K. 1972. Influences of cyclic guanosine 3'5'-monophosphate on the enzymatic hydrolysis of adenosine 3'5'-monophosphate. Naunyn-Schmiedeberg's Arch Pharmakol 274:54–62.

Krinks M, Haiech J, Rhoads A, and Klee C. 1984. Reversible and irreversible activation of cyclic nucleotide phosphodiesterase: Separation of the regulatory and catalytic domains. In: Strada S and Thompson W, eds. Advances in cyclic nucleotide and protein phosphorylation research. Vol 16. New York: Raven Press. 31–47.

Leroy M, Degerman E, Taira M, Murata T, Wong L, Movsesian M, Meacci E, and Manganiello V. 1996. Characterization of two recombinant PDE-3 (cGMP-inhibited cyclic nucleotide phosphodiesterase) isoforms, RcGIP1 and HcGIP2, expressed in NIH3006 murine fibroblasts and SF9 insect cells. Biochem 35:10194–10202.

Li T, Volpp K, and Applebury M. 1990. Bovine cone photoreceptor cGMP phosphodiesterase structure deduced from cDNA clone. Proc Natl Acad Sci 87:293–297.

Liebman P. and Evanczuk A. 1982. Real time assay of rod disk membrane cGMP phosphodiesterase and its controller enzymes. In: Packer L, ed. Methods in enzymology. Vol 81. New York: Academic Press. 532–542.

Lipkin V, Dumler I, Muradov K, Artemyov U, and Etingof R. 1988. Active site of the cyclic GMP phosphodiesterase γ subunit of retinal rod outer segments. FEBS Lett 234:287–290.

Lipkin V, Khramtsov N, Vasilevskaya I, Atabekova N, Muradov K, Gubanov V, Li T, Johnson J, Volpp K, and Applebury M. 1990. Beta subunit of bovine rod photoreceptor cGMP phosphodiesterase. Comparison with the phosphodiesterase family. J Biol Chem 265:12955–12959.

Makino H, Suzuki T, Kajinuma H, Yamazaki M, Ito H, and Yoshida S. 1992. The role of insulin-sensitive phosphodiesterase in insulin action. In: Strada S and Hidaka H, eds. Advances in second messenger and phosphoprotein research. Vol 25. New York: Raven Press. 185–199.

Mandel L and Kuehl F. 1967. Lipolytic action of 3'5'-triiodo-L-thyronine, a cyclic AMP phosphodiesterase inhibitor. Biochem Biophys Res Comm 28:13–18.

Manganiello V, Degerman E, Smith C, Vasta V, Tornquist H, and Belfrage P. 1992. Mechanisms for activation of the rat adipocyte particulate cyclic-GMP-inhibited cyclic AMP phosphodiesterase and its importance in the anti-lipolytic action of insulin. In: Strada S and Hidaka H, eds. Advances in second messenger and phosphoprotein research. Vol. 25. New York: Raven Press. 147–164.

Manganiello V, Tanaka T, and Murashima S. 1990. Cyclic GMP-stimulated cyclic nucleotide phosphodiesterase. In: Beavo J and Houslay M, eds. Isozymes of cyclic nucleotide phosphodiesterases. Sussex England: John Wiley and Sons. 61–85.

Marchmont R, Ayad S, and Houslay M 1981. Purification and properties of the insulin-stimulated cyclic AMP phosphodiesterase from rat liver plasma membranes. Biochem J 195:645–652.

Martins T, Mumby M, and Beavo J 1982. Purification and characterization of a cyclic GMP-stimulated cyclic nucleotide phosphodiesterase from bovine tissues. J Biol Chem 257:1973–1979.

McAllister-Lucas L, Haik T, Colbran J, Sonnenburg W, Seger D, Turko I, Beavo J, Francis S, and Corbin J. 1995. An essential aspartic acid at each of two allosteric cGMP-binding sites of a cGMP-specific phosphodiesterase. J Biol Chem 270:30671–30679.

Meacci E, Taira M, Moos Jr M, Smith C, Mousesian M, Degerman E, Belfrage P, and Manganiello V. 1992. Molecular cloning and expression of human myocardial cGMP-inhibited cAMP phosphodiesterase. Proc Natl Acad Sci 89:3721–3724.

Miki N, Baraban J, Keuins J, Boyce J, and Bitensky M. 1975. Purification and properties of the light-activated cyclic nucleotide phosphodiesterase of rod outer segments. J Biol Chem 250:6320–6327.

Morrison D, Cunnick J, Oppert B, and Takemoto D. 1989. Interaction of the γ-subunit of retinal rod outer segment phosphodiesterase with transducin. J Biol Chem 264:11671–11681.

Morrison D, Rider M, and Takemoto D. 1987. Modulation of retinal transducin and phosphodiesterase activities by synthetic peptides of the phosphodiesterase γ-subunit. FEBS Letts 222:266–270.

Mutus B, Karuppiak N, Sharma R, and Macmanus J. 1985. The different stimulation of brain and heart cyclic AMP phosphodiesterase by oncomodulin. Biochem Biophys Res Comm 131:500–506.

Nair K. 1966. Purification and properties of 3′5′-cyclic nucleotide phosphodiesterase from dog heart. Biochem 5:150–157.

Novack J, Charbonneau H, Bentley J, Walsh K, and Beavo J. 1991. Sequence comparison of the 63,-61-, and 59-kDa calmodulin-dependent cyclic nucleotide phosphodiesterase. Biochem 30:7940–7947.

Oppert B, Cunnick J, Hurt D, and Takemoto D. 1991. Identification of the retinal cyclic GMP phosphodiesterase inhibitory γ-subunit interaction sites on the catalytic α-subunit. J Biol Chem 266:16607–16613.

Ovchinnikov Y, Gubanov V, Khramtosov N, Ischenko V, Zagranichny V, Muradov K, Shuvaeva T, and Lipkin V. 1987. cGMP phosphodiesterase from bovine retina-amino acid sequence of the α-subunit and nucleotide sequence of the corresponding cDNA. FEBS Lett 223:169–173.

Ovchinnikov Y, Lipkin V, Kumarev V, Gubanov V, Zagranichny V, and Muradov K.

1986. Cyclic GMP phosphodiesterase from cattle retina. Amino acid sequence of the γ subunit and nucleotide sequencing of the corresponding cDNA. FEBS Lett 204:288–292.

Pannbacker R, Fleischmann D, and Reed D. 1972. Cyclic nucleotide phosphodiesterase: high activity in a mammalian photoreceptor. Science 175:757–758.

Pennington S. 1971. 3'5'-cyclic adenosine monophosphate phosphodiesterase assay using high speed chromatography. Anal Chem 43:1710–1703.

Pichard A. and Cheung W. 1976. Cyclic 3'5'-nucleotide phosphodiesterase interconversion of multiple forms and their effects on enzyme activity and kinetics. J Biol Chem 251:5726–5737.

Pittler S, Baehr W, Wasmuth J, McConnell D, Champagne M, Van Twinen P, Ledbetter D, and Davis R. 1990. Molecular characterization of human and bovine rod photoreceptor cGMP phosphodiesterase alpha subunit and chromosomal localization of the human gene. Genomics 6:272–283.

Pledger W, Stancel G, Thompson W, and Strada S. 1974. Separation of multiple forms of cyclic nucleotide phosphodiesterase from rat brain by isoelectric focusing. Biochem Biophys Acta 370:242–248.

Poch G. 1971. Assay of phosphodiesterase with radioactivity labeled cyclic 3'5'-AMP as substrate. Naunyn-Schmiedesbergs Archive Fur Pharmakol 268:272–299.

Pozdynyakov N, Yoshida A, Cooper N, Margulis A, Duda T, Sharma R, and Sitaramayya A. 1995. A novel calcium dependent activator of retinal rod outer segment membrane guanylate cyclase. Biochem 34:14279–14283.

Rahn T, Ronnstrandt L, Leroy M, Wernstedt C, Tornquist H, Manganiello V, Belfrage P, and Degermen E. 1996. Identification of the site in the cGMP-inhibited phosphodiesterase phosphorylated in adipocytes in response to insulin and isoproterenol. J Biol Chem 271:11575–11580.

Robison G, Butcher R, and Sutherland E. 1971. Cyclic AMP. New York: Academic Press. 53–110.

Rosen O. 1970. Preparation and properties of a cyclic 3'5'-nucleotide phosphodiesterase isolated from frog erythrocytes. Arch Biochem Biophys 137:435–441.

Russell T, Terasaki W, and Appleman M. 1973. Separate phosphodiesterases for the hydrolysis of cyclic adenosine 3'5'-monophosphate and cyclic guanosine 3'5'-monophosphate in rat liver. J Biol Chem 248:1334–1340.

Russell T, Thompson W, Schneider F, and Appleman M. 1972. 3'5' cyclic adenosine monophosphate phosphodiesterase: negative cooperativity. Proc Natl Acad Sci 69:1791–1795.

Schonhofer P, Skidmore I, Bourne H, and Krishna G. 1972. Cyclic 3'5'-AMP phosphodiesterase in isolated fat cells. Pharmacology 7:65–77.

Schroder J and Plageman P. 1972. Cyclic 3'5'-nucleotide phosphodiesterases of Novikoff rat hepatoma, mouse L and HeLa cells growing in suspension culture. Cancer Res 32:1082–1087.

Senft G, Munske K, Schultz G, and Hoffman M. 1968. The influences of hydrochlorothiazide and other sulfamoyl diuretics on the activity of 3'5' AMP phosphodiesterase in rat kidney. Exp Pathol 259:344–359.

Sette C and Conti M. 1996. Phosphorylation and activation of a cAMP-specific phosphodiesterase by the cAMP-dependent protein kinase. J Biol Chem 271:16526–16534.

Sharma R, Adachi A, Adachi K, and Wang J. 1984. Demonstration of bovine brain calmodulin-dependent cyclic nucleotide phosphodiesterase isozymes by monoclonal antibodies. J Biol Chem 259:9248–9256.

Sharma R and Wang J. 1985. Differential regulation of bovine brain calmodulin-dependent cyclic nucleotide phosphodiesterase isozymes by cyclic AMP-dependent protein kinase and calmodulin-dependent phosphatase. Proc Natl Acad Sci 82:2603–2607.

Sharma R and Wang J. 1986. Calmodulin and Ca^{2+}-dependent phosphorylation and dephosphorylation of 63 kDa subunit-containing bovine brain calmodulin-stimulated cyclic nucleotide phosphodiesterase isozyme. J Biol Chem 261:1322–1328.

Smith K, Scotland G, Beattie J, Trayer I, and Houslay M. 1996. Determination of the structure of the N-terminal splice region of the cyclic-AMP,-specific phosphodiesterase RD1 (RNPDE4A1) by ^{1}H NMR and identification of the membrane association domain using chimeric constructs. J Biol Chem 271:16703–16711.

Sobel B, Dempsey P, and Cooper T. 1968. Adenyl cyclase activity in the chronically denervated cat heart. Biochem Biophys Res Comm 33:758–762.

Stevens F, Walsh M, Ho H, Teo T, and Wang J. 1976. Comparison of calcium-binding proteins. J Biol Chem 251:4495–4500.

Stewart A, Ingebritsen T, Manalan A, Klee C, and Cohen P. 1982. Discovery of a Ca^{2+} and calmodulin-dependent protein phosphatase. FEBS Lett 137:80–84.

Strada S and Thompson W. 1984. Advances in cyclic nucleotide and protein phosphorylation research. Vol 16. New York: Raven Press.

Stroop S and Beavo J. 1992. Sequence homology and structure-function studies of the bovine cyclic-GMP-stimulated and retinal phosphodiesterases. In: Strada S and Hidaka H, eds. Advances in second messenger and phosphoprotein research. Vol 25. New York: Raven Press. 55–71.

Stroop S, Charbonneau H, and Beavo J. 1989. Direct photolabeling of the cGMP-stimulated cyclic nucleotide phosphodiesterase. J Biol Chem 264:13718–13725.

Sutherland E and Rall T. 1958. Fractionization and characterization of a cyclic adenine ribonucleotide formed by tissue particles. J Biol Chem 232:1077–1091.

Takemoto D and Cunnick J. 1992. Structure/function studies of the rod phosphodiesterase subunit. Invest Ophth Vis Sci 33:1005.

Takemoto D, Hansen J, Takemoto L, and Houslay M. 1982. Peptide mapping of multiple forms of cyclic nucleotide phosphodiesterase. J Biol Chem 257:14597–14599.

Takemoto D, Hansen J, Takemoto L, Houslay M, and Marchmont R. 1984. Peptide mapping of cyclic nucleotide phosphodiesterase. In: Strada S and Thompson W, eds. Advances in cyclic nucleotide and protein phosphorylation research. Vol 16. New York: Raven Press. 55–64.

Takemoto D, Hurt D, Oppert B, and Cunnick J. 1992. Domain mapping of the retinal cyclic GMP phosphodiesterase γ-subunit. Biochem J 281:637–643.

Teo T, Wang T, and Wang J. 1973. Purification and properties of the protein activator of bovine heart cyclic adenosine 3'5'-monophosphate phosphodiesterase activities from rat brain. J Biol Chem 248:588–595.

Thompson W and Appleman M. 1971a. Multiple cyclic nucleotide phosphodiesterase activities from rat brain. Biochem 10:311–316.

Thompson W and Appleman M. 1971b. Characterization of cyclic nucleotide phosphodiesterases of rat tissues. J Biol Chem 246:3145–3150.

Thompson W, Little S, and Williams R. 1973. Effect of insulin and growth hormone on rat liver cyclic nucleotide phosphodiesterase. Biochem 12:1889–1894.

Thompson W, Ross C, Pledger W, Strada S, Bonner R, and Hersh E. 1976. Cyclic adenosine 3'5'monophosphate phosphodiesterase: Distinct forms in human lymphocytes and monocytes. J Biol Chem 251:4922–4929.

Trong H, Beier N, Sonnenburg W, Stroop S, Walsh K, Beavo J, and Charbonneau H. 1990. Amino acid sequence of the cyclic GMP stimulated cyclic nucleotide phosphodiesterase from bovine heart. Biochem 29:10280–10288.

Tsuboi S, Matsumoto H, Jackson K, Tsujimoto K, Williams T, and Yamazaki A. 1994. Phosphorylation of an inhibitory subunit of cGMP phosphodiesterase in *Rana catesbiana* rod photoreceptors. J Biol Chem 269:15024–15029.

Turko I, Haiko T, McAllister-Lucas L, Burns F, Francis S, and Corbin J. 1996. Identification of key amino acids in a conserved cGMP-binding site of cGMP-binding phosphodiesterases. J Biol Chem 271:22240–22244.

Udovichenko I, Cunnick J, Gonzalez K, and Takemoto D. 1994. Functional effect of phosphorylation of the photoreceptor phosphodiesterase inhibitory subunit by protein kinase C. J Biol Chem 269:9850–9856.

Uzunov P and Weiss B. 1972. Separation of multiple molecular forms of cyclic adenosine 3'5'-monophosphate phosphodiesterase in rat cerebellum by polyacrylamide gel electrophoresis. Biochem Biophys Acta 284:220–226.

Wachtel H. 1983. Potential antidepressant activity of rolipram and other selective cyclic adenosine 3'5'-monophosphate phosphodiesterase inhibitors. Neuropharmacology 22:267–272.

Wang J, Teo T, Ho H, and Stevens F. 1975. Bovine heart protein activator of cyclic nucleotide phosphodiesterase. In: Greengard P and Robison G, eds. Advances in cyclic nucleotide research. Vol 5. New York: Raven Press. 179–194.

Watterson D, Harrelson W, Keller P, Sharief F, and Vanaman T. 1976. Structural similarities between the Ca^{2+}-dependent regulatory proteins of 3'5' cyclic nucleotide phosphodiesterase and actomyosin ATPase. J Biol Chem 251:4501–4513.

Watterson D and Vanaman T. 1976. Affinity chromatography purification of a cyclic nucleotide phosphodiesterase using immobilized modulator protein, a troponin C-like protein from brain. Biochem Biophys Res Comm 73:40–46.

Weiss B and Costa E. 1968. Regional and subcellular distribution of adenyl cyclase and 3'5'-cyclic nucleotide phosphodiesterase in brain and pineal gland. Biochem Pharmacol 17:2107–2116.

Wheeler G and Bitensky M. 1977. A light-activated GTPase in vertebrate photoreceptors: regulation of light-activated cyclic GMP phosphodiesterase. Proc Natl Acad Sci 72:2320–2324.

Wu Z, Sharma R, and Wang J. 1992. Catalytic and regulatory properties of calmodulin-stimulated phosphodiesterase isozymes. In: Strada S and Hidaka H, eds. Advances in second messenger and phosphoprotein research. Vol 25. New York: Raven Press. 29–43.

Yamamoto T, Manganiello V, and Vaughn M. 1983. Purification and characterization of cyclic GMP-stimulated cyclic nucleotide phosphodiesterase from calf liver—effects of divalent cations on activity. J Biol Chem 258:12526–12533.

Yamazaki A. 1992. The GTP-binding protein-dependent activation and deactivation of cyclic GMP phosphodiesterase in rod photoreceptors. In: Strada S and Hidaka H, eds. Advances in second messenger and phosphoprotein research. Vol 25. New York: Raven Press. 135–145.

Yamazaki A, Bondarenko V, Dua S, Yamazaki M, Usukuwa J, and Hayashi F. 1996. Possible stimulation of retinal rod recovery to dark state by cGMP release from a cGMP phosphodiesterase noncatalytic site. J Biol Chem. 271:32495–32498.

Yan C, Zhao A, Bentley J, and Beavo J. 1996. The calmodulin-dependent phosphodiesterase gene PDE1C encodes several functionally different splice variants in a tissue-specific manner. J Biol Chem 271:25699–25706.

Part IV

NOVEL MESSENGERS

7

Nitric Oxide: Synthesis and Intracellular Actions

Jeffrey L. Garvin

Although the physiological importance of nitric oxide (NO) was not recognized until the 1980s, amyl nitrate, a compound that liberates NO, was first used clinically in 1867 for the treatment of angina pectoris (Brunton 1867). NO can be considered both a paracrine/autocrine hormone and a component of the second messenger cascades of other hormones and physiological signals. This chapter will address NO production, factors that regulate its release and production, the intracellular signaling pathways it activates, and the physiological consequences of such activation.

NITRIC OXIDE SYNTHESIS

Physical and Chemical Properties

NO is a small reactive inorganic molecule that was considered only a toxin and pollutant until its physiological effects were established. NO is a paramagnetic gas with an unpaired electron, making NO a free radical. It is unusual in that most free radicals very rapidly dimerize or dismutate to gain or lose an electron, whereas NO does not. NO reacts rapidly with molecular oxygen and oxygen radicals. This process generates highly reactive compounds, including peroxynitrite ($OONO^-$) (Huie and Padmaja 1993), which are responsible for some of the damaging effects of NO. Alternately, NO can lose an electron to form the nitrosonium cation (NO^+). The half-life of NO has been reported to be as long as 4 minutes in aqueous solutions containing molecular oxygen, but may be as short as 30 seconds in biological systems (Henry et al 1993).

Introduction to Cellular Signal Transduction
A. Sitaramayya, Editor
©1999 Birkhäuser Boston

Production

NO is one of the end products of a reaction in which L-arginine is metabolized to L-citrulline by nitric oxide synthase (NOS). The chemical reaction has been extensively studied, but is not completely understood because of its complexity. It takes place via several steps (Figure 7-1). Completion of the initial steps results in hydroxylation of the nitrogen of the terminal guanidino group of L-arginine to form hydroxy-L-arginine in the presence of molecular oxygen and heme iron, which is reduced from Fe^{+3} to Fe^{+2} in the process. The flavins FAD and FMN are required for this initial reaction and mediate the transfer of electrons from NADPH to molecular oxygen. This reaction is similar to those carried out by other P450 enzymes to which NOS is related (Marletta 1994, Griffith and Stuehr 1995, Schmidt et al 1993).

In subsequent steps, hydroxy-L-arginine is further oxidized to L-citrulline and NO, while at the same time molecular oxygen is reduced to the superoxide free radical, and 0.5 mole of NADPH is oxidized to $NADP^+$. In the end, one mole of elemental oxygen is added to hydroxy-L-arginine, producing L-citrulline and NO, and the other mole of elemental oxygen is reduced to water. These reactions require the cofactors FAD and FMN for the transfer of electrons. Tetrahydrobiopterin is also required, but since NADPH supplies the requisite number of electrons for the reaction, it is not clear whether tetrahydrobiopterin acts as a cofactor (not cycling between dihydro- and tetrahydrobiopterin) or whether NOS possesses dihydrobiopterin reductase activity and the electrons from NADPH are used to return dihydrobiopterin to tetrahydrobiopterin. The tetrahydrobiopterin binding site of NOS appears to be novel in that it has a very high affinity and is not sensitive to competitive inhibitors such as methotrexate. In any event, during the entire reaction five electrons are transferred to L-arginine. Given that NADPH donates electrons in pairs, it is likely that at least two cycles

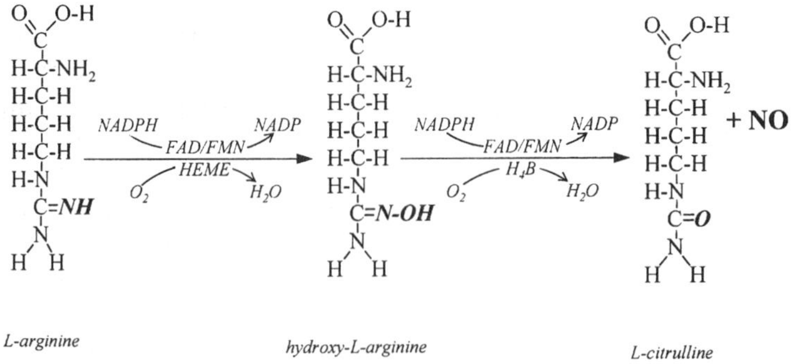

Figure 7-1. Production of nitric oxide from L-arginine.

are necessary in order for NOS to return to its basal state (Marletta 1994, Griffith and Stuehr 1995, Schmidt et al 1993).

The reactions involved in the oxidation of hydroxy-L-arginine by NOS are also responsible for the NADP diaphorase activity of NOS, which is used to detect NOS activity by immunohistochemistry. These reactions can also produce hydrogen peroxide in the absence of hydroxy-L-arginine; thus, in the absence of the NOS substrate L-arginine, or when L-arginine concentrations are low, activation of NOS may produce reactive oxygen species that can damage the cell (Marletta 1994).

NOS Isozymes

There are two separate nomenclature systems for the three major classes of NOS isozymes. The first is numerical, with the three known isozymes being designated NOS I, NOS II, and NOS III. The other is based on the anatomical location or mechanism of activation of the individual isozymes. In this system the NOSs are divided into two groups: "constitutive" and "inducible." There are two constitutive forms of NOS and only one inducible form. The two constitutive forms are referred to as neuronal NOS (nNOS) and endothelial NOS (eNOS), based on the type of cell in which each isozyme was originally identified; under the other system they would be designated NOS I and NOS III. The term "constitutive" refers to the fact that the NOS protein is always present in the cell and NOS activity is allosterically modulated. The term "inducible" refers to the fact that this isozyme of NOS is generally not present in the cell and production of NO by inducible NOS (iNOS) is enhanced by increasing the amount of NOS protein through augmented transcription of mRNA and translation of this mRNA into protein. iNOS was first identified in macrophages but is still referred to as inducible NOS rather than macrophage NOS, since it was thought that only one family member existed. iNOS is also referred to as NOS II. Since neuronal NOS has been found in tissues other than neurons, and inducible NOS has now been found to be constitutively expressed in many types of cells, I will use the designations NOS I, NOS II, and NOS III for nNOS, iNOS, and eNOS, respectively.

Each of the three classes of NOS represents separate gene products. The three NOS genes are complex, with human NOS I having 28 coding sequences (or exons) and 27 spans of DNA between coding sequences (or introns) spread over 100 kilobases (kb) (Fujisawa et al 1994, Nathan and Xie 1994). The mRNA for NOS I codes for 1429 amino acids, with a resulting monomeric protein of 160 kilodaltons (kDa) (Bredt et al 1991). NOS II has 26 exons and 25 introns spread over 37 kb (Marsden et al 1994, Chartrain et al 1994), while NOS III has 26 exons and 25 introns over 21 kb (Marsden et al 1993, Robinson et al 1994). NOS II codes for 1154 amino acids, resulting in a protein of approximately 135 kDa (Chartrain et al 1994), while NOS III mRNA codes for 1205 amino acids, giving a protein with a molecular weight of nearly 135 kDa (Lamas et al 1992). In addition to the complexity of the gene, the native (unmodified) RNAs transcribed from these genes may have their exons combined in different ways to form mature mRNA, and there may be different alleles or families of genes. Two different mRNAs for NOS II have been identified in the kidney (Mohaupt et al 1994), but it is unclear if they are different alleles or are members of a NOS II gene family. It is unclear if the differences in mRNA have an impact on enzymatic activity.

NOS I, II, and III are only 50–60% homologous at the amino acid level, but there is 80–94% homology of a given NOS from different species (Sessa 1994). The enzyme can be divided into several domains: cofactor domains where FAD, FMN, and tetrahydrobiopterin bind, the L-arginine binding site, the NADPH binding site, the calmodulin binding site, and the 5' terminal (Figure 7-2). The homology is greatest in that portion of the protein where the cofactors bind and at the active site. While one might expect the calmodulin binding domain of the three forms of NOS to be highly homologous, this is actually a region of some divergence, resulting in the calmodulin binding domain of NOS I and III having a lower affinity for calmodulin than that of NOS II. Consequently, NOS I and III only bind the Ca/calmodulin complex when its concentration increases in response to an increase in intracellular Ca, whereas NOS II tightly binds calmodulin at normal, unstimulated intracellular Ca concentrations. Thus NOS I and III are activated by increases in intracellular Ca, but NOS II is not (Lincoln et al 1996, Mayer 1994, Schmidt et al 1992b).

NOS is also homologous with cytochrome P450 reductases, another family of enzymes that carry out oxidation/reduction reactions. The homol-

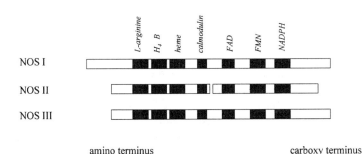

Figure 7-2. Schematic diagram of nitric oxide synthase proteins. Binding sites for substrates and cofactors are shown. The positions of the binding sites for L-arginine, heme, and tetrahydrobiopterin (H_4B) are speculative. These compounds bind at the amino end of the molecule, but the exact locations are unknown. The break in NOS II is an area of deletion as compared to NOS I and III. The break was made so that the cofactor and substrate binding domains would align.

ogy of NOS with cytochrome P450 reductase primarily occurs in those domains responsible for the binding of cofactors and NADPH, since cytochrome P450 reactions also utilize FAD, FMN, tetrahydrobiopterin, NADPH, and molecular oxygen (Dawson and Snyder 1994, Marletta 1994, Griffith and Stuehr 1995).

The active form of NOS is a homodimer, whereas the NOS monomer is inactive. While there are individual binding sites for substrates and cofactors on the monomeric forms, upon dimerization the cofactor binding sites appear to segregate into high- and low-affinity sites (Schmidt et al 1992b). Apparently all sites do not have to be occupied by cofactors in order for NOS to be active; however, tetrahydrobiopterin is required for dimer formation (Stuehr et al 1991). Recently it has been recognized that the availability of other cofactors, especially tetrahydrobiopterin, may limit NOS activity, but the physiological significance is not known and is beyond the scope of this chapter. The availability of the substrate L-arginine may also limit NOS activity, and it is now recognized that some factors that increase NOS activity also increase the transport of L-arginine into the cell (Low and Grigor 1995, Bogle et al 1992, Simmons et al 1996).

NO itself may act in a negative feedback manner to inhibit NOS (Rengasamy and Johns 1993). NO will form nitroso complexes with essentially any enzyme containing a prosthetic heme group, including NOS itself (Abu-Soud et al 1995). Enhanced NO production by NOS II in macrophages has also been shown to reduce the availability of heme moieties for newly synthesized enzymes, thus preventing activation (Albakri and Stuehr 1996). NO may also reduce NOS II production via nitrosylation of nuclear binding factors that regulate NOS II transcription (Colasanti et al 1995, Peng et al 1995, Matthews et al 1996). Interestingly, NOS III has a positive feedback loop involving NO and cGMP (Ravichandran and Johns 1995).

Inhibitors

A number of selective inhibitors have been synthesized, which can be used to study the role of specific isoforms of NOS in the regulation of physiological functions. These are primarily L-arginine analogs and thus inhibit NOS activity competitively. All three classes of NOS can be inhibited by L-monomethyl nitroarginine (LMMNA) (Stenger et al 1995, Rees et al 1989, Ishii et al 1990, Moncada et al 1991). However, the usefulness of this compound in chronic *in vivo* experiments has been called into question, since several investigators have reported that it can be catabolized to NO, albeit at a slow rate. L-nitroarginine (LNA) and its methyl ester analog, L-nitroarginine methyl ester (LNAME) (Moncada et al 1991, Rees et al 1990a, Rees et al 1990b) are now used more widely than LMMNA. LNAME offers an advantage over LNA in that it is more soluble in aqueous solution and may enter the cell via simple diffusion rather than being transported into the cell by a specific carrier. Once inside the cell, LNAME is cleaved by naturally occurring esterases to form LNA (Rees et al 1990b, Moncada et al 1991). In addition to inhibiting NOS, LNAME has been reported to act as a muscarinic antagonist (Buxton et al 1992), among other things.

More recently, a compound reported to selectively inhibit NOS I has been described, 7-nitroindazole (7NI). The basis for describing 7NI as a selective inhibitor of NOS I is data showing that it inhibited NOS activity in the brain when infused *in vivo*, but did not prevent an LNAME-induced increase in blood pressure (Moore et al 1993a, Moore et al 1993b). These data were interpreted to mean that 7NI inhibits NOS I but not NOS III. However, the selectivity of 7NI for NOS I over NOS III is not supported by *in vitro* experiments in which the ability of 7NI to reduce purified NOS I, II, and III activity was measured (Bland-Ward and Moore 1995). While the explanation for this discrepancy is unclear, it may involve *in vitro* modification of 7NI to a more selective metabolite. Nevertheless, until this issue is resolved, data involving 7NI must be interpreted with caution.

NOS II can be selectively inhibited with aminoguanidine (Stenger et al 1995). However, as with other inhibitors, aminoguanidine is not specific, but only selective for NOS II. It is known to have other actions, including inhibition of collagen cross-linking and glycation of proteins, among other things. Recently a more selective inhibitor of NOS II has been reported (Stenger et al 1995). Thus, while inhibitors are useful in studying the physiological roles of the various NOS isozymes, one must keep in mind that inhibitors are only specific until thoroughly studied. Given the importance of NO, it is highly likely that more selective inhibitors of NOS I, II, and III will be developed in the future.

NOS ISOZYMES

Distribution and Regulation of NOS I Activity

NOS I protein is expressed constitutively. It is found in peripheral nerves serving a variety of organs (Rand and Li 1995), including the lung (Asano et al 1994), the central nervous system (Dawson and Dawson 1996, Garthwaite and Boulton 1995), and at least two sites within the kidney: the macula densa (Mundel et al 1992, Bachmann et al 1995) and the inner medullary collecting duct (Terada et al 1992). In the peripheral nervous system NOS I is found in nonadrenergic/noncholinergic nerves (Kelly et al 1996, Dawson and Dawson 1996, Rand and Li 1995) and in some adrenergic nerve terminals (Schwartz et al 1995, Kaye et al 1995). In the central nervous system NOS I is found in the cerebellum, cerebral cortex, dorsal root ganglion, and hippocampus, among others (Dawson and Dawson 1996, Garthwaite and Boulton 1995). NOS I has been described as both a soluble and particulate enzyme. The particulate form associates with cell membranes as a result of posttranslational acylation.

NOS I activity is acutely regulated by changes in intracellular Ca, with increased Ca resulting in increased activity due to binding of Ca/calmodulin. The two primary mechanisms by which Ca is increased in neurons are activation of excitatory neurotransmitter receptors (Garthwaite and Boulton 1995) and action potentials (Rand and Li 1995, Garthwaite and Boulton 1995). Excitatory neurotransmitters such as glutamate bind to their receptors, resulting in depolarization of the cell. This depolarization then opens voltage-gated Ca channels, resulting in increased intracellular Ca and activation of NOS I in the central nervous system (Garthwaite and Boulton 1995). Activation of NOS I in peripheral nerves is similar except that action potentials are the stimulus for cell depolarization (Garthwaite and Boulton 1995, Rand and Li 1995).

While NOS I activity is acutely stimulated by binding of Ca/calmodulin, it can also be regulated by addition or removal of phosphates. Several kinases phosphorylate NOS I, including Ca/calmodulin kinase II, protein kinase C, and protein kinase A (Bredt et al 1992). The addition of phosphates to the enzyme by Ca/calmodulin kinase II and protein kinase C inhibits activity by as much as 75% (Bredt et al 1992, Nakane et al 1991). Early accounts held that protein kinase A did not alter activity of NOS I, although it did phosphorylate the enzyme (Brüne and Lapetina 1991, Bredt et al 1992); however, subsequent data from one of these laboratories show that cyclic nucleotide-dependent kinases, protein kinase A, and cGMP-dependent protein kinase all inhibit NOS I activity by phosphorylating a single site (Dinerman et al 1994). Since both NOS I and Ca/calmodulin kinase II are activated by an increase in Ca, this may provide a means of turning NOS activity off after a physiological stimulus.

Like most proteins, NOS I can also be transcriptionally regulated to induce tissue-specific expression. Transcription can be regulated by other factors as well. In the macula densa, a low-salt diet has been shown to enhance NOS I mRNA expression (Singh et al 1996b). In lung epithelial cells cytokines reduce NOS I activity, although it is not clear whether this is a transcriptional or posttranscriptional effect (Asano et al 1994). Regulation of NOS I transcription is an area needing more investigation.

Distribution and Regulation of NOS II Activity

NOS II protein can be expressed by a wide variety of tissues and cells, including immune cells (Moncada et al 1991, Lyons et al 1992), hepatocytes (Geller et al 1993), fibroblasts, cardiac myocytes (Singh et al 1996a), vascular smooth muscle cells (Moncada et al 1991), Kupffer cells (Moncada et al 1991), renal proximal tubules, thick ascending limbs and inner medullary collecting ducts, and mesangial cells of the glomerulus (Kone et al 1995, Ahn et al 1994). NOS II is primarily found in the cytosol as a soluble enzyme, although there is at least one report showing that it is associated with intracellular vesicles (Vodovotz et al 1995).

NOS II activity is generally considered to be regulated by changes in transcription and translation, since the enzyme binds calmodulin at basal intracellular Ca concentrations and thus is active. The NOS II promoter has more than 22 cis-acting elements, which bind regulatory factors activated by cytokines (such as interferon, tumor necrosis factor, and interleukin), the bacterial toxin lipopolysaccharide, and cAMP (Kunz et al 1994, Geller et al 1993, Mohaupt et al 1995, Mühl et al 1994, Xie et al 1993) and/or protein kinase C (Diaz-Guerra et al 1996) via an AP-1-like site (Xie et al 1993). In addition, increases in cAMP stabilize NOS II mRNA (Kunz et al 1994). The promoter region is also sensitive to glucocorticoids, which depress transcription (Geller et al 1993). Originally it was thought that NOS II activity was Ca-independent, since changes in intracellular Ca did not alter activity. We now know that this Ca independence is due to the high affinity of the enzyme for calmodulin. However, it has recently been reported that transcription of new NOS II mRNA (and consequently activity) can be inhibited by increases in intracellular Ca (Jordan et al 1995).

Although NOS II is thought to be regulated primarily via changes in transcription and translation, other factors help to determine enzyme activity. Availability of substrate and tetrahydrobiopterin (Werner-Felmayer et al 1990) have both been reported to limit the activity of NOS II. Thus, factors that regulate NOS II transcription may also modulate L-arginine influx into cells (Shibazaki et al 1996, Bogle et al 1992, Simmons et al 1996) and tetrahydrobiopterin production, so that NOS II activity is not limited by substrates. In addition to regulation by substrate and cofactor availability, NOS II activity may be regulated by phosphorylation, since it has a consensus sequence for cAMP-dependent protein kinase phosphorylation (Geller et al 1993).

Distribution and Regulation of NOS III Activity

While NOS I and II have been found in many types of cells, NOS III appears to be expressed in relatively few of them, including the endothelial cells lining the vasculature (Moncada et al 1991) and lung (Shaul et al 1994), cardiac myocytes (Seki et al 1996), and brain. NOS III is primarily a particulate enzyme (Boje and Fung 1990, Förstermann et al 1991), associating with the membrane as a result of posttranslation acylation. In the case of NOS III, both palmitate (Liu et al 1996, Robinson et al 1995) and myristate (Liu and Sessa 1994, Busconi and Michel 1993) are added. The latter is irreversible and is required for association with the membrane, while the former is reversible and may be a regulatory factor (Robinson et al 1995, Liu et al 1996).

Like NOS I, NOS III is expressed constitutively and its activity is enhanced by an increase in intracellular Ca (Schmidt et al 1992a, Schmidt et al 1993); however, since NOS III is primarily localized to the endothelial cells lining the blood vessels, the factors that increase Ca in these cells differ from those that increase Ca in cells containing NOS I. The primary activator of NOS III is shear stress on the endothelial cell, generated by the flowing blood. Increasing shear stress results in an increase in intracellular Ca in endothelial cells, although the mechanism involved remains unclear. Depolarization due to K^+ channels (Cooke et al 1991), ATP receptors (Dull and Davies 1991), and changes in the cytoskeleton (Wang et al 1993) may all play a role in the increased intracellular Ca associated with increases in shear stress (Ando et al 1988, Malek and Izumo 1994). Increased intracellular Ca results in the ion being bound to calmodulin. The Ca/calmodulin moiety then binds to and activates NOS III. In contrast to the hypothesis that shear stress induces changes in Ca, Ayajiki et al (1996) have suggested that shear stress activates a tyrosine kinase that in turn alters NOS II activity.

In addition to shear stress, NOS III activity may be regulated by a variety of hormones and factors. Vascular endothelial cells express a number of receptors that, when activated by their ligand, can stimulate NO production and release. These include acetylcholine, bradykinin, vasopressin, and adenosine. These ligands bind to specific receptors coupled to heterotrimeric G-proteins, which activate a number of effector systems including those that result in an increase in intracellular Ca. The increase in Ca then activates NOS III (Moncada et al 1991, Vanhoutte 1989, Griffith and Henderson 1989, Furchgott et al 1981, Lincoln et al 1996, Moncada et al 1991).

In addition to factors that stimulate NO production by NOS III via increases in intracellular Ca, the activity of the enzyme is regulated by phosphorylation and dephosphorylation. Hirata et al (1995) reported that in cultured bovine aortic endothelial cells, protein kinase C decreased both the conversion of L-arginine to citrulline and the release of NO. Protein kinase C has also been shown to phosphorylate NOS III (Michel et al 1993). Interestingly, other factors generally considered to increase NO production also phosphorylate NOS III, including cAMP and cGMP. Phosphorylation by cyclic nucleotide-stimulated kinases activated by these agents is likely to have the same effect on NOS III activity as it does on NOS I, because of the homology of NOS I and III. NOS III is generally associated with the particulate fraction of the cell, as a result of posttranslational acylation. However, ligands that release NO from endothelial cells may also phosphorylate NOS III and remove the palmitate group to make it a soluble protein. Bradykinin has just such an effect, but its role in the regulation of NOS III activity is unclear (Michel et al 1993).

Although NOS III is expressed "constitutively," regulation of transcription is physiologically important. The promoter region of the NOS III gene has several putative and actual *cis*-acting elements, including sterol, cAMP, phorbol ester, NF1, and shear stress regulatory regions (Marsden et al 1993, Robinson et al 1994). Shear stress has been shown to increase not only NOS III activity but also mRNA (Awolesi et al 1994, Shyy et al 1995, Ranjan et al 1995, Uematsu et al 1995). Dexamethasone, which inhibits NOS II transcription, also suppresses transcription of NOS III (Shyy et al 1995). The sterol response element is also active but does not increase transcription; rather, transcription is diminished when sterols are present. Cytokines have been shown to reduce transcription of NOS III (Kelly et al 1996, Lamas et al 1992), perhaps via the NF1 binding site or some other as yet unidentified *cis*-acting element. The promoter of the NOS III gene also contains AP-1 and AP-2 sites that may confer sensitivity to protein kinases A and C, respectively. The AP-2 site appears to be active, since protein kinase C inhibits transcription (Ohara et al 1995). NO produced by NOS III exhibits positive feedback via cGMP to enhance both transcription and translation (Ravichandran and Johns 1995), although the *cis*-acting element responsible is unknown. It may occur via one of the AP sites, since protein kinase A and cGMP-dependent protein kinase have similar motifs for substrate specificity. Finally, the promoter also contains γ-interferon response elements, but they do not appear to be positioned in such a manner as to be active (Yang et al 1990, Pearse et al 1991). Of course, in addition to these elements, the promoter also contains regions that confer cell-specific expression of NOS III (Robinson et al 1994).

INTRACELLULAR ACTIONS OF NO

It is now recognized that NO alters a number of key molecules in prokaryotes and eukaryotes (Figure 7-3). Most of these actions are dependent on NO either altering the redox state of a required metallic cofactor such as Fe, Cu, Zn, or Mn or binding to either thiol or amino groups. The actions of NO may be arbitrarily divided into cytotoxic and regulatory effects. Cytotoxic effects usually require large quantitites of NO and fall into two categories: specific and nonspecific. Specific effects include deamination of DNA, inhibition of aconitases, inhibition of respiratory chain enzymes, interruption of iron metabolism by transferrin, ferritin and other iron-binding proteins, inhibition of ribonucleotide reductase, and inhibition of H-ATPase. Nonspecific effects include generation of reactive species such as nitrosylation of amines and peroxynitrite and nitrosylation of thiol groups. The cytotoxic effects of NO should not be considered entirely negative, for they are the mechanisms by which macrophages destroy invading bacteria and suppress tumor growth. These actions of NO will not be discussed in detail.

Regulatory effects of NO involve activation of guanylate cyclase, ADP-ribosyl transferase and cyclooxygenase (Henry et al 1993; Salvemini et al 1993; Garthwaite and Boulton 1995). Nitrosylation of heme groups and thiols may be regulatory if it occurs at a rate that does not alter enzymatic activity so

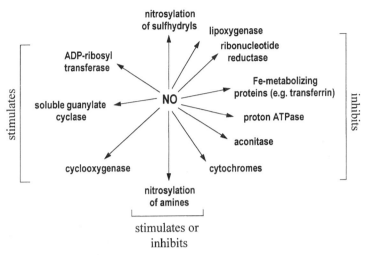

Figure 7-3. Enzymes and proteins affected by nitric oxide. Enzymes stimulated by nitric oxide are depicted on the left. Those inhibited are on the right. Nitrosylation of sulfhydryls and amines may either inhibit or stimulate activity.

drastically that the cell dies. Examples may be ntirosylation of NF-κβ, a nuclear regulatory protein (Matthews et al 1996), respiratory enzymes (Xie and Wolin 1996), and protein phosphatases (Caselli et al 1994). Regulatory effects of NO usually require much lower concentrations of NO than the cytotoxic effects. The precise concentrations for regulatory versus cytotoxic effects depend on cell type and a number of other factors. However, it should be noted that they are not mutually exclusive. For instance, ADP-ribosylation of key proteins may lead to dire consequences for the cell.

NO-Induced Increases in cGMP

Perhaps the best known action of NO is activation of soluble guanylate cyclase. This was the first enzyme shown to be activated rather than inhibited by NO. Soluble guanylate cyclase is a heterodimer with a total molecular weight of 150–160 kDa. The α subunit has a mass of approximately 80 kDa, while the β subunit is slightly smaller with a mass of 70–80 kDa. Each subunit has a terminal catalytic site, but monomers are not active. Currently, two isomers of each subunit have been described, designated α_1, α_2, β_1 and β_2. Interestingly, the β_2 subunit has a carboxy terminal isoprenylation site, which could indicate that this soluble guanylate cyclase may actually be membrane-bound (Koesling et al 1991). Expression of the various isozymes differs from tissue to tissue; however, the physiological significance of this is not understood. The protein contains a heme binding site as well as a binding

site for copper. In the basal state, the iron of the heme moiety lies in a planar
configuration within the porphyrin ring. NO binds iron in soluble guanylate
cyclase, displacing it from its planar position and thereby enhancing the
activity of the molecule 30- to 40-fold by inducing a conformational change.
The mechanism by which NO activates soluble guanylate cyclase was deduced from data showing that the heme-free enzyme could not be stimulated by NO. CO and hydroxyl radicals can activate soluble guanylate
cyclase by a similar mechanism (Waldman and Murad 1987, Drewett and
Garbers 1994, Schultz et al 1991, Koesling et al 1991).

Once activated, soluble guanylate cyclase binds GTP, removes the two
terminal phosphates, and produces an ester linkage between the 3' and 5'
carbons, creating cyclic 3',5' guanylate cyclase. Interestingly, the activated
molecule can also use ATP as a substrate, producing cAMP at 1–10% of the
rate at which it generates cGMP. A metal cofactor (either Mn or Mg) is
required for GTP binding and cleavage, with K_m being 10 and 130 μM,
respectively. Additionally, when soluble guanylate cyclase is activated, it
loses its selectivity for Mg over Mn, and either molecule can maintain
activity. Soluble guanylate cyclase returns to the inactive stage when NO is
removed (Waldman and Murad 1987, Drewett and Garbers 1994, Schultz et
al 1991, Koesling et al 1991).

There are many enzymes whose activity is altered by cGMP, including
cGMP-dependent kinases, cGMP-inhibited phosphodiesterase (phosphodiesterase III), cGMP-stimulated phosphodiesterase (phosphodiesterase II),
and cGMP-gated ion channels (Figure 7-4). cGMP-dependent protein ki-

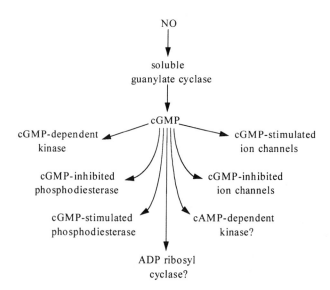

Figure 7-4. Enzymes affected by cGMP.

nase probably accounts for most actions of cGMP, and consequently most of the effects of NO. There are at least three isozymes of cGMP-dependent protein kinase, which have a similar structure and are activated by cGMP in a similar manner. They are divided into two families, type I and type II, because they are products of two genes. There are two forms of type I, α and β, which are the products of alternate splicing. Type I kinases are both soluble and particulate enzymes depending on the type of cell, whereas type II kinase is particulate. The enzyme acts as a dimer, with monomers having a molecular weight of 76–78 kDa in mammals. The amino terminal of cGMP-dependent protein kinase binds cytoskeletal elements with high affinity (Lincoln et al 1996, Lincoln and Corbin 1983).

As the name implies, cGMP-dependent kinase is activated when it binds cGMP, and the activated enzyme transfers a phosphate group from ATP to a protein substrate. In general, this family of kinases is made up of homodimers that bind up to four cGMP moieties. cGMP-dependent protein kinases belong to the larger class of serine/threonine kinases, in that they transfer phosphate moieties to the amino acids serine and threonine. However, in order to be phosphorylated, these amino acids must be part of a consensus sequence, a group of amino acids that provides a binding site for the kinase. The different isozymes of cGMP-dependent protein kinase vary in their affinity for cGMP and substrates. The physiological roles of each isozyme remain to be elucidated (Lincoln et al 1996).

In addition to activating cGMP-dependent protein kinase, NO may also enhance phosphorylation of proteins by all kinases by inhibiting phosphoprotein phosphatases. NO has been shown to inhibit low M_r phosphotyrosine protein phosphatase (Caselli et al 1994). Since all phosphatases have a similar structure in their active sites, including a cysteine residue, NO may alter the activity of the entire family.

cGMP also alters the activity of another major family of enzymes, the phosphodiesterases. Although all phosphodiesterases may be affected by cGMP to some extent, discussion will be limited to those in which cGMP acts as an allosteric modulator, namely cGMP-stimulated and cGMP-inhibited phosphodiesterases. Phosphodiesterases are a family of enzymes that catabolize the cyclic nucleotide compounds cAMP and cGMP to AMP and GMP, respectively. As a whole, the family will cleave both cGMP and cAMP, with various family members having different selectivities for cAMP and cGMP. The phosphodiesterases allosterically modulated by cGMP selectively cleave cAMP over cGMP. Each molecule binds two cyclic nucleotides, with one being bound at a regulatory site and the other at the catalytic site (Thompson 1991, Stroop and Beavo 1991).

cGMP-stimulated phosphodiesterases, or phosphodiesterase II, are a family of enzymes with molecular weights ranging from 102 to 105 kDa (Thompson 1991). They are found in a wide range of tissues, including the adrenal gland, heart, kidney, fibroblasts (Thompson 1991), and brain (Beavo and Reifsnyder 1990), and act as homodimers (Stroop and Beavo 1991). The catalytic site has a K_m for cAMP and cGMP of 30 and 15 µM, respectively. cGMP binds to an allosteric site with an affinity of less than 1 µM. Binding of cGMP to this site stimulates activity 10- to 100-fold. The carboxy terminal contains the active site and has high homology with other phosphodiesterases. A region of 270 amino acids near the middle of the molecule also exhibits high homology and is the allosteric cyclic nucleotide binding site (Stroop and Beavo 1991). Investigation into the physiological importance of this enzyme has been hampered by the lack of selective inhibitors (Conti et al 1991); however, this may change with the release of MEP-1, which selectively inhibits cGMP.

cGMP-inhibited phosphodiesterases, or phosphodiesterase III, are a family of phosphodiesterases with molecular weights ranging from 73 to 135 kDa. Smaller species have been reported, which are likely proteolytic degradation products. They have two binding sites for cyclic nucleotides, one specific for cGMP and the other (the catalytic site) selective for cAMP over cGMP. They are found in a wide range of tissues, including platelets, cardiac myocytes, vascular and brachial smooth muscle, liver, adipocytes, and lymphocytes (Reinhardt et al 1995). The enzyme may be found in either the soluble or particulate fraction. In platelets it is soluble with a molecular weight of 110 kDa, whereas in the liver and adipose tissue it is particulate with molecular weights of 73 and 135 kDa, respectively. The liver enzyme functions as a dimer. cGMP-inhibited phosphodiesterases can be phosphorylated by cAMP-dependent protein kinase, thereby enhancing their catalytic activity. This type of regulation may serve as a form of feedback inhibition in which cGMP inhibits phosphodiesterase activity, cAMP levels and protein kinase A activity increase, cGMP-inhibited phosphodiesterase is phosphorylated, cGMP-inhibited phosphodiesterase catalytic activity is enhanced, and cAMP levels are reduced. K_m for cAMP ranges from 0.15 to 0.4 µM with maximum cleavage rates of 3–9 µmol/min/mg protein. The K_i for cGMP ranges from 0.06 to 0.6 µM. This enzyme will also cleave cGMP with a K_m of 0.02–0.8 µM and a maximum rate of 0.05–2.0 µmol/min/mg protein (Beltman et al 1993). Phosphatidic acid may also regulate particulate forms of cGMP-inhibited phosphodiesterase (Carpenedo and Floreani 1993).

Finally, there are a number of channels whose activity is regulated by cGMP, referred to as cGMP-gated ion channels. Although both activation and inhibition of cGMP-gated channels have been reported, most channels in this category are stimulated by cGMP. The best known is probably the cGMP-gated sodium channel involved in vision and olfaction. Epithelia involved in salt transport also possess cGMP-gated channels. In the kidney, sodium channels that are both inhibited and stimulated by cGMP are expressed, although it is unclear how cGMP-stimulated channels are involved in the regulation of sodium excretion (McCoy et al 1995).

ADP-Ribosylation

Stimulation of ADP-ribosyl transferase has been proposed as a second means whereby NO can regulate cell function, although the significance of this pathway is less well defined than that of the cGMP second messenger cascade. A number of proteins are known to be ADP-ribosylated by NO, including actin, glyceraldehyde phosphate dehydrogenase, and G-proteins, as well as several that have not yet been identified (Brüne et al 1993, Henry et al 1993, Kanagy et al 1995). NO's ability to ADP-ribosylate G-proteins may provide a means whereby NO can influence several cellular processes, since the G-proteins themselves regulate several second messenger cascades. Kanagy et al (1995) have proposed that NO may ADP-ribosylate $G_{\alpha s}$ and $G_{\alpha i}$, thus activating and inactivating these proteins, respectively. This could result in NO-induced cGMP-independent vasodilation due to increased cAMP levels. Furthermore, ADP-ribosylation of $G_{\alpha o}$ and actin would be expected to blunt increases in intracellular Ca induced by vasoconstrictors and to diminish the ability of actin to respond to vasoconstrictor-induced increases in Ca. NO-induced cGMP-independent relaxation of uterine smooth muscle may occur via such a mechanism.

Cyclic ADP Ribose

Although there is little information about this potential second messenger pathway, which was only recently discovered, it may prove to be very important in mammalian systems. In sea urchin eggs cGMP has been found to stimulate ADP ribosyl cyclase, an enzyme that forms cyclic ADP ribose from NAD. Cyclic ADP ribose acts via an intracellular ryanodine receptor to release intracellular stores of Ca. Ryanodine-sensitive stores of Ca have been demonstrated in many tissues, including the brain, but the compound that releases Ca via this mechanism is unknown. It may prove to be NO (Garthwaite and Boulton 1995).

Arachidonic Acid Metabolism

There are at least three families of enzymes that are responsible for the initial metabolism of arachidonic acid to produce biologically active products: cyclooxygenases, cytochrome P450s, and lipoxygenases. All categories are affected by NO. Cyclooxygenase exists in at least two isoforms, referred to as cyclooxygenase I and II (COX I, COX II). These enzymes are responsible for the early steps in the formation of prostaglandins from arachidonic acid. NO has been shown to activate both enzymes in a cGMP-independent manner (Salvemini et al 1993). Further studies have shown that peroxynitrite, which is produced when NO reacts with the superoxide anion, is responsible for this activation (Landino et al 1996). NO also stimulates the activity of prostaglandin H synthase, an enzyme that is responsible for further metabolism of cyclooxygenase products to prostaglandin H_2. The stimulation of prostaglandin H synthase is due to nitrosylation of a sulfhydryl group in the catalytic domain (Haijar et al 1995). Once formed, prostaglandins activate a number of second messenger cascades via G-protein-linked receptors.

In addition to cyclooxygenase, NO also affects lipoxygenase (Maccarrone et al 1996) and cytochrome P450 activity (Wink et al 1993). However, these enzymes are inhibited by NO. Lipoxygenase activity is reduced because NO attacks the iron moiety required for catalytic activity (Maccarrone et al 1996). NO inhibits cytochrome P450 activity via two mechanisms, one that is reversible and the other which is not (Wink et al 1993).

EXAMPLES OF NO-ACTIVATED SECOND MESSENGER CASCADES

In the next section, several examples of the second messenger cascades that activate NO release, as well as those activated by NO, will be addressed. Surprisingly, while we know of many physiological and pathophysiological events in which NO is involved, we have a relatively complete picture in few instances. Representative examples of these are discussed below. The reader must note that the precise second messenger cascade activated by NO and the end physiological response will depend on a variety of parameters, including cell type and local environment. These are too numerous to describe in detail here; consequently, general outlines are given with selected references. For more detailed information, the reader is referred to several excellent reviews in each area.

Cardiovascular System and Endothelium-Derived Relaxing Factor

In 1867, Brunton first used amyl nitrate to treat angina pectoris (Brunton 1867), unknowingly beginning a regimen of treatments whose cellular basis would not be understood for more than 100 years. In what are now classic experiments, Furchgott and Zawadzki first demonstrated in 1980 that the endothelial cells lining the vasculature produce a vasoactive compound. They showed that acetylcholine induced relaxation of aortic strips when the endothelium was intact, but had no effect when the endothelium was removed. These investigators correctly predicted that the endothelium releases a labile factor that induces relaxation of the underlying vascular smooth muscle, and dubbed it endothelium-derived relaxing factor (EDRF) (Furchgott and Zawadzki 1980). These experiments stimulated further investigation into the identity and mechanism of action of this compound, which we now know is NO.

In the vessels of the cardiovascular system, NO has a number of important physiological actions; it serves as a vasodilator, reduces monocyte adherence (Cooke 1996), reduces platelet activation (Pohl and Busse 1990), and prevents neointima formation (Ellenby et al 1996). The first reported physiological effect of NO was vasodilation. *In vivo* NO is released by shear stress or ligands that increase intracellular Ca in endothelial cells. The increase in intracellular Ca activates NOS III by promoting the binding of Ca/calmodulin to the enzyme. Once activated, NOS III catabolizes L-arginine to NO, which diffuses out of the cell. On the blood side of the endothelium, NO is rapidly scavenged by hemoglobin. On the vascular smooth muscle side, NO diffuses into the muscle and activates soluble guanylate cyclase. This increases cGMP, which in turn activates cGMP-dependent protein kinase. From this point it is not precisely known how cGMP-dependent protein kinase induces relaxation; however, this is thought to involve a decrease in intracellular calcium (Johnson and Lincoln 1985). At least three mechanisms are possible (Figure 7-5). First, cGMP-dependent kinase may increase Ca uptake by the sarcoplasmic reticulum Ca ATPase, removing it from the cytoplasm, by phosphorylating a regulatory protein such as phospholamban (Raeymaekers et al 1988, Sarcevic et al

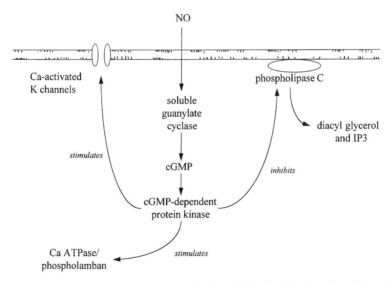

Figure 7-5. Mechanisms responsible for nitric oxide-induced relaxation of vascular smooth muscle. NO diffuses into the muscle and activates soluble guanylate cyclase, increases cGMP, and activates cGMP-dependent protein kinase. cGMP-dependent protein kinase-induced relaxation may involve three mechanisms: increased Ca uptake by the sarcoplasmic reticulum Ca ATPase due to phosphorylation of phospholamban; activation of Ca-activated K channels, causing hyperpolarization of the cell membrane and decreased Ca influx due to closure of Ca channels; and inhibition of vasoconstrictor-stimulated phospholipase C. IP3, inositol triphosphate.

1989). Next, cGMP-dependent protein kinase may stimulate Ca-activated K channels, causing hyperpolarization of the cell membrane (Archer et al 1994) and decreased Ca influx into the cell due to closure of Ca channels. Finally, cGMP-dependent protein kinase may inhibit vasoconstrictor-stimulated phospholipase C, a key enzyme in the second messenger cascade of vasoconstrictors (Rapoport 1986, Ruth et al 1993).

In addition to cGMP-dependent protein kinase, other pathways may be involved in NO-induced dilation. For instance, cGMP may directly activate ion channels, high levels of cGMP may directly activate cAMP-dependent protein kinase (Lincoln et al 1996), or cGMP-inhibited phosphodiesterase may increase cAMP levels (Haynes et al 1991) with consequent activation of cAMP-dependent protein kinase. The mechanisms of vasorelaxation are reviewed by Lincoln et al (1996).

In spite of the rapid scavenging of NO by hemoglobin in the circulation, sufficient NO escapes to prevent platelet aggregation on the vessel wall. Ligands such as thrombin and adenosine stimulate platelet aggregation via receptor-mediated release of stored granules. NO from endothelial cells diffuses into the platelets, where it increases cGMP and activates cGMP-dependent protein kinase (Pohl and Busse 1990). Phosphorylation of an unknown substrate, or substrates, blocks the increase in intracellular Ca that signals aggregation and secretion. One such substrate may be a component of the cytoskeleton. Phosphorylation of a 46–50-kDa actin-binding protein disrupts the ability of platelets to aggregate by preventing the effects of other kinases activated by agents that stimulate aggregation (Schmidt et al 1993). Additionally, NO-induced increases in cGMP reduce the activity of cGMP-inhibited phosphodiesterase. Inactivation of this enzyme increases cAMP and cAMP-dependent protein kinase activity. Activation of cAMP-dependent protein kinase prevents aggregation and secretion (Beltman et al 1993).

Platelets not only respond to NO but also produce NO via NOS III (Moncada et al 1991). NO blockade of platelet aggregation may be very important in preventing clot formation, since many of the agents that can induce aggregation, such as collagen, ADP, or arachidonate, do not do so when physiological concentrations of L-arginine are present. Agents that induce aggregation work by increasing intracellular Ca; thus, enhanced production of NO by this increase in Ca may be a means of feedback inhibition. However, feedback inhibition can be overridden in some cases. Thrombin-induced aggregation is due to an increase in intracellular Ca, but it is not inhibited by NO except when cGMP is pharmacologically enhanced by phosphodiesterase inhibition.

In the heart, NO obviously plays an important role in keeping the vessels patent via vasodilation and prevention of platelet aggregation. However, it also plays an important role in regulating the force and rate of contraction. The effects of NO on the heart are reviewed by Kelly et al (1996).

NOS I is found in adrenergic, cholinergic, and nonadrenergic/noncholinergic nerves supplying the heart (Kelly et al 1996). In sympathetic neurons NO appears to reduce release of norepinephrine (Schwartz et al 1995) as well as block its uptake (Kaye et al 1995). Such an action would tend to decrease heart rate and contractility (Figure 7-6).

NOS III has been found in a variety of cells within the heart, including the myocytes. In general, it decreases contractility and has a negative chronotropic effect. NOS III may be activated by at least three mechanisms. First, normal pacing can activate NOS and induce the release of NO. In paced cardiocytes, the addition of a NOS inhibitor increases the rate of

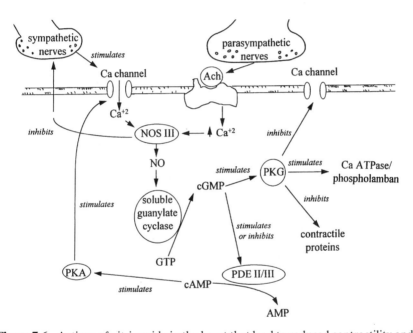

Figure 7-6. Actions of nitric oxide in the heart that lead to reduced contractility and pacing. NO stimulates soluble guanylate cyclase and increases cGMP levels. cGMP may activate cGMP-dependent protein kinase (PKG), cGMP-inhibited phosphodiesterase (PDE III), and cGMP-stimulated phosphodiesterase (PDE II). PKG may reduce the force and rate of contraction, possibly by phosphorylating troponin I or by phosphorylating phospholamban. PDE III is inhibited by the increases in cGMP brought about by NO. This may result in an increase in cAMP and cAMP-dependent protein kinase (PKA). PKA in turn activates Ca channels, countering the effects of PKG. In contrast, cGMP may stimulate PDE II, reduce cAMP levels and PKA activity, and thereby reduce Ca channel activity. Nitric oxide inhibits the release of neurotransmitters from sympathetic nerves, as well as their reuptake. Ach, acetylcholine.

contraction (Finkel et al 1995). β-adrenergic stimulation also stimulates the release of NO. β-adrenergic stimulation ultimately results in the activation of Ca channels, and the resultant increase in intracellular Ca activates NOS III. This fact was demonstrated by showing that inhibition of NOS increases the chronotropic and ionotropic effects of β-adrenergic stimulation (Keaney et al 1996). Finally, cholinergic nerves will activate NOS III in the cardiomyocytes. Activation of the muscarinic receptor by acetylcholine also results in increased intracellular Ca (Balligand et al 1993).

Once formed, the NO generated by NOS III, as well as other sources of NOS, has a variety of effects within the myocyte. NO stimulates soluble guanylate cyclase and increases cGMP levels. cGMP has several possible effects, including activation of cGMP-dependent protein kinase and cGMP-stimulated phosphodiesterase (PDE II), and blocking cGMP-inhibited phosphodiesterase (PDE III). cGMP-dependent protein kinase reduces the force and rate of contraction, possibly by phosphorylating Ca channels thereby inactivating them (Sumii and Sperelakis 1995). This kinase has also been reported to phosphorylate troponin I in the contractile machinery, reducing its affinity for calcium (Shah et al 1994). Finally, cGMP-dependent protein kinase may also phosphorylate phospholamban (Raeymaekers et al 1988), enhancing Ca uptake by the Ca ATPase.

The role of the phosphodiesterases is more interesting and controversial. PDE III is inhibited by the increases in cGMP brought about by NO. This results in an increase in cAMP and cAMP-dependent protein kinase. cAMP-dependent protein kinase in turn activates Ca channels. Thus, inhibition of PDE III by cGMP tends to counter the effects of cGMP-dependent protein kinase (Wang and Lipsius 1995). However, the effects of the kinase appear to predominate. In contrast, cGMP stimulates PDE II and reduces cAMP levels. The decrease in cAMP diminishes cAMP-dependent protein kinase activity and Ca channel activity (Hartzell and Fishmeister 1986, Han et al 1996). Consequently, stimulation of PDE II by NO-induced increases in cGMP enhances the effects of cGMP-dependent protein kinase. However, the effects of PDE II may only be important when cAMP levels are elevated. Whether or not the effects of PDE II or III are important may also be species-dependent. Additionally, they depend on the levels of cGMP in an individual cell. Low levels of cGMP have been reported to first inhibit PDE III; then, as levels increase, PDE II is stimulated (Méry et al 1993).

NO also inhibits atherogenesis in animal models by reducing platelet adherence, monocyte adhesion, and neointima formation (Cooke 1996). However, the role of NO in pathological states is beyond the scope of this chapter.

Central and Peripheral Nervous System

NO is produced in nonadrenergic/noncholinergic nerves, where it mediates cerebral blood flow, pain perception, gastrointestinal motility, urogenital function, and penile erection, as well as other functions. Several of the physiological functions of these nerves, including gastrointestinal motility (Dawson and Dawson 1996) and penile erection (Burnett et al 1992), are due to relaxation of smooth muscle. Release of NO from nerve terminals is the result of an increase in Ca caused by action potentials. The increase in Ca stimulates activity of NOS I, resulting in the production of NO, which activates soluble guanylate cyclase and enhances cGMP levels in the target tissue. The second messenger cascades are thought to be similar to those described above for relaxation of vascular smooth muscle.

In addition to the peripheral nervous system, NO also plays a key role in the central nervous system, including the hippocampus, cerebellum, corpus striatum, dorsal root ganglia, and hypothalamic-pituitary axis. NO has been implicated in long-term potentiation and depression in the hippocampus and cerebellum, respectively, and also in neurotransmission. The central nervous system effects of NO are reviewed by Garthwaite and Boulton (1995) and Dawson and Dawson (1996).

The rudimentary steps through which NO alters neurotransmitter release are beginning to be outlined, although we do not have a complete picture. In the cerebellum, glutamate binds to NMDA receptors on granular and basket cells and increases intracellular Ca, which activates NOS I. The resulting NO diffuses to the Purkinje cells, where it activates soluble guanylate cyclase; this increases cGMP, which in turn enhances the activity of cGMP-dependent protein kinase. The kinase is then thought to phosphorylate a regulatory protein, DARPP-32 (Tsou et al 1993). When phosphorylated, DARPP-32 acts as a protein phosphatase inhibitor. Consequently, protein phosphatases are inhibited and the effects of other kinases are enhanced (Garthwaite and Boulton 1995). Alternately, cGMP-dependent protein kinase may directly phosphorylate proteins associated with vesicle fusion (Dawson and Snyder 1994), which enhances neurotransmitter release (Figure 7-7).

In regions of the brain such as the hippocampus, cerebral cortex, and basal ganglia, all of which possess high levels of cGMP-stimulated phosphodiesterase, NO may act via a different mechanism. In the hippocampus, cGMP suppresses at least one type of Ca current. Since the hippocampus possesses both NOS I and cGMP-stimulated phosphodiesterase, it has been postulated that NO activates soluble guanylate cyclase, increases cGMP, and activates cGMP-stimulated phosphodiesterase. Activation of this phosphodiesterase decreases cAMP, resulting in diminished cAMP-dependent protein kinase and reduced Ca channel activity (Doerner and Alger 1988). This mechanism may be the basis for long-term depression (Figure 7-7).

Finally, peroxynitrite appears to have important actions in the central nervous system. NMDA receptor activity is altered by peroxynitrite. This has direct effects on neurotransmission. Cyclooxygenase is also activated by peroxynitrite in the brain. This results in an increase in prostaglandin syn-

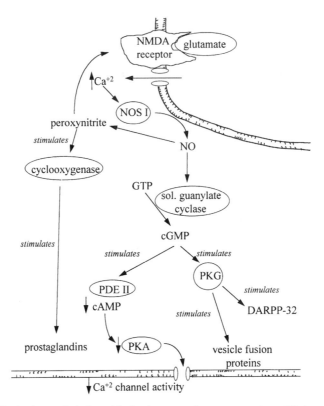

Figure 7-7. Actions of nitric oxide in the central nervous system. Glutamate binds to NMDA receptors on granular and basket cells and increases intracellular Ca, which activates NOS I. The resulting NO diffuses to the Purkinje cells, where it activates soluble guanylate cyclase; this increases cGMP, which in turn enhances the activity of cGMP-dependent protein kinase (PKG). PKG may phosphorylate regulatory proteins such as DARPP-32 or proteins associated with vesicle fusion. In other regions of the brain, NO may activate cGMP-stimulated phosphodiesterase (PDE II). Activation of this phosphodiesterase decreases cAMP, resulting in diminished cAMP-dependent protein kinase (PKA) and reduced Ca channel activity.

thesis. The prostaglandins thus produced have a variety of effects (Yamagata et al 1993).

NO is also involved in excitatory neurotransmitter toxicity. Excess amounts of excitatory neurotransmitters such as glutamate result in massive release of NO via the NMDA receptor. The magnitude of the NO release causes DNA damage, which activates poly-ADP-ribose synthase; this in turn converts NAD to nicotinamide while consuming ATP (Zhang et al 1994, Lautier et al 1993). This process, coupled with the inhibition of ATP production due to NO-inhibiting aconitases and mitochondrial cytochromes, results in depletion of ATP and NAD with resultant cell death (Dawson and Snyder 1994).

Kidney

NO has many roles in the kidney. Originally it was thought that NO caused salt and water retention, because animals treated with LNAME to inhibit NOS showed pronounced natriuresis and diuresis. However, more recent studies have shown that the natriuresis and diuresis induced by systemic NOS inhibition are the result of increased blood pressure. When NO synthesis is stimulated only in the kidney, it increases salt and water excretion. Thus it appears that the regulatory role of NO is to promote diuresis and natriuresis. NO affects many types of cells within the kidney in a concerted manner to produce these effects.

NO has been shown to increase renal blood flow and decrease renal vascular resistance. Detailed *in vitro* studies have shown that NO is released from and also dilates the afferent arteriole, the major resistance vessel in the kidney (Juncos et al 1995, Navar et al 1996). The mechanism of vasodilation is presumed to be similar to that described above for the vasculature in general. Such dilation increases blood flow to the glomerulus, which tends to enhance glomerular filtration rate.

NO derived from NOS I in the macula densa has also been shown to inhibit tubuloglomerular feedback, a process that promotes salt and water retention by reducing glomerular filtration when the NaCl concentration of luminal fluid increases (Wilcox et al 1992, Navar et al 1996). Although NOS I expression in the macula densa is altered by changes in dietary salt content, it is unclear how NO inhibits tubuloglomerular feedback. Our laboratory has proposed that NO may act as an autacoid in the macula densa, inhibiting Na/K/2 Cl cotransport, which is essential for tubuloglomerular feedback.

Although NO's actions in the renal vasculature all tend to enhance salt and water excretion, NO can induce natriuresis and diuresis without altering glomerular filtration or renal blood flow (Lahera et al 1991, Lahera et al 1993, Majid and Navar 1994). This led to the hypothesis that NO inhibits Na and water absorption by the nephron. NO inhibits salt and water reabsorption by the proximal tubule, but only at high concentrations (Guzman et al 1995, Roczniak and Burns 1996), and thus the importance of NO as a physiological regulator of proximal nephron function is unclear. However, the proximal nephron expresses NOS II after ischemic injury, so that the effects of NO may become important in pathological states or during periods of ischemia in this segment.

NO also inhibits NaCl absorption by medullary and cortical thick ascending limbs of the loop of Henle. Interestingly, NOS II is expressed in these segments, and the NO it produces has been shown to act as an autacoid, inhibiting salt absorption (Stoos et al 1996). Since production of NO by NOS II in the thick ascending limb increases with the NaCl content of the diet, we have proposed that the ability of NO to inhibit NaCl absorption in the thick ascending limb is important in eliminating a salt load. The mechanism by which the salt content of the diet is transmitted to the thick ascending limb is unclear, but may involve changes in osmolality or angiotensin.

The most distal portion of the nephron, the collecting duct system, was the first nephron segment shown to be sensitive to NO. Zeidel et al (1986) demonstrated that NO donors inhibit sodium transport by the inner medullary collecting duct by blocking apical membrane Na channels via an increase in intracellular cGMP. Light et al (1990), using the patch clamp technique, showed that cGMP reduces the probability of apical sodium channels being open. Interestingly, cGMP regulates open channel probability in the inner medullary collecting duct via a mechanism dependent on cGMP-dependent protein kinase as well as by direct interaction (Figure 7-8).

Subsequently our laboratory has shown that in the cortical collecting duct NO inhibits both basal (Stoos et al 1995) and vasopressin-stimulated sodium and water reabsorption (Garcia et al 1996a). The latter process

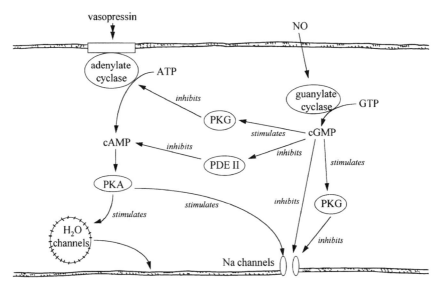

Figure 7-8. Actions of nitric oxide in the renal collecting duct. NO activates soluble guanylate cyclase and increases cGMP. cGMP regulates open channel probability via a mechanism dependent on cGMP-dependent protein kinase (PKG), as well as by direct interaction. cGMP inhibits vasopressin-stimulated water reabsorption by activating both PKG and cGMP-stimulated phosphodiesterase (PDE II), resulting in decreased cAMP levels. PKA, cAMP-dependent protein kinase.

involves activation of soluble guanylate cyclase, increases in cGMP, and activation of both cGMP-dependent protein kinase and cGMP-stimulated phosphodiesterase. Stimulation of these cGMP-dependent enzymes results in decreased cAMP levels. The fall in cAMP and consequent decrease in cAMP-dependent protein kinase activity diminish transport. If the decrease in cAMP is blocked, NO has no effect on vasopressin-stimulated water absorption (Garcia et al 1996b) (Figure 7-8).

Immune Responses

NO is produced by NOS II in macrophages. As such, its production is regulated primarily by factors that stimulate gene transcription, including bacterial lipopolysaccharide and cytokines. Foreign, cancerous, and invading cells induce the release of cytokines, whereas lipopolysaccharide is a product of the bacteria themselves. These factors dramatically enhance NOS II transcription and translation in macrophages. Given that NOS II avidly binds calmodulin, NOS II is active at basal Ca concentrations. Consequently, the large increase in NOS II transcription and translation leads to generation of large, cytotoxic amounts of NO. The NO thus produced acts at a variety of levels. First, NO inhibits energy production by binding cytochromes (Welter et al 1996) and aconitases (Castro et al 1994, Hausladen and Fridovich 1994). It may also reduce ATP production by ADP-ribosylating key enzymes such as glyceraldehyde phosphate dehydrogenase (Tao et al 1994). NO blocks DNA synthesis and damages DNA (Henry et al 1993, Moncada et al 1991). Finally, many cell processes are disrupted as a result of the nonspecific nitrosylation of amines and sulfhydryl groups. These actions result in either the death or slow reproduction of the cancerous/foreign cells so that phagocytic immune cells can purge the body of the invaders.

Other Tissues

NO plays an important role in the physiology of several other tissues, including the gastrointestinal tract, the respiratory system, skeletal muscle, and bone. In the gastrointestinal tract it regulates motility via cGMP-independent mechanisms (Takeuchi et al 1996) and water and solute absorption (Schmidt et al 1993, Barry et al 1994). In the lung it controls ion absorption (Schmidt et al 1993) and bronchial smooth muscle tone (Rand and Li 1995, Gaston et al 1994). In bone it is involved in the regulation of reabsorption via a cGMP-independent process (MacIntyre et al 1991). NO plays an important role in the reproductive system, being involved in penile erection (Rand and Li 1995) and uterine contraction (Kuenzli et al 1996). The number of physiological processes NO is known to be involved in is increasing rapidly.

SUMMARY

NO plays a wide variety of roles in the regulation of physiological function. It plays a role in the regulation of blood pressure, cardiac function, central nervous system function, renal function, and immune responses. NO is produced by three classes of enzymes (NOS I, II, and III), which have different physiological stimulators. NOS I and III are activated by Ca, whereas NOS II is not. NO stimulates several second messenger cascades. The precise effects of NO depend on the type of cell and the amount produced. The primary second messenger cascade activated by NO in the regulation of physiological processes is the cGMP cascade; however, cGMP-independent mechansims are important in some tissues and circumstances. Diminished or enhanced production of NO may lead to several pathological conditions.

REFERENCES

Abu-Soud HM, Wang J, Rousseau DL, Fukuto JM, and Ignarro LJ. 1995. Neuronal nitric oxide synthase self-inactivates by forming a ferrous-nitrosyl complex during aerobic catalysis. J Biol Chem 270:22997–23006.
Ahn KY, Mohaupt MG, Madsen KM, and Kone BC. 1994. In situ hybridization localization of mRNA encoding inducible nitric oxide synthase in rat kidney. Am J Physiol 267:F748–F757.
Albakri QA and Stuehr DJ. 1996. Intracellular assembly of inducible NO synthase is limited by nitric oxide-mediated changes in heme insertion and availability. J Biol Chem 271:5414–5421.
Ando J, Komatsuda T, and Kamiya A. 1988. Cytoplasmic calcium response to fluid shear stress in cultured vascular endothelial cells. In Vitro Cell Dev Biol 24:871–877.
Archer SL, Huang JMC, Hampl V, Nelson DP, Shultz PJ, and Weir EK. 1994. Nitric oxide and cGMP cause vasorelaxation by activation of a charybdotoxin-sensitive K channel by cGMP-dependent protein kinase. Proc Natl Acad Sci USA 91:7583–7587.
Asano K, Chee CB, Gaston B, Lilly CM, Gerard C, Drazen JM, and Stamler JS. 1994. Constitutive and inducible nitric oxide synthase gene expression, regulation, and activity in human lung epithelial cells. Proc Natl Acad Sci USA 91:10089–10093.
Awolesi MA, Widmann MD, Sessa WC, and Sumpio BE. 1994. Cyclic strain increases endothelial nitric oxide synthase activity. Surgery 116:439–445.
Ayajiki K, Kindermann M, Hecker M, Fleming I, and Busse R. 1996. Intracellular pH and tyrosine phosphorylation but not calcium determine shear stress-induced nitric oxide production in native endothelial cells. Circ Res 78:750–758.
Bachmann S, Bose HM, and Mundel P. 1995. Topography of nitric oxide synthesis by localizing constitutive NO synthase in mammalian kidney. Am J Physiol 268:F885–F898.

Balligand J-L, Kelly RA, Marsden PA, Smith TW, and Michel T. 1993. Control of cardiac muscle cell function by an endogenous nitric oxide signalling system. Proc Natl Acad Sci USA 90:347–351.

Barry MK, Aloisi JD, Pickering SP, and Yeo CJ. 1994. Nitric oxide modulates water and electrolyte transport in the ileum. Ann Surg 219:382–388.

Beavo JA and Reifsnyder DH. 1990. Primary sequence of cyclic nucleotide phosphodiesterase isozymes and the design of selective inhibitors. Trends Pharmacol Sci 11:150–155.

Beltman J, Sonnenburg WK, and Beavo JA. 1993. The role of protein phosphorylation in the regulation of cyclic nucleotide phosphodiesterases. Mol Cell Biochem 127/128:239–253.

Bland-Ward PA and Moore PK. 1995. 7-Nitroindazole derivatives are potent inhibitors of brain, endothelium and inducible isoforms of nitric oxide synthase. Life Sci 57:131–135.

Bogle RG, Baydoun AR, Pearson JD, Moncada S, and Mann GE. 1992. L-arginine transport is increased in macrophages generating nitric oxide. Biochem J 284:15–18.

Boje KM and Fung H-L. 1990. Endothelial nitric oxide generating enzyme(s) in the bovine aorta: subcellular location and metabolic characterization. J Pharmacol Exp Ther 253:20–26.

Bredt DS, Ferris CD, and Snyder SH. 1992. Nitric oxide synthase regulatory sites. J Biol Chem 267:10976–10981.

Bredt DS, Hwang PM, Glatt CE, Lowenstein C, Reed RR, and Snyder SH. 1991. Cloned and expressed nitric oxide synthase structurally resembles cytochrome P-450 reductase. Nature 351:714–718.

Brüne B, Dimmeler S, Vedia LM, and Lapetina EG. 1993. Nitric oxide: a signal for ADP-ribosylation of proteins. Life Sci 54:61–70.

Brüne B and Lapetina EG. 1991. Phosphorylation of nitric oxide synthase by protein kinase A. Biochem Biophys Res Commun 181:921–926.

Brunton TL. 1867. Amyl nitrate in angina pectoris. Lancet ii:97.

Burnett AL, Lowenstein CJ, Bredt DS, Chang TSK, and Snyder SH. 1992. Nitric oxide: a physiologic mediator of penile erection. Science 257:401–403.

Busconi L and Michel T. 1993. Endothelial nitric oxide synthase: N-terminal myristoylation determines subcellular localization. J Biol Chem 270:995–998.

Buxton ILO, Cheek DJ, Eckman D, Westfall DP, Sanders KM, and Keef KD. 1992. N^G-Nitro L-arginine methyl ester and other alkyl esters of arginine are muscarinic receptor antagonists. Circ Res 72:387–395.

Carpenedo F and Floreani M. 1993. Stimulation of rat liver microsomal cGMP-inhibited cAMP phosphodiesterase (PDE III) by phospholipase C and D. Biochem Biophys Res Commun 190:609–615.

Caselli A, Camici G, Manao G, Moneti G, Pazzagli L, Cappugi G, and Ramponi G. 1994. Nitric oxide causes inactivation of the low molecular weight phosphotyrosine protein phosphatase. J Biol Chem 269:24878–24882.

Castro L, Rodriguez M, and Radi R. 1994. Aconitase is readily inactivated by peroxynitrite, but not by its precursor, nitric oxide. J Biol Chem 269:29409–29415.

Chartrain NA, Geller DA, Koty PP, Sitrin NF, Nussler AK, Hoffman EP, Billiar TR, Hutchinson NI, and Mudgett JS. 1994. Molecular cloning, structure and chromosomal localization of the human inducible nitric oxide synthase gene. J Biol Chem 269:6765–6772.

Colasanti M, Persichini T, Menegazzi M, Mariotto S, Giordano E, Calderera CM, Sogos V, Lauro GM, and Suzuki H. 1995. Induction of nitric oxide synthase mRNA expression: suppression by exogenous nitric oxide. J Biol Chem 270:26731–26733.

Conti M, Jin S-LC, Monaco L, Repaske DR, and Swinnen JV. 1991. Hormonal regulation of cyclic nucleotide phosphodiesterases. Endocr Rev 12:218–234.

Cooke JP. 1996. Role of nitric oxide in progression and regression of atherosclerosis. West J Med 164:419–424.

Cooke JP, Rossitch E Jr, Andon NA, Loscalzo J, and Dzau VJ. 1991. Flow activates an endothelial potassium channel to release an endogenous nitrovasodilator. J Clin Invest 88:1663–1671.

Dawson TM and Dawson VL. 1996. Nitric oxide synthase: role as a transmitter/mediator in brain and endocrine system. Annu Rev Med 47:219–227.

Dawson TM and Snyder SH. 1994. Gases as biological messengers: nitric oxide and carbon monoxide in the brain. J Neurosci 14:5147–5159.

Diaz-Guerra MJM, Bodelon OG, Velasco M, Whelan R, Parker PJ, and Bosca L. 1996. Up-regulation of protein kinase C-ε promotes the expression of cytokine-induced nitric oxide synthase in RAW 264.7 cells. J Biol Chem 271:32028–32033.

Dinerman JL, Steiner JP, Dawson TM, Dawson V, and Snyder SH. 1994. Cyclic nucleotide dependent phosphorylation of neuronal nitric oxide synthase inhibits catalytic activity. Neuropharmacology 33:1245–1251.

Doerner D and Alger BE. 1988. Cyclic GMP depresses hippocampal Ca^{2+} current through a mechanism independent of cGMP-dependent protein kinase. Neuron 1:693–699.

Drewett JG and Garbers DL. 1994. The family of guanylyl cyclase receptors and their ligands. Endocr Rev 15:135–162.

Dull RO and Davies PF. 1991. Flow modulation of agonist (ATP)-response (Ca^{2+}) coupling in vascular endothelial cells. Am J Physiol 261:H149–H154.

Ellenby MI, Ernst CB, Carretero OA, and Scicli AG. 1996. Role of nitric oxide in the effect of blood flow on neointima formation. J Vasc Surg 23:314–322.

Finkel MS, Oddis CV, Mayer OH, Hattler BG, and Simmons RL. 1995. Nitric oxide synthase inhibitor alters papillary force-frequency relationship. J Pharmacol Exp Ther 272:945–952.

Förstermann U, Pollock JS, Schmidt HHHW, Heller M, and Murad F. 1991. Calmodulin-dependent endothelium-derived relaxing factor/nitric oxide synthase is present in the particulate and cytosolic fractions of bovine aortic endothelial cells. Proc Natl Acad Sci USA 88:1788–1792.

Fujisawa H, Ogura T, Kurashima Y, Yokoyama T, Yamashita J, and Esumi H. 1994.

Expression of two types of nitric oxide synthase mRNA in human neuroblas-toma cell lines. J Neurochem 63:140–145.

Furchgott RF and Zawadzki JV. 1980. The obligatory role of endothelial cells in the relaxation of arterial smooth muscle by acetylcholine. Nature 288:373–376.

Furchgott RF, Zawadzki JV, and Cherry PD. 1981. Role of endothelium in the vasodilator response to acetylcholine. In: Vanhoutte PM and Leusen I, eds. Vasodilatation. New York: Raven Press. 49–65.

Garcia NH, Pomposiello SI, and Garvin JL. 1996a. Nitric oxide inhibits ADH-stimu-lated osmotic water permeability in the cortical collecting duct. Am J Physiol 270:F206–F210.

Garcia NH, Stoos BA, Carretero OA, and Garvin JL. 1996b. Mechanism of the nitric oxide-induced blockade of collecting duct water permeability. Hypertension 27:679–683.

Garthwaite J and Boulton CL. 1995. Nitric oxide signaling in the central nervous system. Annu Rev Physiol 57:683–706.

Gaston B, Drazen JM, Loscalzo J, and Stamler JS. 1994. The biology of nitrogen oxides in the airways. Am J Respir Crit Care Med 149:538–551.

Geller DA, Nussler AK, DiSilvio M, Lowenstein CJ, Shapiro RA, Wang SC, Simmons RL, and Billiar TR. 1993. Cytokines, endotoxin and glucocorticoids regulate the expression of inducible nitric oxide synthase in hepatocytes. Proc Natl Acad Sci USA 90:522–526.

Griffith OW and Stuehr DJ. 1995. Nitric oxide synthases: properties and catalytic mechanism. Annu Rev Physiol 57:707–736.

Griffith TM and Henderson AH. 1989. EDRF and the regulation of vascular tone. Int J Microcirc Clin Exp 8:383–396.

Guzman NJ, Fang M-Z, Tang S-S, Ingelfinger JR, and Garg LC. 1995. Autocrine inhibition of Na^+/K^+-ATPase by nitric oxide in mouse proximal tubule epi-thelial cells. J Clin Invest 95:2083–2088.

Haijar DP, Lander HM, Pearce FA, Upmacis RK, and Pomerantz KB. 1995. Nitric oxide enhances prostaglandin-H synthase-1 activity by a heme-independent mechanism: evidence implicating nitrosothiols. J Am Chem Soc 117:3340–3346.

Han X, Kobzik L, Balligand J-L, Kelly RA, and Smith TW. 1996. Nitric oxide synthase (NOS3)-mediated cholinergic modulation of Ca^{2+} current in adult rabbit atrioventricular nodal cells. Circ Res 78:998–1008.

Hartzell HC and Fishmeister R. 1986. Opposite effects of cyclic GMP and cAMP on Ca^{2+} current in single heart cells. Nature 323:273–275.

Hausladen A and Fridovich I. 1994. Superoxide and peroxynitrite inactivate aconi-tases, but nitric oxide does not. J Biol Chem 269:29405–29408.

Haynes J, Kithas PA, Taylor AE, and Strada SJ. 1991. Selective inhibition of cGMP-inhibitable cAMP phosphodiesterase decreases pulmonary vasoreactivity. Am J Physiol 261:H487–H492.

Henry Y, Lepoivre M, Drapier J-C, Ducrocq C, Boucher J-L, and Guissani A. 1993. EPR characterization of molecular targets for NO in mammalian cells and organelles. FASEB J 7:1124–1134.

Hirata K-i, Kuroda R, Sakoda T, Katayama M, Inoue N, Suematsu M, Kawashima S, and Yokoyama M. 1995. Inhibition of endothelial nitric oxide synthase activity by protein kinase C. Hypertension 25:180–185.

Huie RE and Padmaja S. 1993. The reaction of NO with superoxide. Free Radic Res Comm 18:195–199.

Ishii K, Chang B, Kerwin JF, Huang Z-J, and Murad F. 1990. N^ω-nitro-L-arginine: a potent inhibitor of endothelium-derived relaxing factor formation. Eur J Pharmacol 176:219–223.

Johnson RM and Lincoln TM. 1985. Effects of nitroprusside, glyceryl trinitrate and 8-bromo-cyclic GMP on phosphorylase a formation and myosin light chain phosphorylation in rat aorta. Mol Pharmacol 27:333–342.

Jordan ML, Rominski B, Jaquins-Gerstl A, Geller D, and Hoffman RA. 1995. Regulation of inducible nitric oxide production by intracellular calcium. Surgery 118:138–146.

Juncos LA, Garvin JL, Carretero OA, and Ito S. 1995. Flow modulates myogenic responses in isolated microperfused rabbit afferent arterioles via endothelium-derived nitric oxide. J Clin Invest 95:2741–2748.

Kanagy NL, Charpie JR, and Webb RC. 1995. Nitric oxide regulation of ADP-ribosylation of G proteins in hypertension. Med Hypotheses 44:159–164.

Kaye DM, Wiviott SD, Balligand J-L, and Smith TW. 1995. Nitric oxide inhibits norepinephrine uptake into cardiac sympathetic neurons. Circulation 92:I-507 (abstract).

Keaney JF Jr, Hare JM, Balligand J-L, Kelly RA, Loscalzo J, Smith TW, and Colucci WS. 1996. Inhibition of nitric oxide synthase augments myocardial contractile responses to β-adrenergic stimulation. Am J Physiol 271:H2646–H2652.

Kelly RA, Balligand J-L, and Smith TW. 1996. Nitric oxide and cardiac function. Circ Res 79:363–380.

Koesling D, Schultz G, and Böhme E. 1991. Sequence homologies between guanylyl cyclases and structural analogies to other signal-transducing proteins. FEBS Lett 280:301–306.

Kone BC, Schwöbel J, Turner P, Mohaupt MG, and Cangro CB. 1995. Role of NF-κB in the regulation of inducible nitric oxide synthase in an MTHAL cell line. Am J Physiol 269:F718–F729.

Kuenzli KA, Bradley ME, and Buxton ILO. 1996. Cyclic GMP-independent effects of nitric oxide on guinea-pig uterine contractility. Br J Pharmacol 119:737–743.

Kunz D, Mühl H, Walker G, and Pfeilschifter J. 1994. Two distinct signaling pathways trigger the expression of inducible nitric oxide synthase in rat renal mesangial cells. Proc Natl Acad Sci USA 91:5387–5391.

Lahera V, Navarro J, Biondi ML, Ruilope LM, and Romero JC. 1993. Exogenous cGMP prevents decrease in diuresis and natriuresis induced by inhibition of NO synthesis. Am J Physiol 264:F344–F347.

Lahera V, Salom MG, Fiksen-Olsen MJ, and Romero JC. 1991. Mediatory role of endothelium-derived nitric oxide in renal vasodilatory and excretory effects of bradykinin. Am J Hypertens 4:260–262.

Lamas S, Marsden PA, Li GK, Tempst P, and Michel T. 1992. Endothelial nitric oxide synthase: molecular cloning and characterization of a distinct constitutive enzyme isoform. Proc Natl Acad Sci USA 89:6348–6352.

Landino LM, Crews BC, Timmons MD, Morrow JD, and Marnett LJ. 1996. Peroxynitrite, the coupling product of nitric oxide and superoxide, activates prostaglandin biosynthesis. Proc Natl Acad Sci USA 93:15069–15074.

Lautier D, Lagueux J, Thibodeau J, Menard L, and Poirier GG. 1993. Molecular and biochemical features of poly(ADP-ribose) synthetase in neurotoxicity. Mol Cell Biochem 122:171–193.

Light DB, Corbin JD, and Stanton BA. 1990. Dual ion-channel regulation by cyclic GMP and cyclic GMP-dependent protein kinase. Nature 344:336–339.

Lincoln TM and Corbin JD. 1983. Characterization and biological role of the cGMP-dependent protein kinase. In: Greengard P and Robison GA, eds. Advances in cyclic nucleotide research. New York: Raven Press. 139–192.

Lincoln TM, Cornwell TL, Komalavilas P, MacMillan-Crow LA, and Boerth N. 1996. The nitric oxide-cyclic GMP signaling system. In: Barney M, ed. Biochemistry of smooth muscle contraction. Orlando: Academic Press. 257–268.

Liu J, Garcia-Cardena G, and Sessa WC. 1996. Palmitoylation of endothelial nitric oxide synthase is necessary for optimal stimulated release of nitric oxide: implications for caveolae localization. Biochemistry 35:13277–13281.

Liu J and Sessa WC. 1994. Identification of covalently bound amino-terminal myristic acid in endothelial nitric oxide synthase. J Biol Chem 269:11691–11694.

Low BC and Grigor MR. 1995. Angiotensin II stimulates system y^+ and cationic amino acid transporter gene expression in cultured vascular smooth muscle cells. J Biol Chem 270:27577–27583.

Lyons CR, Orloff GJ, and Cunningham JM. 1992. Molecular cloning and functional expression of an inducible nitric oxide synthase from a murine macrophage cell line. J Biol Chem 267:6370–6374.

Maccarrone M, Corasaniti MT, Guerrieri P, Nisticò G and Agrò AF. 1996. Nitric oxide-donor compounds inhibit lipoxygenase activity. Biochem Biophys Res Commun 219:128–133.

MacIntyre I, Zaidi M, Towhidul Alam ASM, Datta HK, Moonga BS, Lidbury PS, Hecker M, and Vane JR. 1991. Osteoclast inhibition: an action of nitric oxide not mediated by cyclic GMP. Proc Natl Acad Sci USA 88:2936–2940.

Majid DSA and Navar LG. 1994. Blockade of distal nephron sodium transport attenuates pressure natriuresis in dogs. Hypertension 23:1040–1045.

Malek AM and Izumo S. 1994. Molecular aspects of signal transduction of shear stress in the endothelial cell. J Hypertens 12:989–999.

Marletta MA. 1994. Nitric oxide synthase: aspects concerning structure and catalysis. Cell 78:927–930.

Marsden PA, Heng HHQ, Duff CL, Shi X-M, Tsui L-C, and Hall AV. 1994. Localization of the human gene for inducible nitric oxide synthase (NOS 2) to chromosome 17q11.2–q12. Genomics 19:183–185.

Marsden PA, Heng HHQ, Scherer SW, Stewart RJ, Hall AV, Shi X-M, Tsui L-C, and

Schappert KT. 1993. Structure and chromosomal localization of the human constitutive endothelial nitric oxide synthase gene. J Biol Chem 268:17478–17488.

Matthews JR, Botting CH, Panico M, Morris HR, and Hay RT. 1996. Inhibition of NF-κB binding by nitric oxide. Nucleic Acids Res 24:2236–2242.

Mayer B. 1994. Regulation of nitric oxide synthase and soluble guanylyl cyclase. Cell Biochem Funct 12:167–177.

McCoy DE, Guggino S, and Stanton BA. 1995. The renal cGMP-gated cation channel: its molecular structure and physiological role. Kidney Int 48:1125–1133.

Méry P-F, Pavoine C, Belhassen L, Pecker F, and Fischmeister R. 1993. Nitric oxide regulates cardiac Ca^{2+} current. J Biol Chem 268:26286–26295.

Michel T, Li GK, and Busconi L. 1993. Phosphorylation and subcellular translocation of endothelial nitric oxide synthase. Proc Natl Acad Sci USA 90:6252–6256.

Mohaupt MG, Elzie JL, Ahn KY, Clapp WL, Wilcox CS, and Kone BC. 1994. Differential expression and induction of mRNAs encoding two inducible nitric oxide synthases in rat kidney. Kidney Int 46:653–665.

Mohaupt MG, Schwobel J, Elzie JL, Kannan GS, and Kone BC. 1995. Cytokines activate inducible nitric oxide synthase gene transcription in inner medullary collecting duct cells. Am J Physiol 268: F770–F777.

Moncada S, Palmer RMJ, and Higgs EA. 1991. Nitric oxide: physiology, pathophysiology, and pharmacology. Pharmacol Rev 43:109–142.

Moore PK, Babbedge RC, Wallace P, Gaffen ZA, and Hart SL. 1993a. 7-Nitro indazole, an inhibitor of nitric oxide synthase, exhibits anti-nociceptive activity in the mouse without increasing blood pressure. Br J Pharmacol 108:296–297.

Moore PK, Wallace P, Gaffen Z, Hart SL, and Babbedge RC. 1993b. Characterization of the novel nitric oxide synthase inhibitor 7-nitro indazole and related indazoles: antinociceptive and cardiovascular effects. Br J Pharmacol 110:219–224.

Mühl H, Kunz D, and Pfeilschifter J. 1994. Expression of nitric oxide synthase in rat glomerular mesangial cells mediated by cyclic AMP. Br J Pharmacol 112:1–8.

Mundel P, Bachmann S, Bader M, Fischer A, Kummer W, Mayer B, and Kritz W. 1992. Expression of nitric oxide synthase in kidney macula densa cells. Kidney Int 42:1017–1019.

Nakane M, Mitchell J, Förstermann U, and Murad F. 1991. Phosphorylation by calcium calmodulin-dependent protein kinase II and protein kinase C modulates activity of nitric oxide synthase. Biochem Biophys Res Commun 180:1396–1402.

Nathan C and Xie Q. 1994. Regulation of biosynthesis of nitric oxide. J Biol Chem 269:13725–13728.

Navar LG, Inscho EW, Majid DSA, and Immig JD. 1996. Paracrine regulation of the renal microcirculation. Physiol Rev 76:426–536.

Ohara Y, Sayegh HS, Yamin JJ, and Harrison DG. 1995. Regulation of endothelial constitutive nitric oxide synthase by protein kinase C. Hypertension 25:415–420.

Pearse RN, Feinman R, and Ravetch JV. 1991. Characterization of the promoter of the human gene encoding the high-affinity IgG receptor. Transcriptional induction by γ-interferon is mediated through common DNA response elements. Proc Natl Acad Sci USA 88:11305–11309.

Peng H-B, Libby P, and Liao JK. 1995. Induction and stabilization of IκBα by nitric oxide mediates inhibition of NF-κB. J Biol Chem 270:14214–14219.

Pohl U and Busse R. 1990. Endothelium-dependent modulation of vascular tone and platelet function. Eur Heart J 11:35–42.

Raeymaekers L, Hofmann F, and Casteels R. 1988. Cyclic GMP-dependent protein kinase phosphorylates phospholamban from cardiac and smooth muscle. Biochem J 252:269–273.

Rand MJ and Li CG. 1995. Nitric oxide as a neurotransmitter in peripheral nerves: nature of transmitter and mechanism of transmission. Annu Rev Physiol 57:659–682.

Ranjan V, Xiao Z, and Diamond SL. 1995. Constitutive NOS expression in cultured endothelial cells is elevated by fluid shear stress. Am J Physiol 269:H550–H555.

Rapoport RM. 1986. Cyclic guanosine monophosphate inhibition of contraction may be mediated through inhibition of phosphatidyl inositol hydrolysis in rat aorta. Circ Res 58:407–410.

Ravichandran LV and Johns RA. 1995. Up-regulation of endothelial nitric oxide synthase expression by cyclic guanosine 3′, 5′-monophosphate. FEBS Lett 374:295–298.

Rees DD, Palmer RMJ, Hodson HF, and Moncada S. 1989. A specific inhibitor of nitric oxide formation from L-arginine attenuates endothelium-dependent relaxation. Br J Pharmacol 96:418–424.

Rees DD, Palmer RMJ, Schulz R, Hodson HF, and Moncada S. 1990a. Characterization of three inhibitors of endothelial nitric oxide synthase in vitro and in vivo. Br J Pharmacol 101:746–752.

Rees DD, Schulz R, Hodson HF, Palmer RMJ, and Moncada S. 1990b. Identification of some novel inhibitors of the vascular nitric oxide synthase in vitro and in vivo. In: Moncada S and Higgs EA, eds. Nitric oxide from L-arginine: a bioregulatory system. Amsterdam: Elsevier Science Publishing. 485–487.

Reinhardt RR, Chin E, Zhou J, Taira M, Murata T, Manganiello VC, and Bondy CA. 1995. Distinctive anatomical patterns of gene expression for cGMP-inhibited cyclic nucleotide phosphodiesterases. J Clin Invest 95:1528–1538.

Rengasamy A and Johns RA. 1993. Regulation of nitric oxide synthase by nitric oxide. Mol Pharmacol 44:124–128.

Robinson LJ, Busconi L, and Michel T. 1995. Agonist-modulated palmitoylation of endothelial nitric oxide synthase. J Biol Chem 270:995–998.

Robinson LJ, Weremowicz S, Morton CC, and Michel T. 1994. Isolation and chromosomal localization of the human endothelial nitric oxide synthase (NOS3) gene. Genomics 19:350–357.

Roczniak A and Burns KD. 1996. Nitric oxide stimulates guanylate cyclase and regulates sodium transport in rabbit proximal tubule. Am J Physiol 270:F106–F115.

Ruth P, Wang G-X, Boekhoff I, May B, Pfeifer A, Penner R, Korth M, Breer H, and Hofmann F. 1993. Transfected cGMP-dependent protein kinase suppresses calcium transients by inhibition of inositol 1, 4, 5-trisphosphate production. Proc Natl Acad Sci USA 90:2623–2627.

Salvemini D, Misko TP, Masferrer JL, Seibert K, Currie MG, and Needleman P. 1993. Nitric oxide activates cyclooxygenase enzymes. Proc Natl Acad Sci USA 90:7240–7244.

Sarcevic B, Brookes V, Martin TJ, Kemp BE, and Robinson PJ. 1989. Atrial natriuretic peptide-dependent phosphorylation of smooth muscle cell particulate fraction proteins is mediated by cGMP-dependent protein kinase. J Biol Chem 264:20648–20654.

Schmidt HHHW, Lohman SM, and Walter U. 1993. The nitric oxide and cGMP signal transduction system: regulation and mechanism of action. Biochim Biophys Acta 1178:153–178.

Schmidt HHHW, Pollock JS, Nakane M, Förstermann U, and Murad F. 1992b. Ca^{2+}/calmodulin-regulated nitric oxide synthases. Cell Calcium 13:427–434.

Schmidt HHHW, Smith RM, Nakane M, and Murad F. 1992a. Ca^{2+}/calmodulin-dependent NO synthase type I: a biopteroflavinprotein with Ca^{2+}/calmodulin-independent diaphorase and reductase activities. Biochemistry 31:3243–3249.

Schultz S, Yuen PST, and Garbers DL. 1991. The expanding family of guanylyl cyclases. Trends Pharmacol Sci 12:116–120.

Schwartz P, Diem R, Dun NJ, and Förstermann U. 1995. Endogenous and exogenous nitric oxide inhibits norepinephrine release from rat heart sympathetic nerves. Circ Res 77:841–848.

Seki T, Hagiwara H, Naruse K, Kadowaki M, Kashiwagi M, Demura H, Hirose S, and Naruse M. 1996. In situ identification of messenger RNA of endothelial type nitric oxide synthase in rat cardiac myocytes. Biochem Biophys Res Commun 218:601–605.

Sessa WC. 1994. The nitric oxide synthase family of proteins. J Vasc Res 31:131–143.

Shah AM, Spurgeon HA, Sollott SJ, Talo A, and Lakatta EG. 1994. 8-Bromo-cGMP reduces the myofilament response to Ca^{2+} in intact cardiac myocytes. Circ Res 74:970–978.

Shaul PW, North AJ, Wu LC, Wells LB, Brannon TS, Lau KS, Michel T, Margraf LR, and Star R. 1994. Endothelial nitric oxide synthase is expressed in cultured human bronchiolar epithelium. J Clin Invest 94:2231–2236.

Shibazaki T, Fujiwara M, Sato H, Fujiwara K, Abe K, and Bannai S. 1996. Relevance of the arginine transport activity to the nitric oxide synthesis in mouse peritoneal macrophages stimulated with bacterial lipopolysaccharide. Biochim Biophys Acta 1311:150–154.

Shyy JY-J, Li Y-S, Lin M-C, Chen W, Yuan S, Usami S, and Chien S. 1995. Multiple cis-elements mediate shear stress-induced gene expression. J Biomech 28:1451–1457.

Simmons WW, Closs EI, Cunningham JM, Smith TW, and Kelly RA. 1996. Cytokines and insulin induce cationic amino acid transporter expression in cardiac myocytes: regulation of L-arginine transport and NO production by CAT-1, CAT-2A and CAT-2B. J Biol Chem 271:11694–11702.

Singh K, Balligand J-L, Fischer TA, Smith TW, and Kelly RA. 1996. Regulation of cytokine-inducible nitric oxide synthase in cardiac myocytes and microvascular endothelial cells. J Biol Chem 271:1111–1117.

Singh I, Grams M, Wang W-H, Yang T, Killen P, Smart A, Schnermann J, and Briggs JP. 1996. Coordinate regulation of renal expression of nitric oxide synthase, renin, and angiotensinogen mRNA by dietary salt. Am J Physiol 270:F1027–F1037.

Stenger S, Thüring H, Röllinghof M, Manning P, and Bogdan C. 1995. L-N^6-(1-iminoethyl)-lysine potently inhibits inducible nitric oxide synthase and is superior to N^G-monomethyl-arginine in vitro and in vivo. Eur J Pharmacol 294:703–712.

Stoos BA, Garcia NH, and Garvin JL. 1995. Nitric oxide inhibits sodium reabsorption in the isolated perfused cortical collecting duct. J Am Soc Nephrol 6:89–92.

Stoos BA, Nahhas F, and Garvin JL. 1996. Inhibition of Cl absorption (J_{cl}) by NO produced in the thick ascending limb (THAL) by endogenous nitric oxide synthase (NOS) increases during salt loading. J Am Soc Nephrol 7:1291 (abstract)

Stroop SD and Beavo JA. 1991. Structure and function studies of the cGMP-stimulated phosphodiesterase. J Biol Chem 266:23802–23809.

Stuehr DJ, Cho HJ, Kwon NS, and Nathan CF. 1991. Purification and characterization of the cytokine-induced macrophase nitric oxide synthase: and FAD- and FMN-containing flavoprotein. Proc Natl Acad Sci USA 88:7773–7777.

Sumii K and Sperelakis N. 1995. cGMP-dependent protein kinase regulation of the L-type Ca^{2+} current in rat ventricular myocytes. Circ Res 77:803–812.

Takeuchi T, Kishi M, Ishii T, Nishio H, and Hata F. 1996. Nitric oxide-mediated relaxation without concomitant changes in cyclic GMP content of rat proximal colon. Br J Pharmacol 117:1204–1208.

Tao Y, Howlett A, and Klein C. 1994. Nitric oxide regulation of glyceraldehyde-3-phosphate dehydrogenase activity in *Dictyostelium discoideum* cells and lysate. Eur J Biochem 224:447–454.

Terada Y, Tomita K, Nonoguchi H, and Marumo F. 1992. Polymerase chain reaction localization of constitutive nitric oxide synthase and soluble guanylate cyclase messenger RNAs in microsdissected rat nephron segments. J Clin Invest 90:659–665.

Thompson WJ. 1991. Cyclic nucleotide phosphodiesterases: pharmacology, biochemistry and function. Pharmacol Ther 51:13–33.

Tsou K, Snyder GL, and Greengard P. 1993. Nitric oxide/cGMP pathway stimulates phosphorylation of DARPP-32, a dopamine- and cAMP-regulated phosphoprotein in the substantia nigra. Proc Natl Acad Sci USA 90:3462–3465.

Uematsu M, Ohara Y, Navas JP, Nishida K, Murphy TJ, Alexander RW, Nerem RM, and Harrison DG. 1995. Regulation of endothelial cell nitric oxide synthase mRNA expression by shear stress. Am J Physiol 269:C1371–C1378.

Vanhoutte PM. 1989. Endothelium and control of vascular function. Hypertension 13:658–667.

Vodovotz Y, Russell D, Xie Q-w, Bogdan C, and Nathan C. 1995. Vesicle membrane association of nitric oxide synthase in primary mouse macrophages. J Immunol 154:2914–2925.

Waldman SA and Murad F. 1987. Cyclic GMP synthesis and function. Pharmacol Rev 39:163–196.

Wang N, Butler JP, and Ingber DE. 1993. Mechanotransduction across the cell surface and through the cytoskeleton. Science 260:1124–1127.

Wang YG and Lipsius SL. 1995. Acetylcholine elicits a rebound stimulation of Ca^{2+} current mediated by pertussis toxin-sensitive G-protein and cAMP-dependent protein kinase A in atrial myocytes. Circ Res 76:634–644.

Welter RW, Yu L, and Yu C-A. 1996. The effects of nitric oxide on electron transport complexes. Arch Biochem Biophys 331:9–14.

Werner-Felmayer G, Werner ER, Fuchs D, Hausen A, Reibnegger G, and Wachter H. 1990. Tetrahydrobiopterin-dependent formation of nitrite and nitrate in murine fibroblasts. J Exp Med 172:1599–1607.

Wilcox CS, Welch WJ, Murad F, Gross SS, Taylor G, Levi R, and Schmidt HHW. 1992. Nitric oxide synthase in macula densa regulates glomerular capillary pressure. Proc Natl Acad Sci USA 89:11993–11997.

Wink DA, Osawa Y, Darbyshire JF, and Jones CR. 1993. Inhibition of cytochromes P450 by nitric oxide and a nitric oxide-releasing agent. Arch Biochem Biophys 300:115–123.

Xie Q-w, Whisnant R, and Nathan C. 1993. Promoter of the mouse gene encoding calcium-independent nitric oxide synthase confers inducibility by interferon γ and bacterial lipopolysaccharide. J Exp Med 177:1779–1784.

Xie Y-W and Wolin MS. 1996. Role of nitric oxide and its interaction with superoxide in the suppression of cardiac muscle mitochondrial respiration. Circulation 94:2580–2586.

Yamagata K, Andreasson KI, Kaufmann WE, Barnes CA, and Worley PF. 1993. Expression of a mitogen-inducible cyclooxygenase in brain neurons: regulation by synaptic activity and glucocorticoids. Neuron 11:371–386.

Yang Z, Sugawara M, Ponath PD, Wessendorf L, Banerji J, Li Y, and Strominger JL. 1990. Interferon γ response region in the promoter of the human DPA gene. Proc Natl Acad Sci USA 87:9226–9230.

Zeidel ML, Seifter JL, Lear S, Brenner BM, and Siva P. 1986. Atrial natriuretic peptides inhibit oxygen consumption in kidney medullary collecting duct cells. Am J Physiol 251:F379–F383.

Zhang J, Dawson VL, Dawson TM, and Snyder SH. 1994. Nitric oxide activation of poly (ADP-ribose) synthetase in neurotoxicity. Science 263:687–689.

Part V

REGULATORY MECHANISMS AND ION TRANSPORT IN SIGNAL TRANSDUCTION

8

Ionic Channels Mediating
Sensory Transduction

TAKASHI KURAHASHI AND GEOFFREY H. GOLD

INTRODUCTION

Fast signaling in the nervous system is mediated by changes in the voltage across the neuronal cell membrane. These changes in membrane voltage are caused by ion channels. Ion channels are membrane proteins that selectively allow small ions, such as Na^+, K^+, Ca^{2+}, and Cl^-, to cross the cell membrane. Ion transporters generate gradients in the concentration of these ions across the cell membrane, such that most ions have different gradients. For example, the extracellular Na^+ concentration is typically 130 mM in mammals, but the intracellular Na^+ concentration is typically only about 10 mM, because of active transport by Na^+/K^+-ATPase. Therefore, there is an inward gradient for Na^+, causing Na^+ to diffuse down its concentration gradient into the cell through ion channels that conduct Na^+. On the other hand, K^+ is high inside the cell and low outside, so that K^+ flows out of the cell through ion channels that conduct K^+. In the resting state, most neurons have a membrane potential of between -60 and -70 mV. This negative resting membrane potential is caused by K^+-selective ion channels.

The ability of an ion channel to control the membrane potential can be thought of as follows. If a cell has an ionic gradient for K^+ only, and the cell membrane is permeable only to K^+, then K^+ will flow out of the cell (down its concentration gradient), until enough charge has built up on the outside of the cell membrane to prevent further efflux of K^+. At this point the cell has reached an equilibrium state in which there is no further net movement of K^+ across the cell membrane. This equilibrium state is referred to as

Introduction to Cellular Signal Transduction
A. Sitaramayya, Editor
©1999 Birkhäuser Boston

Donnan equilibrium, and the voltage created by the K^+ gradient is defined by the Nernst potential,

$$E_m = -\frac{RT}{F} \ln \frac{[K^+]_i}{[K^+]_o},$$

where E_m is the equilibrium potential, $[K^+]_i$ is the cytoplasmic potassium concentration, $[K^+]_o$ is the external K concentration, and R, T, and F are usual constants respectively. Remember that K^+ was initially high inside the cell, so this equilibrium will result in an accumulation of positive charge on the outside of the cell. This separation of charge across the cell membrane creates an electric field within the membrane. This electric field causes a difference in the voltage across the cell membrane, such that the voltage on the inside of the cell is negative with respect to the outside. If this cell has a Na^+ gradient as well, and the cell membrane becomes permeable to Na^+, Na^+ will flow down its concentration gradient and into the cell. Because this inward movement of Na^+ will reduce the charge separation caused by K^+, the inward movement of Na^+ will cause a positive shift in the voltage across the membrane (Figure 8-1). Negative shifts in membrane potential are commonly called hyperpolarization, and positive shifts in membrane potential are called depolarization.

The Nernst potential for an ion is commonly referred to as the reversal potential for that ion. This is because at potentials more positive than the Nernst potential, current flows out of the cell. On the other hand, at potentials more negative than the Nernst potential, current flows into the cell. Therefore, at the Nernst potential, the current reverses from inward to outward. Measurements of the reversal potential for a stimulus-evoked current often provide information about which ion is carrying that current, because the reversal potentials for most ions are well known. In addition, electrophysiologists may vary the intracellular or extracellular ion concentrations to vary the reversal potential for a particular ion. This manipulation can provide additional information about which ions carry the current.

Up until the 1980s, neurophysiologists knew only that stimulating cells caused an increase or decrease in the flow of ions across the cell membrane. It was hypothesized that this flow of ions was carried by ion channels, consisting of membrane-spanning protein molecules that allowed ions to flow across the cell membrane. The lipids of biological membranes are hydrophobic, so ions cannot cross the membrane without the help of ion channels or transporters. This hypothesis received confirmation in the early 1980s, with the development of the patch clamp technique by Neher, Sakmann, and colleagues (Hamill et al 1981, Sakmann and Neher 1995). With this technique, a saline-filled glass micropipette forms a high-resistance seal on the surface of a cell membrane. Once this seal is formed, the current across the patch of mem-

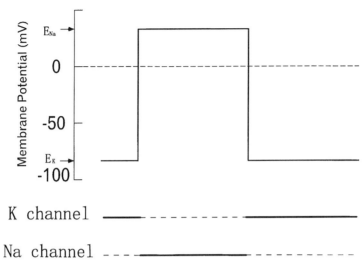

Figure 8-1. A scheme explaining the relation between the membrane potential and openings of ionic channels. E_K and E_{Na} indicate equilibrium potentials for K and Na ions, respectively. When K channels are open (indicated by a solid line), the membrane potential becomes identical to E_K. However, when Na channels become functional, instead of K channels, the membrane potential shifts to E_{Na} value.

brane enclosed by the tip of the micropipette can be measured with high resolution. In this way, Neher and Sakmann were able to measure the current flow through single ion channels. They found that ion channels existed in two states, open and closed. Transitions between the open and closed states occur stochastically. Therefore, when an ion channel is opened by voltage or by a neurotransmitter, there is an increased probability of finding the ion channel in the open state, but individual ion channels do not open and close at the same time. More recently, some ion channels have been found to exist in intermediate states, which allow currents to flow that are intermediate in magnitude between the fully open and fully closed states. The conductance of the membrane is defined as the current that flows across it, divided by the voltage difference across it. Therefore, the more ion channels that are open, the more current will flow across the membrane, because current flows through each ion channel independently. Consequently, the opening of ion channels increases the conductance of the membrane.

Ion channels can be divided into two categories: those that are opened or closed by voltage and those that are opened or closed by the binding of small molecules, such as neurotransmitters (on the outside of the cell) and intracellular messengers (on the inside of the cell). In nerves, voltage-dependent Na^+ and K^+ ion channels conduct signals over long distances by generating transient voltage spikes called action potentials. The nerve membrane (neuronal axons) contains voltage-dependent Na channels that are opened by membrane depolarization. However, these ion channels only remain open for a few milliseconds. In addition, there are voltage-dependent K channels that are opened rapidly by membrane depolarization and accelerate the recovery of the membrane potential back to its resting value. As the depolarization caused by the Na channels spreads through the neuron, this activates other Na channels, causing the action potential to propagate. This is how signals are propagated along nerves. In most nerves, the intensity of the signal is proportional to the frequency of the action potentials. Sensory receptor cells, for example, cause membrane depolarization or hyperpolarization, and therefore increase or decrease the frequency of action potentials that are sent to the brain. This is how the brain receives information about the occurrence and intensity of sensory stimuli. In some nerves, action potentials can occur at rates of several hundred per second and propagate at speeds of up to 10 m/sec (unmyelinated axon). However, the speed of propagation increases with nerve fiber diameter. Some fibers that transmit pain signals have relatively small diameters and propagate action potentials relatively slowly. This is why there is usually a delay between stubbing one's toe and feeling it begin to hurt.

Ion channels that are opened or closed by neurotransmitters allow signals to spread from one neuron to another. At specific sites called synapses, one neuron releases a neurotransmitter, such as acetylcholine or glutamate, in response to membrane depolarization. This released neurotransmitter then diffuses a very small distance (usually only a few hundred angstroms) to an adjacent neuron, which contains ion channels that are opened or closed when the neurotransmitter binds to the ion channel. The opening or closing of the ion channels in the second (or postsynaptic) neuron, causes a change in the membrane voltage, thus spreading the signal to that cell. In addition, intracellular second messengers, such as cyclic nucleotides and inositol-1,4,5-trisphosphate (IP3), can open ion channels from intracellular binding sites.

This very brief introduction is intended to provide the reader with a general idea of how ion channels function in the nervous system. There are a large number of different ion channels in the nervous system, with different ion selectivities and different opening and closing rates. The reader may wish to read introductory neurobiology texts for a more complete and rigorous introduction to this subject (e.g., Hagiwara 1983, Hille 1984, Sakmann and Neher 1995). The remainder of this chapter will focus on the role of ion channels in mediating sensory transduction in photoreceptor and olfactory receptor cells. However, the principles outlined for these two types of sensory receptor cells apply equally well to other neurons.

SENSORY TRANSDUCTION

Human beings and animals are able to perceive many kinds of stimuli. Human beings can sense light, sound, odors, the taste of foods, chemical

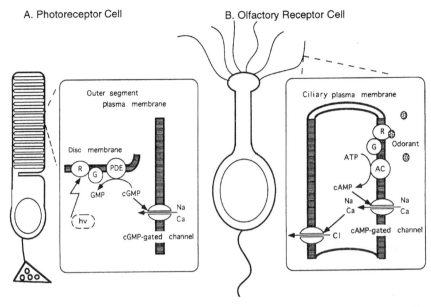

Figure 8-2. Transduction cascades in rod photoreceptor and olfactory receptor cells. Left panel, rod photoreceptor cell. Signal transduction is carried out in the outer segment. R, rhodopsin; G, transducin; PDE, phosphodiesterase. Right panel, olfactory cilia are the site for olfactory transduction. R, odorant receptor protein; G, olfactory-specific G-protein; AC, adenylyl cyclase. Note that principal cascades are very similar between photoreceptor and olfactory receptor cells, except for the presence of Ca^{2+} -activated Cl channels in the olfactory system.

irritation, heat, cold, and touch. Many fish can sense weak electric fields, and can use this ability to locate other fish and obstacles in darkness.

How is it that we can detect so many different kinds of stimuli? All information in the nervous system is encoded in changes in the voltage across neuronal cell membranes. Therefore, animals have evolved mechanisms for converting (transducing) different stimuli into electrical signals. The remainder of this chapter will summarize what is known about the visual and olfactory transduction mechanisms. We chose these two sensory modalities because, despite the very different natures of light and chemical stimuli, these two transduction mechanisms are remarkably similar (Fig 8-2).

VISUAL TRANSDUCTION

We are able to see because of photoreceptors in the eye, which transduce light into an electrical signal. Photoreceptors are found in the retina, which lines the back of the eye. The retina contains two types of photoreceptors:

rods and cones. Rods are more sensitive and mediate vision in dim light, whereas cones are less sensitive and mediate vision in bright light. In order to maximize their sensitivity, rods have more pigment, allowing them to absorb more of the incident light. The pigment and the transduction mechanism are located in a part of the rods called the outer segment. Within the outer segment there is a dense stack of membrane disks that contain the light-absorbing visual pigment rhodopsin at very high density. Prior to the 1970s, electrophysiological measurements had shown that the response originates in the outer segment plasma membrane. Therefore, the outer segment contains the complete transduction mechanism. This has aided research on the visual transduction mechanism, because outer segments are relatively easy to purify. Therefore, biochemists could isolate enough photosensitive material to study the biochemistry of visual transduction. In the early 1970s, it was discovered that light activated a phosphodiesterase that breaks down the intracellular messenger cyclic GMP (for review, see e.g., Stryer 1986). Rhodopsin activates the phosphodiesterase by means of a rod-specific G-protein, called transducin. These biochemical results suggested that cyclic GMP might play a role in generating an electrophysiological response. However, at that time, protein phosphorylation was the only mechanism by which cyclic GMP was known to affect ion channels. Phosphorylation is a relatively slow reaction, whereas photoreceptors are known to detect light within milliseconds. This discrepancy prevented the acceptance of the cyclic GMP model.

The proof that cyclic GMP does in fact mediate visual transduction came from the discovery of ion channels that are directly opened by intracellular cyclic GMP. This type of ion channel was discovered by using the patch clamp technique to excise an inside-out patch of membrane from the rod outer segment (Figure 8-3, Fesenko et al 1985). In this configuration, cyclic GMP could be applied to the internal surface of the membrane patch. It was found that cyclic GMP rapidly and reversibly opens ion channels in the plasma membrane. This effect occurs in the absence of ATP and GTP, which indicates that phosphorylation, which requires ATP, is not involved in the opening or closing of these ion channels. Therefore, it appears that these ion channels are opened by the binding of cyclic GMP to the channel protein. These ion channels have been called cyclic-GMP-gated channels.

But the existence of these ion channels did not prove that a decrease in cyclic GMP concentration actually mediates the light response. Evidence that a decrease in cyclic GMP concentration does cause the light response was provided in the following studies: (1) Light-sensitive and cGMP-gated channels show very similar ionic permeabilities to all alkali metal ions, suggesting that both ionic channels are the same (Hodgkin et al 1985, Fesenko et al 1985, Menini 1990). (2) Activities of unitary events (single-channel current) display essentially the same properties for the light-suppressive and cGMP-activated currents (Matthews 1986, Matthews and

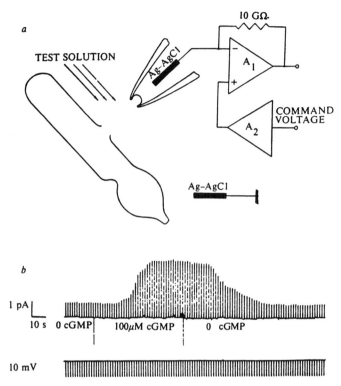

Figure 8-3. Direct measurement of current through cGMP-gated channel located in the rod outer segment. Reproduced from Fesenko el al (1985) with permission.

Watanabe 1987). (3) Light stimulation can actually suppress the current induced by cytoplasmic cGMP, when it is loaded into rods from a whole-cell recording pipette (Cobbs and Pugh 1985). Similar light-sensitive suppression of cGMP-induced currents was confirmed in a "truncated outer segment" preparation (Yau and Nakatani 1985). In this recording method, rod outer segment is inserted into a suction pipette, and the external parts are truncated, so that one can freely superfuse the cytoplasmic side of the rod outer segment under the recording pipette. Since this preparation also maintains the whole transduction system included in the outer segment, light stimulation activates a reaction of the transduction machinery. In the absence of cyclic GMP, the truncated preparation displays no current, but addition of cyclic GMP induces a large inward current, which is suppressed by light. This suppression requires the presence of GTP, suggesting an involvement of GTP-binding protein (transducin). ATP is required for the response recovery to inactivate (by phosphorylation) the light-activated rhodopsin.

The reversal potential for the cyclic-GMP-gated channels is close to zero millivolts. Hence, these are depolarizing ion channels. However, light causes a decrease in the cyclic GMP concentration through increased hydrolysis of cyclic GMP. Therefore, light causes a decrease in a depolarizing conductance, thus causing hyperpolarization of rod and cone photoreceptors. Hyperpolarizing response to light of single photoreceptor cells *in situ* has been revealed by an earlier study by Tomita (1970).

The discovery of cyclic-GMP-gated ion channels in rods provided a mechanism whereby activation of the phosphodiesterase could generate an immediate change in membrane conductance, and established that visual transduction is mediated by changes in cyclic GMP concentration. The discovery of this ion channel also revealed a new way by which cyclic nucleotides could control electrical activity in the nervous system. Related ion channels have since been discovered in a variety of other tissues.

One of the key features of transduction in rods is their high sensitivity. In fact, rods can reliably detect single photons of light, which bleach just one molecule of rhodopsin, out of about a billion per rod. How is this great sensitivity achieved? All of the components of the transduction mechanism are associated with intracellular membranes. Rhodopsin is an integral membrane protein, which means that it is embedded in the membrane. The G-protein transducin and the phosphodiesterase are peripheral membrane proteins, which means that they are bound to the surface of the membrane. The membrane is composed of a bilayer of phospholipid molecules, which have an oil-like consistency. Therefore, all of the protein molecules involved in transduction can diffuse within the plane of the membrane as a result of thermal vibrations. This motion is very important, because it is the cause of the rod's high sensitivity. When a rhodopsin molecule is bleached, it remains in an active state for several tens (up to several hundreds) of milliseconds. During this time, it can collide with more than five hundred G-proteins and activate them (Stryer 1986). Therefore, we can think of the transduction mechanisms as an enzymatic cascade, in which rhodopsin activates many G-proteins, and each G-protein activates a phosphodiesterase, which hydrolyzes many molecules of cyclic GMP. In this way, a single bleached rhodopsin molecule generates a highly amplified response. Similar enzymatic cascades mediate other forms of signal transduction throughout the body.

OLFACTORY TRANSDUCTION

Odors are transduced into an electrophysiological response by specialized receptor cells in the nasal cavity, called olfactory receptor cells. In contrast to photoreceptors, which hyperpolarize in response to light, olfactory receptor cells depolarize in response to odors. The question of how odors cause this membrane depolarization began to be studied after 1985, when an odor-stimulated adenylyl cyclase, which produces the intracellular messenger cyclic AMP, was discovered in olfactory receptor cells (Pace et al 1985). In addition, it was shown that GTP was required for activation of the cyclase, which suggested a receptor protein (G-protein) cascade, such as mediates visual transduction.

In olfactory receptor cells, transduction occurs in thin (0.25 micron diameter) cylinders called cilia (Kurahashi 1989a, Lowe and Gold 1991). Thin cylinders maximize the membrane area for binding odorant molecules, thereby increasing the probability that an odorant molecule will bind to a receptor protein in the ciliary membrane. This contrasts with photoreceptor outer segments, which have a high-density of membranes, to maximize the concentration of the visual pigment. Photoreceptor outer segments are also based on a ciliary structure. Therefore, it was proposed that there is an evolutionary relationship between visual and olfactory receptors. If so, this suggests that there may be some parallels between the visual and olfactory transduction mechanism. It was already known that odorants cause an increase in cyclic AMP concentration in the cilia. Therefore, Nakamura and Gold (1987) hypothesized that there might be cyclic-AMP-gated ion channels in olfactory cilia that would serve to convert an increase in cyclic AMP concentration into membrane depolarization. Therefore, they took the same approach as Fesenko et al. (1985) did, which was to excise patches from the ciliary membrane. In this way, they observed cyclic-AMP-gated ion channels in the ciliary membrane. However, cyclic GMP also activates this ion channel, so these channels have been termed cyclic-nucleotide-gated (CNG) ion channels.

After the discovery of the CNG channels in olfactory cilia, it was still necessary to show that the CNG channels actually generate the odorant-induced depolarization. First, odorant-activated and cyclic-AMP-induced currents exhibit the same properties: both exhibit Ca^{2+}-mediated desensitization (Figure 8-4) (Kurahashi 1990, Kurahashi and Shibuya 1990, Zufall et al 1991), both are cation-selective (Figure 8-5) (Kurahashi 1989a, 1990, Frings et al 1992), both exhibit the same single-channel conductance (Firestein et al 1993), and both are concentrated in the ciliary membrane (Kurahashi 1989a, 1990, Lowe and Gold 1991, Kurahashi and Kaneko 1993, see also below). Second, the currents evoked by simultaneous odorant- and cyclic-AMP-stimulation sum nonlinearly, as expected if the odorant response is mediated by the CNG channels (Lowe and Gold 1993a). Recently, Brunet et al (1996) demonstrated a causal relationship between the CNG channels and odorant responses by showing that a knockout mouse that lacks CNG channels does not exhibit odorant responses to any odorants tested. This established that the odorant response is actually mediated by the CNG channels.

The electrophysiological properties of olfactory CNG channels are very similar to those described for the cyclic-GMP-gated ion channels found in photoreceptors. Indeed, these molecules must have a common evolutionary

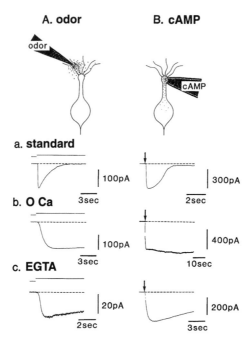

Figure 8-4. Comparison between odor- and cAMP-induced response (Ca^{2+} dependence). A. odorant (amyl acetate) induced responses. B. cAMP-induced responses. From the top traces, normal Ringer's solution, 0 Ca solution, and EGTA-loaded cells. Note that the current responses show very similar time courses between A and B.

origin, because they have similar amino acid sequences. Perhaps the biggest difference between the visual and olfactory channels is their sensitivity to cyclic nucleotide: the photoreceptor channels have high sensitivity to cyclic GMP with very low sensitivity to cyclic AMP, whereas the olfactory channels have similar sensitivity to both cyclic AMP and cyclic GMP. *In situ*, however, it is believed that cyclic AMP plays the dominant role in generating odorant responses.

Detailed reviews on the properties of cGMP-gated channels in rods and CNG channels in olfactory cells have already been written. Here, we present a brief summary of the channel properties, focusing on the olfactory channel and its role in mediating olfactory perception. Readers interested in more information may wish to read reviews, such as Yau and Baylor (1989), Kaupp and Koch (1992), Menini (1995).

The olfactory CNG channel is cation-selective (Figure 8-5), so the main ions mediating membrane depolarization are Na^+ and Ca^{2+}. But divalent cations, such as Ca^{2+} and Mg^{2+}, have an additional effect: they bind more

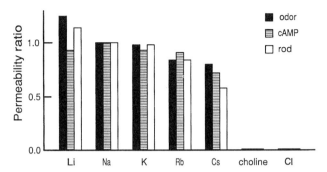

Figure 8-5. Ionic selectivity of odor-activated conductance, olfactory cAMP channel, and rod cGMP-gated channel. Data from Kurahashi (1989), Kurahashi (1990), and Menini (1990).

strongly than monovalent cations to a negatively charged site in the pore of the channel, and briefly prevent the flow of ions through the pore. This effect of divalent cations causes the current through the open channel to fluctuate at a very high frequency, which is called "flickery block" of the channel. A similar phenomenon has been observed for the rod channel (Haynes et al 1986). In the presence of physiological concentrations of Ca^{2+} and Mg^{2+}, this "flickery block" reduces the mean current through the open channel one to two orders of magnitude. This decreases the effective single-channel conductance, increasing the amplitude resolution of current generation by cyclic AMP. This phenomenon is important for allowing small signals to be generated by both photoreceptors and olfactory receptors, without being limited by the discreteness of the single-channel conductance. Indeed, when divalent cations are removed, the single-channel conductance is quite large, generating current steps of about 10^{-12} amps (Nakamura and Gold 1988, Suzuki 1990, Kurahashi and Kaneko 1991, Zufall et al 1991). Such large current steps would generate voltage steps of about 5 millvolts in mammalian olfactory receptor cells. This value is calculated by using Ohm's law (voltage = current × resistance, where the membrane resistance of mammalian receptor cells is around 5×10^{10} ohms). If olfactory receptor cells could generate responses only in 5 millivolt steps, this would limit the resolution for detecting weak odorant stimuli.

As mentioned earlier, the olfactory transduction mechanism is localized to the cilia. Therefore, it is expected that all components of the mechanism will be concentrated in the cilia. For example, the putative odorant receptor proteins (Buck and Axel 1991), the olfactory-specific G-protein(G_{olf}) (Jones and Reed 1989), and the odorant-stimulated adenylyl cyclase (Pace et al 1985, Bakalyar and Reed 1990) are all concentrated in the cilia. In addition, the CNG channels are also concentrated in the cilia, as revealed by a density of 200–1000 μm^{-2} (Kurahashi and Kaneko 1993, Kleene 1994), while in the dendrite or cell body the channel density is only a few μm^{-2} (Kurahashi and Kaneko 1993).

The CNG channels are both thought to consist of four subunits, including two subunits called "alpha" and two called "beta" (Chen et al 1993). The amino acid sequence (primary structure) of the olfactory alpha subunit is similar to that of the alpha subunit of the photoreceptor channel (60% identical, see Kaupp et al 1989, Dhallan et al 1990). On the basis this amino acid sequence similarity, the three-dimensional structure of these channels is also expected to be similar. Furthermore, this three-dimensional structure is similar to that of a voltage-gated potassium channel, called *Shaker*, both having six transmembrane domains and a pore-forming region that is the site for ion permeation (Goulding et al 1993). The cyclic AMP binding site is located in the carboxy-terminal (C-terminal) region that projects into the cytoplasm. Each subunit has only one cyclic nulceotide binding site (Dhallan et al 1990); it is reasonable to think that there are four cAMP binding sites in one CNG channel molecule.

Olfactory Adaptation and Channel Modulation by Ca^{2+}-Calmodulin

It has long been known that olfactory receptor cells desensitize or adapt in response to prolonged stimulation (Ottson 1956, Getchell and Shepherd 1978). Adaptation occurs in all sensory receptors, and allows the cells to respond over a wider range of stimulus magnitudes than would be possible if sensitivity could not decrease in response to strong stimuli. In 1990, Kurahashi and Shibuya demonstrated in solitary preparations that the odorant-induced current also shows adaptation to prolonged stimuli. They also showed that adaptation was abolished when Ca^{2+} was removed from the extracellular solution. The odorant-activated ion channels (CNG channels) have a sixfold higher permeability to Ca^{2+} than to Na^+ (Kurahashi and Shibuya 1990). This implies that Ca^{2+} influx is the cause of adaptation. However, this experiment did not show what the intracellular target for Ca^{2+} was. Adaptation could occur at several points in the transduction mechanism, including phosphorylation of the receptor proteins, inhibition of the adenylyl cyclase, or stimulation of the phosphodiesterase. It was also possible that Ca^{2+} desensitizes the CNG channels to cyclic AMP.

Recently, Kurahashi and Menini (1997) showed that adaptation is due almost entirely to desensitization of the CNG channels. They accomplished this by using a photoactivatable form of cyclic AMP to produce cyclic AMP within the cell, independently of the adenylyl cyclase. They showed that an adapting stimulus reduced the current evoked by the photoactivated cyclic AMP to the same extent as the response to an odorant stimulus. This showed that adaptation occurs after the production of cyclic AMP. They also showed that the decay of the current evoked by the photoactivated cyclic AMP was unchanged, which showed that activation of the phosphodiesterase does not play a role in adaptation. This implied that desensitization of the CNG channels is the dominant mechanism of adaptation.

The ability of Ca^{2+} to desensitize the CNG channel had already been demonstrated by Chen and Yau (1994). They measured the sensitivity of the channel to cyclic nucleotides as a function of Ca^{2+}. Calmodulin, a protein that mediates Ca^{2+} regulation of many proteins, was required for this effect. Thus, it is calmodulin with Ca^{2+} bound to it (Ca-calmodulin) that actually changes the affinity of the channel to cAMP by 20-fold. A putative Ca-calmodulin binding site is thought be located near the N-terminal cytoplasmic side of the CNG channel. Balasubramanian et al (1996) have reported the presence of another, as yet unidentified, cytoplasmic factor(s) that also desensitizes the CNG channel to Ca^{2+}. There is also evidence that the CNG channel can be desensitized by Ca^{2+} without a requirement for any intracellular protein (Lynch and Lindemann, 1994). However, this effect is relatively small compared to that mediated by Ca^{2+}-calmodulin.

Ca^{2+}-activated Cl Channel

Because the presence of Ca^{2+} reduces the current through the CNG channels, almost all measurements of the odorant-induced and cGMP-activated current were carried out in the absence of Ca^{2+}. This experimental condition had the advantage of increasing the odorant-induced current, but also prevented the Ca^{2+} influx that would occur under normal conditions. However, in 1991, Kleene and Gesteland (1991a) recorded currents across the ciliary membrane by a new method: rather than pulling small membrane patches off the side of a cilium, they sucked most of a cilium into a patch clamp electrode. This configuration was equivalent to an excised inside-out cilium. In a later paper, Kleene and Gesteland (1991b) showed that this conductance was Cl^- selective. Thus, the ciliary membrane contains Ca^{2+}-dependent Cl channels, in addition to the CNG channels. However, it was not known what the Cl^- reversal potential was in olfactory receptor cells. In most cells, the Cl^- reversal potential is more negative than the resting membrane potential, so that Cl^- conductances generally hyperpolarize cells. Also in 1993, Kurahashi and Yau demonstrated that when Ca^{2+} is present outside the cell, the odorant-induced inward current is a mixture of a cationic current through the CNG channels and a Cl^- current through the Ca^{2+}-dependent Cl channels. They also demonstrated, using a recording procedure that would minimally perturb the intracellular Cl^- concentration, that the reversal potential for Cl^- was more positive than the resting potential, i.e., very close to the reversal potential for the CNG channels, and that about half of the odorant-induced current was carried by Cl^-. However, it is possible that the Cl^- reversal potential may have been altered in these experiments, so Zhainazarov and Ache (1995) repeated this measurement by a different method that would not allow the intracellular Cl$^-$ concentration to be altered during the recording. They confirmed the conclusion of Kurahashi and Yau that the Cl$^-$ reversal potential is near zero millivolts in olfactory receptor cells. Thus, the generation of the odorant-induced current occurs in two steps: first cyclic AMP activates the CNG channels, causing membrane depolarization. Second Ca^{2+} influx via the CNG channels activates the Ca^{2+}-dependent Cl channels, causing additional depolarization. Recently, Kleene et al (personal communication) have shown that the Cl channels have a very small single-channel current, so they do not increase the noise of the transduction mechanism.

The function of this two-stage mechanism of current generation is not known with certainty, but a possible role has been suggested by Lowe and Gold (1993b). They used photoactivated cyclic AMP to measure the cyclic AMP concentration dependence of the current in the presence of extracellular Ca^{2+}, and observed a 4th or higher power dependence on cyclic AMP concentration. This high power dependence can be explained by the sequential activation of the CNG and Cl channels, because CNG channels open in proportion to the square of cyclic AMP concentration, and the Cl channels open in proportion to the square of the Ca^{2+} concentration. Putting these two processes in sequence can explain the 4th or higher power dependence of the total current on cyclic AMP concentration. Ca^{2+} and cyclic AMP buffering within the cell may account for higher than 4th power dependence on cyclic AMP concentration. An interesting property of this high power dependence on cyclic AMP concentration is that current is not proportional to cyclic AMP concentration at low concentrations. Indeed, no current is generated until the cyclic AMP concentration exceeds a threshold. Thus, the combination of the CNG and Cl channels can be thought of as introducing a threshold into the transduction mechanism: odorant concentrations below a certain threshold cannot generate a current. This property was surprising, because it reduces the sensitivity of olfactory receptor cells. Therefore, Lowe and Gold (1995) proposed that the function of this ionic cascade was to prevent fluctuations in the basal cyclic AMP concentration from generating a current. In support of this hypothesis, they demonstrated a high basal adenylyl cyclase activity in the absence of odorants, and showed that there are large fluctuations in the basal cyclic AMP concentration.

Another function for the coexistence of the CNG and Cl channels was proposed by Kurahashi and Yau (1993). The olfactory cilia are embedded in a thin layer of mucus that covers the olfactory epithelium. Therefore, any changes in the ionic composition of the mucus should alter the magnitude of the odorant-induced current. However, the odorant-induced current is the sum of two currents (depolarization by the CNG channels is due to cation influx into the cell, whereas depolarization by the Cl channels is caused by Cl$^-$ flowing out of the cell). Therefore, if the ionic concentration of the mucus is concentrated by drying, or diluted by water entering the

nose, this would have opposite effects on the currents through the CNG and
Cl channels. Therefore, this may be a mechanism for stabilizing the odorant-
induced current. Indeed, it has been reported that olfactory responses are
relatively insensitive to changes in the composition of the mucus *in situ*
(Tucker and Shibuya 1965, Suzuki 1981, Shoji et al 1993), or to the change
in external Na reduction in solitary preparation (Kurahashi and Yau 1993).

Voltage-gated Ionic Channels

In addition to the ionic channels that generate the odorant-induced current,
olfactory receptor cells contain a variety of other ionic channels that generate action potentials and modulate the signal that is sent to the brain
(Miyamoto et al 1992, Nevitt and Moody 1992, Schild 1989, Firestein and
Werblin 1987, Kawai et al 1996, Rajendra et al 1992, Trombley and Westbrook 1991). These other ion channels are located in dendrosomatic membrane rather than in ciliary membrane, and play important roles in
determining the response that is transmitted to the brain.

Patch clamp recordings from single olfactory receptor cells have revealed a variety of ionic channels in the plasma membrane. It is now known
that olfactory receptor cells express at least five types of voltage-activated
channels: (1) Na channel, (2) L-type Ca channel, (3) T-type Ca channel, (4)
delayed rectifier K channel, and (5) Ca^{2+}-activated K channels. As explained in the Introduction, membrane depolarization generates depolarizing transient voltage spikes in the receptor cell axon, called action potentials.
In a sense, you can think of the generation of action potentials as an analog-
to-digital conversion of information. This is because the membrane depolarization in the receptor cell can attain any level, and is therefore like an
analog signal. However, the amplitude of the action potentials is fixed, so it
is like a digital signal. Therefore, the timing of the frequency of occurrence
of action potentials is the only information carried by the receptor cell
axons. These action potentials propagate along the receptor cell axon to the
olfactory bulb, which is the first stage of olfactory processing in the brain. It
has long been assumed that olfactory receptor cells utilize only voltage-activated Na and K channels to generate action potentials. However, detailed
voltage-clamp experiments by Kawai et al (1996) revealed that the transient
inward current activated by the depolarizing voltage step is actually a mixture of both Na^+ currents and transient Ca^{2+} (called T-type) currents. Moreover, T- type Ca channels can be activated at more negative potentials than
Na^+ channels can be. Therefore, the function of these Ca channels appears
to be to decrease the depolarization that is necessary to generate action
potentials. This also increases the sensitivity of olfactory receptor cells.

In the olfactory cells, surprisingly, there are no detectable K channels
that are open around the resting membrane potential ($-70mV$ see e.g.,
Kawai et al 1996), while in rod photoreceptor cells, or in other neurons in

general, K channels mediating inward rectification are open. This lack of K channels that are open around the resting membrane potential is an important specialization for olfactory receptor cells. Keeping the resting membrane conductance to a minimum creates a high input resistance, exceeding 5×10^9 ohms in amphibians and 3×10^{10} ohms in mammals. Because of this very high membrane resistance in the unstimulated state, very small currents can generate significant membrane depolarization. This increases the sensitivity of olfactory receptor cells.

Finally, the decay time course of the action potential is determined by the two types of K channels (delayed rectifier and Ca-activated K channels; Kurahashi 1989, thesis).

CONCLUSION

Prior to 1985, the mechanisms of visual and olfactory transduction were uncertain. However, the discovery of the cyclic GMP-gated ionic channels in photoreceptors established that cyclic GMP is the intracellular messenger that mediates visual transduction. In addition, this work on the visual transduction mechanism provided precedents that stimulated work on the olfactory transduction mechanism. Today, both transduction mechanisms are understood in considerable detail, and future work will provide additional information about the molecular mechanisms that determine sensitivity and the response kinetics of these two transduction mechanisms.

The development of the patch clamp technique has made much of this progress possible. For example, because of the small sizes of photoreceptors and olfactory receptors, it was difficult to obtain recordings of single cells, without the patch clamp technique. Prior to the development of the patch clamp technique, single cell recordings required impaling single cells with sharp microelectrodes. However, this method of recording invariably introduced a large leakage conductance, which altered the properties of the cell. In addition, the use of excised patches to study second-messenger-gated ionic channels enables the discovery of cyclic-nucleotide-gated ionic channels. Recent work has demonstrated that the same types of ionic channels are expressed in a variety of cells besides photoreceptors and olfactory receptors. Therefore, cyclic-nucleotide-gated ionic channels now constitute a major class of ionic channels.

REFERENCES

Bakalyar HA and Reed RR. 1990. Identification of a specialized adenylyl cyclase that may mediate odorant detection. *Science* 250:1403–1406.

Balasubramanian S, Lynch JW and Barry PH. 1996. Calcium-dependent modulation of the agonist affinity of the mammalian olfactory cyclic nucleotide-gated channel by calmodulin and a novel endogenous factor. *J. Membrane Biol.* 15:12–23.

Boekoff I and Breer H. 1992. Termination of second messenger signaling in olfaction. *Proc Natl Acad Sci* 89:471–474.

Brunet LJ, Gold GH and Ngai J. 1996. General anosmia caused by a targeted disruption of the mouse olfactory cyclic nucleotide-gated cation channel. *Neuron* 17:681–693.

Buck L and Axel R. 1991. A novel multigene family may encode odorant receptors: A molecular basis for odor recognition. *Cell* 65:175–187.

Chen T-Y et al. 1993. A new subunit of the cyclic nucleotide-gated cation channel in retinal rods. *Nature* 362:764–767.

Chen T-Y and Yau K-W. 1994. Direct modulation by Ca^{2+}-calmodulin of cyclic nucleotide-activated channel of rat olfactory receptor neurons. *Nature* 368:545–548.

Cobbs WH and Pugh Jr EN. 1985. Cyclic GMP can increase rod outer-segment light-sensitive current 10-fold without delay of excitation. *Nature* 313:585–587.

Dhallan RS, Yau K-W, Schrader KA Reed RR. 1990. Primary structure and functional expression of a cyclic nucleotide-activated channel from olfactory neurons. *Nature* 347:184–187.

Fesenko EE, Kolesnikov SS and Lyubarsky A L. 1985. Induction by cyclic GMP of cationic conductance in plasma membrane of retinal rod outer segment. *Nature* 313:310–313.

Firestein S et al. 1991. Single odor-sensitive channels in olfactory receptor neurons are also gated by cyclic nucleotides. J Neurosci 11:3565–3572.

Frings S et al. 1992. Properties of cyclic nucleotide-gated channels mediating olfactory transduction. *J Gen Physiol* 100:45–67.

Getchell TV and Shepherd GM. 1978. Adaptive properties of olfactory receptors analysed with odour pulses of varying durations. *J Physiol* 282:541–560.

Goulding EH et al. 1993. Role of H5 domain in determining pore diameter and ion permeation through cyclic nucleotide-gated channels. *Nature* 364:61–64.

Hagiwara S. 1983. Membrane potential-dependent ion channels in cell membrane. New York: Raven Press.

Hamill OP, Marty A, Neher E, Sakmann B, and Sigworth FJ. 1981. Improved patch-clamp techniques for high resolution current recording from cells and cell-free membrane patches. *Pfluegers Arch* 391:85–100.

Haynes LW, Kay AR and Yau K-W. 1986. Single cyclic GMP-activated channel activity in excised patches of rod outer segment membrane. *Nature* 321:66–70.

Hille B. 1984. Ionic channels of excitable membranes. Massachusetts: Sinauer Associates Inc.

Hodgkin AL, McNaughton PA and Nunn BJ. 1985. The ionic selectivity and calcium dependence of the light-sensitive pathway in toad rods. *J Physiol* 358:447–468.

Jones DT and Reed RR. 1989. G_{olf}: an olfactory neuron specific-G-protein involved in odorant signal transduction. *Science* 244:790–795.

Kaupp UB and Koch K-W. 1992. Role of cGMP and Ca^{2+} in vertebrate photoreceptor excitation and adaptation. *Annu. Rev. Physiol.* 54:153–175.

Kaupp UB, Niidome T, Tanabe T, Terada S, Bonigk W, Stuhmer W, Cook NJ, Kanagawa K, Matsuo H, Hirose T, Miyata T and Numa S. 1989. Primary structure and functional expression from complementary DNA of the rod photoreceptor cyclic GMP-gated channel. *Nature* 342:762–766.

Kawai F, Kurahashi T, and Kaneko A. 1996. T-type Ca^{2+} channel lowers the threshold of spike generation in the newt olfactory receptor cell. *J Gen Physiol* 108:525–536.

Kleene SJ and Gesteland RC. 1991a. Transmembrane currents in frog olfactory cilia. *J Membrane Biol* 120:75–81.

Kleene SJ, and Gesteland RC, 1991b. Calcium-activated Chloride Conductance in frog olfactory cilia. *J Neurosci* 11:3624–3629.

Kleene SJ, Gesteland RC and Bryant SH. 1994. An electrophysiological survey of frog olfactory cilia. *J Exp Biol* 195:307–328.

Kurahashi T. 1989a. Activation by odorants of cation-selective conductance in the olfactory receptor cell isolated from the newt. *J Physiol* 419:177–192.

Kurahashi T. 1989b. Molecular mechanisms of olfactory transduction. Ph.D. thesis. Tsukuba University, Japan.

Kurahashi T. 1990. The response induced by intracellular cyclic AMP in isolated olfactory receptors of the newt. *J Physiol* 430:355–371.

Kurahashi T and Kaneko A. 1991. High density cAMP-gated channels at the ciliary membrane in the olfactory receptor cell. *NeuroReport* 2:5–8.

Kurahashi T and Kaneko A. 1993. Gating properties of the cAMP-gated channels in toad olfactory receptor cells. *J. Physiol.* Vol 466:287–302.

Kurahashi T and Menini A. 1997. Mechanism of odorant adaptation in the olfactory receptor cell. *Nature* 385:725–729.

Kurahashi T and Shibuya T. 1990. Ca^{2+}-dependent adaptive properties in the isolated olfactory receptor cell of the newt. *Brain Res* 515:261–268.

Kurahashi T and Yau K-W. 1993. Co-existence of cationic and chloride components in odorant-induced current of vertebrate olfactory receptor cells. *Nature* 363:71–74.

Lowe G and Gold GH. 1991. The spatial distribution of odorant sensitivity and odorant-induced currents in salamander olfactory receptor cells. *J Physiol* 442:147–168.

Lowe G and Gold GH. 1993a. Contribution of the ciliary cyclic nucleotide-gated conductance to olfactory transduction in the salamander. *J Physiol* 462:175–196.

Lowe G and Gold GH. 1993b. Nonlinear amplification by calcium-dependent chloride channels in olfactory receptor cells. *Nature* 366:283–286.

Lowe G and Gold GH. 1995. Olfactory transduction is intrinsically noisy. *Proc Natl Acad Sci* 92:7864–7868

Lynch JW and Lindemann B. 1994. Cyclic nucleotide-gated channels of rat olfactory receptor cells: divalent cations control the sensitivity to cAMP. *J Gen Physiol* 103:87–106.

Matthews G. 1986. Comparison of the light-sensitive and cyclic GMP-sensitive conductance of the rod photoreceptor: noise characteristics. *J Neurosci* 6:2521–2526.

Matthews G and Watanabe S-I. 1987. Properties of ion channels closed by light and opened by guanosine 3′,5′-cyclic monophosphate in toad retinal rods. *J Physiol* 389:691–715.

Menini A. 1990. Currents carried by monovalent cations through cyclic GMP-activated channels in excised patches from salamander rods. *J Physiol* 424:167–185.

Menini A. 1995. Cyclic nucleotide-gated channels in visual and olfactory transduction. *Biophys Chem* 55:185–196.

Miyamoto T, Restrepo D, and Teeter JH. 1992. Voltage-dependent and odorant-regulated currents in isolated olfactory receptor neurons of the channel catfish. *J Gen Physiol* 99:505–530.

Nakamura T and Gold GH. 1987. A cyclic nucleotide-gated conductance in olfactory receptor cilia. *Nature* 325:442–444.

Nakamura T and Gold GH. 1988. Single channel properties of the ciliary cyclic nucleotide-gated conductance. *Chem Senses* 13:723–724.

Nevitt GA and Moody WJ. 1992. An electrophysiological characterization of ciliated olfactory receptor cells of the coho salmon *Oncorhynchus kisutch. J Exp Biol* 166:1–17.

Ottoson D. 1956. Analysis of the electrical activity of the olfactory epithelium. *Acta Physiologica Scandinavica* 35:7–83.

Pace U, Hanski E, Salomon Y, and Lancet D. 1985. Odorant-sensitive adenylate cyclase may mediate olfactory reception. *Nature* 316:255–258.

Rajendra S Lynch JW and Barry PH. 1992. An analysis of Na$^+$ currents in rat olfactory receptor neurons. *Pfluegers Arch* 420:342–346.

Sakmann B and Neher E. 1995. Single-channel recording. 2nd edition. New York: Plenum Press.

Schild D. 1989. Whole-cell currents in olfactory receptor cells on *Xenopus laevis. Brain Res* 78:223–232.

Shoji et al. 1993. Transduction mechanisms in the olfactory and vomeronasal organs of turtles. *Brain Behav Evol* 41:192–197.

Stryer L. 1986. Cyclic GMP cascade of vision. *Ann Rev Neurosci* 9:87–119.

Suzuki N. 1978. Effects of different ionic environments on the responses of single olfactory receptors in the lamprey. Comp Biochem Physiol 61A:461–467.

Suzuki N. 1990. Single cyclic nucleotide-activated ion channel activity in olfactory receptor soma membrane. Neurosci Res Suppl 12:S113–126.

Tomita T. 1970. Electrical activity of vertebrate photoreceptors. Q Rev Biophys 3:179–222.

Trombley PQ and Westbrook GL. 1991. Voltage-gated currents in identified rat olfactory receptor neurons. *J Neurosci* 11(2):435–444.

Tucker D and Shibuya T. 1965. A physiologic and pharmacologic study of olfactory receptors. In: Cold Spring Harbor Symposia on Quantitative Biology 30:207–215.

Yau K-W and Baylor DA. 1989. Cyclic GMP-activated conductance of retinal photoreceptor cells. *Annu Rev Neurosci* 12:289–327.

Yau K-W and Nakatani K. 1985. Light-suppressible, cyclic GMP-sensitive conductance in the plasma membrane of a truncated rod outer segment. *Nature* 317:252–255.

Zhainazarov AB and Ache BW. 1995. Odor-induced currents in Xenopus olfactory receptor cells measured with perforated-patch recording. *J Neurophysiol* 74:479–483.

Zufall F, Shepherd GM and Firestein S. 1991. Inhibition of the olfactory cyclic nucleotide gated ion channels by intracellular calcium. *Proc R Soc Lond B* 246:225–230.

9

Bits for an Organic Microprocessor:
Protein Phosphorylation/Dephosphorylation

PETER J. KENNELLY

1. DECISIONS, DECISIONS: INTEGRATING MULTIPLE SIGNALS IN THE BIOLOGICAL WORLD

Survival in a dynamic external environment demands the ability to monitor and respond to a wide range of internal and external variables, or signals. The binding of a signal to its receptor initiates a sequence or "cascade" of molecular events inside the cell that modulate relevant metabolic, nuclear, motile, or other processes. The transmission of an often extracellular receptor binding event into the interior of the cell and its translation into a catalytic or other response is called signal transduction. For an organism to be consistently successful in nature's continual competition for scarce resources, such responses must be rapid and efficient. Efficiency demands that the response be comprehensive in scope to ensure against wastage of either materials or energy. For example, the activation of the contractile machinery in a muscle cell is accompanied by several secondary and tertiary effects. To meet the requirements of initiating and sustaining contractile activity, a concomitant reconfiguration of the metabolic processes that generate cellular energy and mobilize stored energy sources takes place. The diversion of cellular resources to meet the overriding demands of the contractile process, in turn, requires compensatory fine-tuning of other pathways.

Complicating matters is the cacophony of competing and contradictory signals bombarding the cell at any one instant. A microbe floating in a puddle is continually subjected to variations in temperature, sunlight, external pH, oxygen tension, and in the availability of the compounds that supply

Introduction to Cellular Signal Transduction
A. Sitaramayya, Editor
©1999 Birkhäuser Boston

the carbon, nitrogen, sulfur, phosphorus, etc. needed to sustain life. Mammalian cells are bathed in a churning humoral sea containing scores of hormones whose shifting levels are punctuated by flashes of neuronal excitation. Cells need mechanisms for integrating information and making decisions, for selecting the response(s) most appropriate for a particular combination of signal inputs. In other words, signal transduction demands much more than the construction of molecular switches for turning particular cellular processes "on" and "off." Simple linear connections between a receptor and its primary site of action are not enough. Successful signal transduction demands the creation of highly branched, tightly interwoven, integrated command and control networks—organic microprocessors—in which the signaling pathways triggered by particular effectors actively interact with and influence one another, a process oftentimes referred to as "cross-talk."

Nature creates sophisticated microprocessors from the simple organic components that reside within cells in essentially the same way that man builds computers with inorganic materials. The universal building blocks for all computers are simple binary elements called bits. A bit can be anything with the capability to exist in two freely interchangeable, functionally distinct states. This function, or output, is the ability to influence other bits in the microprocessor and the objects whose activities the microprocessor directs. By cleverly arranging these simple bits in complex arrays, powerful computers can be constructed. Simplicity is transformed into sophistication and power by the manner in which these building blocks are linked and layered together into interactive networks.

In man-made computers, bits are electromagnetic in nature, as are their functional outputs. In the biological world, bits are chemical in nature. Proteins form the building blocks for most, but not all, biomolecular bits (Bray 1995, Marks 1996). This is the natural consequence of the broad spectrum of chemical, catalytic, and conformational capabilities of proteins, a functional versatility that renders them the predominant source of chemical activity and the principal agents of physical motion in cells. The preeminence of protein-based bits also reflects the importance of time as a variable. The use of prepositioned component proteins imbues cellular microprocessors with the ability to react to signaling inputs in as little as a fraction of a second. Gene expression, by contrast, requires minutes to hours to take effect. Fast reaction times are a must if an organism is to take advantage of a fleeting opportunity or to effectively neutralize an impending threat.

Among the cellular functions that can be affected by the functional outputs of protein bits is the pattern of gene expression in the cell. The expression of genes and the control thereof are absolutely dependent upon extant proteins and their regulation. Gene expression works in partnership with protein-based bits to confer a powerful adaptive plasticity upon the cell's organic microprocessor by providing the means to reconfigure its composition and architecture over time. This ability to continuously "customize"—down to the level of selecting among subtly different isozymic forms of a particular protein—provides an efficient and economical mechanism for meeting the specific, long-term demands introduced by gradually unfolding processes such as cellular differentiation, seasonal transitions in the external environment, etc.

Nature utilizes two general molecular mechanisms for creating proteinaceous bits: reversible ligand binding and covalent modification. The binding of a protein to a second molecular entity, whether by noncovalent or covalent means, produces a new species that is chemically distinct with the potential to exhibit new functional properties. Numbered among the cell's regulatory ligands are hormonal second messengers such as cyclic nucleotides and Ca^{2+}, allosteric effectors such as AMP and hydroxysterols, and auxiliary subunits that serve to inhibit or activate catalytic efficiency or to target proteins to specific subcellular locations. Numbered among the scores of covalent modification processes that take place in living cells are oxidation, hydroxylation, methylation, and phosphorylation (Graves et al 1994). While some mechanisms, such as protein phosphorylation-dephosphorylation, are featured as regulatory modalities more prominently than others, all are intimately interwoven to form the fabric of an interdependent, interactive network.

2. WHY PHOSPHORYLATION-DEPHOSPHORYLATION?

Protein phosphorylation-dephosphorylation represents one of the preeminent molecular mechanisms for modulating the functional properties of proteins. It takes place in virtually every living cell and has been employed, in one organism or another, for the regulation of virtually every aspect of cellular existence. What are the factors responsible for its popularity?

2.1. Physicochemical impact

Protein phosphorylation offers a lot of "bang for your buck" (Westheimer 1987). The ability to produce two chemically distinguishable protein forms is only of value if these forms differ in some functional output. Since function follows structure, if you wish to alter the former then you must affect the latter. While the addition of an 80-Da phosphoryl group increases a 40,000-Da polypeptide's molecular mass by only 0.2%, its physicochemical properties render it a potent perturber of local and global protein structure. Phosphoryl groups possess two dissociable protons, the first of which has a pKa < 2, and thus is dissociated under virtually all circumstances. The second proton has a pK_a that is a unit or more below neutrality: ≈ 5.8 in the free phosphoamino acids phosphoserine, phosphothreonine, and phosphotyrosine; 4.8 in phosphoaspartate; and 4.3 in phosphoarginine (Vogel 1989).

Thus, unless the nature of the local microenvironment dictates otherwise, the second proton also will tend to be dissociated at physiological pH, and the phosphoryl group will bear a charge of -2. Studies with phosphoproteins appear to confirm this. The phosphoseryl residues in phosvitin (Sanchez-Ruiz and Martinez-Carrion 1988, Graves and Luo 1994), ovalbumin (Sanchez-Ruiz and Martinez-Carrion 1988), and cardiac troponin I (Jacquet et al 1993) exist predominantly as dianions at pH 7, as does the phosphohistidine in enzyme I of the bacterial phosphoenolpyruvate:sugar phosphotransferase system (Rajagopal et al 1994).

A phosphoryl group's high charge density, coupled with its strong propensity to form multiple hydrogen bonds, can transform a hydrophobic region on a protein into a hydrophilic one, a net positively charged microenvironment into a net negatively charged one, a basic one to an acidic one, etc. Such changes in local charge character, chemistry, and hydrophilicity can lead to dramatic changes in the functional characteristics of the affected protein. In bacteria, for example, isocitrate dehydrogenase is inactivated by the phosphorylation of a threonine residue that resides near the entrance to the active site (Hurley et al 1990). The introduction of a negatively charged phosphoryl group blocks binding of the polyanionic substrate isocitric acid via simple electrostatic repulsion, shutting down catalytic activity (Thorsness and Koshland 1987).

A second example of how local physicochemical effects can alter an enzyme's functional properties is illustrated by 3-hydroxy-3-methylglutaryl-CoA (HMG-CoA) reductase (Omkumar and Rodwell 1994, Omkumar et al 1994). In mammals, this enzyme's catalytic activity is greatly attenuated by phosphorylation of Ser[871]. This seryl residue lies near a catalytically important histidyl residue, His[865]. Normally, this histidine donates a proton to the coenzyme A anion (CoAS-) that is formed when HMG-CoA is reduced by NADPH. Once protonated, the CoASH dissociates from the enzyme. When Ser[871] is phosphorylated, however, the phosphate group forms a hydrogen bond with the catalytic histidine that effectively sequesters the needed proton, which locks the CoAS− into the active site, where it blocks substrate binding.

While a covalent phosphorylation event sometimes exerts its influence on protein functional properties in the highly localized manner described above, usually its perturbing effects ripple throughout complete domains or even the entire molecule to affect sites far distant from the phosphoryl group. The best-studied example of the effectiveness of phosphorylation as a modulator of protein secondary and tertiary structure is glycogen phosphorylase (Johnson and Barford 1993). In its dephosphorylated or b state, glycogen phosphorylase exists as a homodimer sensitive to allosteric regulation by a variety of metabolites, including ATP, AMP, and glucose 6-P. The phosphorylation of Ser[14] by glycogen phosphorylase-b kinase (phosphorylase kinase) to form glycogen phosphorylase-a has several conformational and functional consequences. A network of new hydrogen bonds is formed, linking phosphoserine[14] to Arg[69] from the same polypeptide and Arg[43] from the neighboring one. The amino-terminal tail dramatically shifts position—more than 36 angstroms using Ser[14] as a reference point—and the locations of the aforementioned arginines shift under the pull of their new hydrogen bonding partner. The sweeping conformational changes that result are accompanied by a marked shift in the sensitivity of the enzyme to allosteric regulation. Whereas glycogen phosphorylase-b is activated by 5'-AMP and inhibited by glucose 6-P and ATP, glycogen phosphorylase-a is largely insensitive to these allosteric effectors. (In addition, skeletal muscle, but not liver, glycogen phosphorylase-a dimerizes to form homotetramers.) In this way phosphorylation serves to override the constitutive allosteric mechanisms modulating glycogen metabolism when appropriate hormonal or neuronal stimuli demand.

2.2 Enzymatic "reversibility"

In addition to its physicochemical impact, phosphorylation-dephosphorylation plays a prominent role in cellular signal transduction and information processing because of its ready "reversibility." In order to create molecular bits, it is necessary that the binary elements used be capable of rapidly and controllably interchanging between states. Many covalent modification processes, e.g. partial proteolysis or hydroxylation, are difficult to "undo" under the conditions of temperature, solvent composition, pH, reactant concentration, etc., that exist inside living cells. (Chemical stability may be highly desirable for covalent modifications that serve a purely structural role.) However, the hydrolysis of phosphoester, phosphoramide, and acyl-phosphate bonds, to regenerate the original, unmodified protein, is a thermodynamically favorable process that is readily catalyzed by protein phosphatases.

It is important to notice that the dephosphorylation of phosphoproteins through the action of protein phosphatases does not represent the direct reversal of the protein phosphorylation process (Figure 9-1). The phosphorylation of proteins using a high-energy compound such as ATP as the source of the phosphoryl group is a thermodynamically favorable process. This necessarily renders the reciprocal process energetically unfavorable. Nature has escaped the tyranny of the equilibrium constant by using a distinct, thermodynamically favorable pathway to regenerate the dephosphorylated protein, the hydrolysis of the phosphate-protein bond by water.

The use of different chemical reactions involving separate protein catalysts for the interconversion of (phospho)protein bits constitutes a key source of the extraordinary ability of signal transduction networks to integrate information. Since the protein kinase(s) responsible for phosphorylating a protein and the protein phosphatase(s) that dephosphorylates it

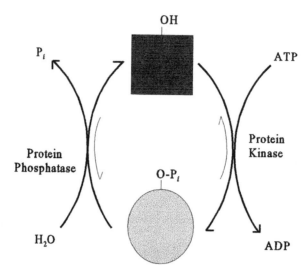

Figure 9-1. The interconversion of proteins between their phosphorylated and dephosphorylated states is carried out by two distinct chemical reactions requiring their own unique enzyme catalysts.

oftentimes respond to different signals that may be synergistic or antagonistic, the phosphoprotein forms a decision node in the cellular microprocessor. Its net state of phosphorylation is determined by the balance between countervailing protein kinase and protein phosphatase activities. The resulting functional output of the phosphoprotein constitutes a decision between two or, if multiple protein kinases and/or protein phosphatases target a particular site, more inputs. In addition, this dynamic competition permits steady states to be achieved in which only a portion of the targeted protein molecules are phosphorylated, creating the potential for a graded, as opposed to a simply all-or-none, response.

3. THE SIGNAL TRANSDUCERS: PROTEIN KINASES AND PROTEIN PHOSPHATASES

3.1. Introduction

How many proteins are targeted for the receipt of (de)phosphorylation-mediated signals? Using radioactive phosphate and two-dimensional gel electrophoresis, 1200 distinct phosphoprotein species have been detected in a human fibroblast cell line (Levenson and Blackshear 1989) and 150 in *E. coli* (Cozzone 1991). In higher eukaryotes, hundreds of protein kinases (Hanks and Hunter 1995, Hunter 1987) and protein phosphatases (Charbonneau and Tonks 1992, Shenolikar and Nairn 1991) are required to interconnect this vast number of phosphoprotein targets with the array of receptors that monitor a broad spectrum of intra- and extracellular signals. It is estimated that up to 5% of a eukaryotic genome encodes protein kinases (Hunter 1995), and a similar proportion likely is devoted to the protein phosphatases—a clear reflection of the demanding nature of the complex task of biological control. Ironically, this vast spectrum of protein kinases and protein phosphatases appears to have been constructed using a mere handful of standard modules as their foundation.

3.2. Definitions and Nomenclature

Protein kinases catalyze the transfer of a phosphoryl group from ATP or other high-energy compound to a nucleophilic acceptor group located on one or more of the amino acid side chains in a protein. Candidate acceptors include the hydroxyl groups of serine, threonine, and tyrosine to form phosphomonoesters; the carboxylate groups of aspartic and glutamic acid to form acyl phosphates; and the nitrogen atoms on the side chains of histidine, lysine, and arginine to form what are oftentimes referred to as phosphoramides. While the hydroxyl groups of hydroxyproline and hydroxylysine and the thiol group of cysteine possess the chemical potential to function as phosphoacceptors, no evidence for their modification *in vivo* via phosphorylation has been encountered to date.

Protein phosphatases catalyze the hydrolysis of protein-phosphate bonds. In general, protein kinases and protein phosphatases are classified on the basis of their (phospho)amino acid specificity, e.g. protein-tyrosine kinases, protein-serine/threonine phosphatases, etc. In the case of the histidine kinases this nomenclature is somewhat misleading as the histidine in question resides on the protein kinase itself, while the enzyme ultimately phosphorylates an aspartyl residue on its exogenous target proteins. For many years it was thought that the protein kinases and protein phosphatases that acted on the aliphatic hydroxyl groups of serine and threonine did not act upon the aromatic (i.e., phenolic) hydroxyl group of tyrosine, and vice-versa. Recently, several exceptions to this pattern have been encountered. These are referred to as dual-specific or dual-specificity protein kinases and protein phosphatases.

There is no standard system for naming protein kinases. The electric methods employed to date reflect their multifaceted nature. Many protein kinases are named for the molecular mechanism by which they are controlled, as in the cyclic-AMP-dependent protein kinase (cAPK), the cyclin-dependent protein kinases, or Ca^{2+}/calmodulin-dependent protein kinase II (CaM kinase II); the extracellular signals to which they respond, as in the stress-activated protein kinases (SAP kinases) or extracellular-signal regulated protein kinases (ERKs), also known as the mitogen-activated protein (MAP) kinases; the genes that encode them, as in src, CheA, HOG1, or cdc2; or the reactions that they catalyze, as in phosphorylase kinase, myosin light chain kinase, and the c-Jun N-terminal kinase. A complicating factor behind this last, substrate-based, nomenclature is the tendency of many protein kinases to target multiple proteins in the cell and an even wider spectrum of sometimes physiologically irrelevant ones in the test tube, where it is possible to achieve supraphysiological concentrations of both catalyst and substrate. Thus, although casein kinase I and casein kinase II both avidly phosphorylate casein *in vitro*, we now know that this only takes place under artificial conditions created in the laboratory. A number of protein kinases have been identified that phosphorylate other protein kinases. These are sometimes referred to as "kinase kinases," as in MAP kinase kinase—which itself is the target for a MAP kinase kinase kinase!

Because of the length of many of these names, abbreviations and acronyms are frequently used: MLCK for myosin light chain kinase, PKC for the calcium-activated and phospholipid-dependent protein kinase, JNK for c-Jun N-terminal kinase, etc. Further complicating the nomenclature picture is the propensity of many organisms, particularly higher eukaryotes, to elaborate multiple variants of their protein kinases. The most striking example of this is PKC, for which twelve isozymic forms have been identified thus far. In some cases these isozymes are expressed in a tissue-specific manner and are so designated, e.g. smooth muscle MLCK. More often isozymic forms are designated by Greek letters, as in PKC-α, -β, -γ, -δ, . . .

The protein phosphatase nomenclature is slightly less complicated. A major share of the credit for this belongs to Cohen and coworkers (Ingebritsen and Cohen 1983), who devised a systematic scheme for naming the protein-serine/threonine phosphatases in the early 1980s. Under this system protein-serine/threonine phosphatases were all designated by the abbreviation PP, to which alphanumeric designations were added to differentiate the enzymes on the basis of their functional characteristics. Type 1 protein phosphatases (PP1) preferentially dephosphorylate the β subunit of phosphorylase kinase and are sensitive to the heat-stable inhibitor proteins I-1 and I-2, while the type 2 enzymes preferentially dephosphorylate the α subunit of phosphorylase kinase and are insensitive to these compounds. The type 2 protein phosphatases were further subdivided according to their dependence on divalent metal ions. PP2A is metal-ion-independent, PP2B is Ca^{2+}-dependent, and PP2C is Mg^{2+}-dependent. Recently discovered protein-serine/threonine phosphatases include PP5, PPT, PPX, etc. With the discovery of various isozymic and oligomeric forms of these proteins, subscripts have been added, as in $PP2A_0$.

The protein-tyrosine phosphatases and dual-specific protein phosphatases are often referred to as PTPs and DSPs, respectively. However, while the nomenclature in this area is largely acronym/abbreviation-driven, it is not systematic. While many of the protein-tyrosine phosphatases are designated by names such as PTP-1B, PTP-PEST, or PTP-H1, this is not universal practice, as evidence by names such as VH-1 and VHR. The DSPs use designations ranging from MKP-1, for MAP kinase phosphatase-1, to cdc25, which stands for cell division cycle mutant 25.

3.3. The Major Protein Kinase Superfamilies

Although literally hundreds of different protein kinases exist in nature, the overwhelming majority of those characterized to date segregate into two structurally homologous superfamilies: the histidine kinases and the cAPK-like protein kinases. Not only are these two protein kinase superfamilies distinguishable at the level of amino acid sequence, they feature strikingly different catalytic mechanisms.

3.3.1. HISTIDINE KINASES.   The histidine kinases were first discovered in bacteria as part of a regulatory module called the two-component system (Stock et al 1992, Swanson et al 1994). A "typical" bacterium possesses up to forty such modules, which participate in the regulation of chemotaxis, phototaxis, osmoregulation, etc. Recently, the two-component system has been discovered in a variety of eukaryotic organisms including yeast (Maeda et al 1994), where it participates in osmosensing, and *Arabidopsis thaliana* (Chang et al 1993), where it participates in the ethylene-response cascade. Interestingly, in the former, the two-component module acts as the lead, or upstream, element of a signal transduction cascade featuring downstream cAPK-like protein kinases of the MAP kinase family.

The two components to which the name refers consist of a protein kinase domain approximately 250 amino acids in extent and a response-regulator domain about 120 amino acids long. These domains are found in a number of different molecular arrangements. In some cases the histidine kinase is autonomous, and complexes with the receptor molecule responsible for sensing the signal of interest. In many cases the protein kinase module is directly incorporated into the polypeptide encoding the signal-sensing domain, forming a receptor kinase. Similarly, the response regulator containing the aspartyl residue that constitutes the ultimate target of the histidine kinase has been found both as a separate, autonomous protein and fused with the histidine kinase domain in a contiguous polypeptide unit.

Histidine kinases carry out phosphorylation of response regulator molecules via a two-step process. The catalytically competent form of the enzyme is a dimer. When activated, each of the two protein kinase domains phosphorylates a conserved histidyl residue on its partner. This event is called transphosphorylation or intermolecular autophosphorylation. In the second step, the phosphoryl group on the histidine is transferred to the aspartyl residue on the response regulator.

3.3.2. cAPK-LIKE PROTEIN KINASES. For many years this group of enzymes was simply referred to as the "eukaryotic" protein kinases, since it was believed that the protein kinases utilizing this molecular architecture were only found in the members of this domain. Today we know that these protein kinases are found in many bacteria and, perhaps, in the archaea as well (Kennelly and Potts 1996). In this chapter we will refer to them as the cAPK-like protein kinases, because the catalytic subunit of cAPK forms the basic template for the family (Taylor et al 1993).

The cAPK-like protein kinases are characterized by a 250–300 amino acid catalytic domain recognizable by the presence, order, and spacing of 12 conserved "subdomains" (Hanks and Hunter, 1995). These subdomains range from short stretches of sequence such as Gly-Xaa-Gly-Xaa-Xaa-Gly (subdomain I) to the single conserved glutamate of subdomain III. X-ray structures have been solved for several family members (Knighton et al 1991). They reveal a catalytic domain consisting of an N-terminal nucleotide binding lobe hinged to a C-terminal protein binding lobe. The cAPK-like protein kinases catalyze the direct, one-step transfer of the γ-phosphoryl group of a nucleotide triphosphate to a substrate protein in a ternary protein kinase-substrate protein-NTP complex. It should be noted that most cAPK-like protein kinases catalyze their inter- or intramolecular autophosphorylation. However, this represents a self-modification process. It is not obligatory for the phosphorylation of exogenous proteins, as is the case with the phosphohistidine formed as a transient catalytic intermediate in the histidine kinases and many other phosphotransfer enzymes.

3.3.3. OTHER PROTEIN KINASES. A few protein kinases have been discovered that do not fit neatly into the two superfamilies above. The isocitrate dehydrogenase kinase/phosphatase from bacteria is a protein-serine/threonine kinase and phosphatase responsible for the phosphorylation and inactivation of isocitrate dehydrogenase. This protein kinase is unique in amino acid sequence among the known protein kinases (Klumpp et al 1988), and is further distinguished by the presence within the same polypeptide of the protein phosphatase that counterbalances its actions. Pyruvate dehydrogenase kinase and the branched chain α-ketoacid dehydrogenase kinase form a second set of protein kinases that bear weak, but recognizable, homology to the histidine kinases of the two-component system (Harris et al 1995). Unlike "conventional" histidine kinases, these enzymes function as monomers and phosphorylate serine and threonine residues on their protein substrates. No phosphohistidine has been detected in these enzymes; however, their autophosphorylation on seryl residues has been reported (Davie et al 1995).

Consequently some classify them as a separate, "mitochondrial" protein kinase family (Harris et al 1995). Finally, the phosphorylation of proteins, particularly chromatin proteins such as histones, on histidyl, lysyl, and arginyl
residues, has been reported sporadically in the literature for many years.
However, very little is known regarding the protein kinases responsible for
these events.

3.4. The Major Protein Phosphatase Superfamilies

As is the case with the protein kinases, more than one hundred known or
potential protein phosphatases have been described in molecular detail. The
vast majority fall into three structurally distinguishable superfamilies, the
PP1/2A/2B superfamily, the PP2C superfamily, and the HAT phosphatases.
As was the case with the protein kinases, these superfamilies display differences not only in structure but also in catalytic mechanism.

3.4.1. THE PP1/2A/2B SUPERFAMILY. The protein serine/threonine phosphatases were first separated into two major types on the basis of their
substrate specificity *in vitro* and their sensitivity to inhibitor proteins (Ingebritsen and Cohen 1983). However, with the acquisition of a library of amino
acid/gene sequence data, it was discovered that the primary division in this
scheme did not reflect the structural/evolutionary relationships between
these enzymes (Barton et al 1994). PP1 was found to be part of a larger
genetic superfamily that includes two type 2 protein phosphatases, PP2A
and PP2B. The other type 2 protein phosphatase, PP2C, constitutes a completely distinct class.

The members of the PP1/2A/2B superfamily share a common catalytic
core domain of roughly 300 amino acids (Barton et al 1994). In recent years,
the membership of this family has been expanded to include new forms such
as PP5 and PPT, as well as representatives from bacteriophages (Cohen and
Cohen 1989) and bacterial (Missiakas and Raina 1997) and archaeal (Leng
et al 1995) organisms. This catalytic core domain is highly conserved through
phylogeny, and its N-terminal half exhibits significant sequence homology
with diadenosine tetraphosphatase from *E. coli* (Koonin 1993). These observations suggest that the members of this enzyme superfamily employ a
general mechanism for catalysis that is shared by phosphohydrolases of
many types.

In eukaryotes, PP1 and PP2A represent the most quantitatively significant source of protein-serine/threonine phosphatase activity (Cohen 1991).
Their catalytic core domain is contained in a catalytic subunit of approximately 35 kDa that forms part of a hetero-oligomeric complex. The auxiliary
subunits in these complexes are thought to provide for the regulation of
activity and to impart the ability to target or discriminate substrates. The
catalytic domain of PP1, for example, binds both to a regulatory subunit

called inhibitor-2 (I-2), which serves to control its catalytic efficiency, and to a variety of targeting subunits that serve to recruit it to specific locations within the cell (Hubbard and Cohen 1993). The catalytic subunit of PP2A complexes with two types of subunits called A and B. The first acts as a structural scaffold that remains complexed to the catalytic subunit at all times. The B subunits come in a wide variety of sizes and sequences. Consequently, PP2A is not a single enzyme but rather a family of 20 or more protein phosphatases each characterized by its individual B subunit, which confers upon it its own unique substrate specificity and/or subcellular localization (Sontag et al 1995, Zhao et al 1997). The catalytic core domain of Ca^{2+}-activated protein phosphatase PP2B, also known as calcineurin, is part of a 61-kDa polypeptide—the A subunit—that is stably complexed with a B subunit almost identical in sequence with the Ca^{2+}-binding protein calmodulin (CaM).

The members of the PP1/2A/2B superfamily are metalloenzymes that bind a pair of divalent metal ions (Chu et al 1996). It is postulated that one or more of the metals may serve to activate a water molecule for direct attack on the phosphoprotein substrate—catalysis proceeds via a single step. X-ray crystallography reveals that these metals reside together near the intersection of three grooves in the enzyme's surface (Goldberg et al 1995). It is postulated that these grooves can be used alone or in combination to bind phosphoproteins, imparting to the catalytic subunits of PP1 and PP2A the potential to dephosphorylate a wide range of potential substrates.

3.4.2. PP2C.   PP2C was first identified as a Mg^{2+}-dependent protein phosphatase activity in eukaryotes. Although it is suspected that the divalent metal participates in catalysis, and that dephosphorylation takes place via a one-step mechanism involving direct attack by an activated water molecule, data confirming this remains lacking. The mammalian enzyme is a monomer of 42–44 kDa whose sequence bears no discernible resemblance to other protein phosphatases (Tamura et al 1989). Recently, a bacterial form of PP2C, the SpoIIE gene product, has been identified in *Bacillus subtilis*, where it functions in the control of sporulation (Duncan et al 1995).

3.4.3. HAT PHOSPHATASES.   The third, and potentially most numerically prolific (Charbonneau and Tonks 1992), major family of protein phosphatases is those that dephosphorylate tyrosine residues—the HAT phosphatases. HAT stands for His-Arg-Thiolate, an acronym derived from the active site signature sequence for these enzymes: His-Cys-Xaa$_5$-Arg. Unlike the protein-serine/threonine phosphatases, the HAT phosphatases employ a two-step catalytic mechanism (Guan and Dixon 1991). In the first step, the thiol group of the active site cysteinyl residue, which has been deprotonated by the adjacent histidine, carries out a nucleophilic attack on the protein-bound phosphoryl group. This results in the transfer of the phosphoryl group from the substrate protein to the enzyme. The dephosphorylated protein dissociates from the enzyme, to be replaced by a water molecule that in the second step hydrolyses the cysteinyl phosphate group to generate inorganic phosphate and the free enzyme.

HAT phosphatases come in two forms, those that specifically act on tyrosyl residues, the PTPs, and those that will dephosphorylate both seryl/threonyl and tyrosyl residues, the DSPs. X-ray crystallographic analysis suggests that the major determinant as to whether a HAT phosphatase exclusively dephosphorylates tyrosyl residues is, to a large degree, the depth of the catalytic cleft (Jiang et al 1995). The side chain of tyrosine is much longer than that of serine or threonine, so in "deep cleft" enzymes only tyrosine can project sufficiently far into the active site to be accessible to the catalytic machinery residing therein.

The membership of the HAT family of protein phosphatases is much more numerous than that of the PP1/2A/2B superfamily, suggesting a greater degree of specialization at the amino acid sequence level. Although some recognizable subgroups exist among the HAT phosphatases, such as the MKPs and the VHR-like phosphatases, as a group the members of the family exhibit negligible sequence conservation beyond their active site signature motif.

3.4.4. OTHER PROTEIN PHOSPHATASES.   A second set of potential protein-tyrosine phosphatases is the low molecular weight acidic PTPs. First discovered in mammals (Wo et al 1992), they have recently been found in bacteria as well (Li and Strohl 1996). These enzymes contain an active site signature sequence very similar to that of the HAT phosphatases, Val/Leu-Cys-Xaa$_5$-Arg. Their general catalytic mechanism is thought to be similar to that of the HAT enzymes, an initial transfer of the protein-bound phosphate to the active site cysteine of the phosphatase, followed by hydrolysis of the phosphocysteinyl enzyme to regenerate the free phosphatase. While these enzymes can act as protein-tyrosine phosphatases *in vitro*, no clear consensus has emerged as to whether they do so under physiological circumstances. As was the case with the protein kinase domain of AceK, the isocitrate dehydrogenase kinase/phosphatase, this unusual bifunctional enzyme's protein phosphatase also remains unique (Klumpp et al 1988).

It is unclear as to whether specialized protein phosphatases exist for the dephosphorylation of histidine (and other phosphoramides) or aspartate (and other acyl phosphates) (Wong et al 1993). At least some of the response regulator proteins of the two-component system contain an endogenous phosphatase activity that acts on their aspartyl phosphate groups, and some evidence indicates that binding of other proteins to the response regulator can modulate this activity (Swanson et al 1994). However, we have only just

begun to explore the molecular mechanisms responsible for the dephosphorylation of phosphoramides and acylphosphates on proteins.

4. ACTIVATING AND DIRECTING THE SIGNAL: REGULATION AND SUBSTRATE SPECIFICITY OF PROTEIN KINASES AND PROTEIN PHOSPHATASES

4.1. Introduction

Because they are the intermediaries responsible for communication between the receptors that sense extra- and intracellular signals and the proteins that carry out cellular responses, the when and where of protein kinase
and protein phosphatase action must be carefully regulated. Mechanisms
must exist (1) to activate and inactivate the catalytic potential of these
enzymes upon the receipt of appropriate signals by cellular receptors and
(2) to select protein targets for their catalytic actions. While most descrip-
tions of signal transduction cascades confine themselves to describing an
apparently linear sequence of events leading directly from receptor to protein kinase or protein phosphatase to the primary target protein, in reality
most signaling pathways branch. Some of these branches target protein
kinases and protein phosphatases in "other" signaling pathways or cascades.
Their dual nature as both the receivers and transducers of signals is a key
mechanism for creating the complex, interactive molecular circuitry of the
cellular microprocessor.

The need for postreceptor signal branching, both as a mechanism for
cross-talk with other signaling pathways and as a means to create comprehensive cellular responses, imposes a set of paradoxical requirements for
protein kinases and protein phosphatases. On the one hand they must be
exquisitely specific, able to target not just a specific protein from among the
thousand(s) present in the cell, but a specific amino acid from among the
many chemically equivalent serines, threonines, tyrosines, and aspartates
that dot their surfaces. On the other hand, they oftentimes must be capable
of recognizing two or more proteins whose size, shape, charge character, and
catalytic functions differ radically. As will be seen below, nature employs
multiple mechanisms, both direct and indirect, to control the where and
when of protein-kinase- and protein-phosphatase-mediated signal transduction.

4.2. Regulation by Catalytic Activation

One way to turn a particular signal transduction cascade "on" or "off" is to
activate or inactivate the relevant protein kinase or protein phosphatase.
This is accomplished by the same means as for any other enzyme, via
covalent modification, binding of allosteric and other regulatory ligands, etc.

4.2.1. DIRECT ACTIVATION BY THE SIGNAL LIGAND, RECEPTOR KINASES, AND RECEPTOR PHOSPHATASES.   One of the simplest means for controlling protein kinase or protein phosphatase activity is to incorporate their functions directly into the signal receptor molecule itself, to create receptor kinases and receptor phosphatases. The signal molecule acts as an allosteric regulator, turning on or, sometimes, off the receptor kinase/phosphatase's activity for as long as the ligand remains bound.

Receptor kinases exist that sense both intra- and extracellular inputs. An example of the former is the AMP-activated protein kinase, AMP-PK (Hardie 1992). In the cell, acetyl-CoA forms an important intermediate in the conversion of food or catabolically recycled biomaterials into energy. This energy can be made immediately available as ATP or placed in long-term storage by incorporation into fatty acids and sterols. The AMP-PK helps control the distribution of acetyl-CoA between the generation of immediately available cellular energy and the biosynthesis of lipid storage forms. As its name implies, AMP-PK is activated through the binding of the allosteric ligand 5'-AMP. When activated, the enzyme phosphorylates and attenuates the activity of the rate-controlling enzymes for the biosynthesis of fatty acids and sterols: acetyl-CoA carboxylase and HMG-CoA reductase, respectively (Figure 9-2). When ATP levels drop, such as when nutrient carbohydrate becomes depleted, AMP-PK is activated. Through its ability to target multiple substrates, it shuts down the biosynthesis of energy-storing

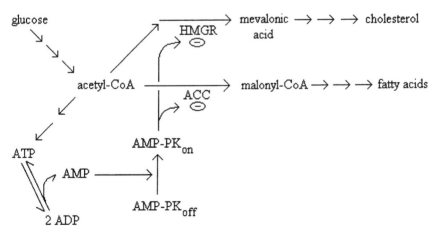

Figure 9-2. Regulation of the incorporation of acetyl-CoA to lipids via the AMP-activated protein kinase. Abbreviations used include AMP-PK, AMP-activated protein kinase; HMGR, 3-hydroxy-3-methylglutaryl Coenzyme A reductase; ACC, acetyl-CoA carboxylase. The minus signs indicate the occurrence of a protein phosphorylation event that inhibits the activity of the target phosphoprotein.

lipids in a single comprehensive action, thus increasing the pool of acetyl-CoA available for the generation of ATP.

Why use AMP as the activating ligand? Why not monitor the level of ADP, the immediate hydrolysis product of ATP? The most readily sensed signals are those that undergo dramatic, i.e. manifold, changes in concentration. ADP levels are highly buffered in the cell. Even in a well-nourished cell, the proportion of the total adenine nucleotide pool represented by ADP is fairly high—a consequence of the high turnover of ATP. Moreover, ADP levels change relatively little as ATP levels start to fall, because the enzyme adenylate kinase attempts to meet the ATP shortfall by catalyzing phosphoryl transfer between two molecules of ADP to create one molecule of ATP and another of AMP. However, as a consequence, AMP levels—which are normally very low—quickly rise several-fold to provide a timely and readily detected indicator of impending metabolic distress.

Some hints of the complex, interlocking nature of signal transduction processes also are provided by the AMP-PK. First, inactivation of acetyl-CoA carboxylase by AMP-PK leads to a drop in the level of its reaction product, malonyl-CoA. In addition to its role as an intermediate in fatty acid biosynthesis, malonyl-CoA blocks fatty acid oxidation by inhibiting fatty acid transport into mitochondria. When AMP-PK inactivates acetyl-CoA carboxylase, it releases this malonyl-CoA block, turning on ATP generation via fatty acid oxidation. In addition, the AMP-PK is itself activated by phosphorylation, but only when AMP is bound to it. Phosphorylation both stimulates the activity of AMP-PK markedly above that seen in the presence of AMP alone, and renders its activity largely AMP-independent. In addition, the binding of AMP decreases the susceptibility of AMP-PK to dephosphorylation by protein phosphatases and the signals to which they respond.

Another receptor kinase that senses an intracellular signal is the DNA-PK. It binds to and is activated by double-stranded (ds) ends in DNA and other helix discontinuities. It is postulated that DNA-PK binds to double-strand breaks, where it helps recruit and activate DNA repair proteins (Gottlieb and Jackson 1994). Other intracellular receptor kinases are activated by dsRNA from invading viruses (Samuel 1993). However, most receptor kinases and phosphatases are found as part of the cytoplasmic domains of transmembrane proteins that sense extracellular inputs, such as hormones, cytokines, nutrients, cell-surface antigens, etc. Receptors can also bind to protein kinases and protein phosphatases noncovalently, to form signal-sensitive receptor: transducer complexes.

4.2.2. SECOND MESSENGERS.   A second mechanism for catalytic activation is the employment of intracellular second messengers as activators of protein kinases and protein phosphatases. In this case, the binding of ligand to receptor leads to a catalytic or other event that creates or releases an allosteric effector molecule into the cell. This allosteric effector functions as

an intracellular hormone surrogate, or second messenger. Binding of epinephrine to its extracellular receptor leads to the G-protein-mediated activation of the enzyme adenylate cyclase (Tang and Gilman 1992). This
enzyme synthesizes cAMP using ATP. The major intracellular target, or
receptor, for cAMP is cAPK. In the absence of cAMP, cAPK exists as a
tetramer consisting of two catalytic subunits bound to a regulatory subunit
dimer (Figure 9-3). The regulatory subunit inhibits the enzyme's catalytic
activity and interacts with anchoring proteins that hold the tetramer at

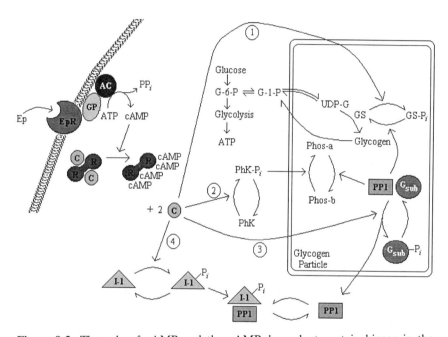

Figure 9-3. The role of cAMP and the cAMP-dependent protein kinase in the
mobilization of glycogen in response to hormones. Binding of epinephrine (Ep) to
its cell surface receptor (EpR) activates the enzyme adenylate cyclase (AC) via a
mechanism involving G-proteins (GP). Adenylate cyclase produces the second messenger cAMP from ATP. Two molecules of this second messenger bind to each of the
regulatory subunits (R) present in the inactive, heterotetrameric form of the cAMP-
dependent protein kinase (cAPK), triggering release of active catalytic subunits (C).
The C subunits target four proteins for phosphorylation: 1. glycogen synthase (GS),
2. phosphorylase kinase (PhK), 3. the glycogen targeting subunit (G_{sub}) of protein
phosphatase-1 (PP-1), and 4. the regulatory subunit of PP1, I-2. Collectively, these
actions trigger increased phosphorylation of glycogen synthase and glycogen phosphorylase (Phos) via a synergistic mechanism that couples the activation of protein
kinase activity (cAPK and PhK) with the concomitant inactivation of the opposing
protein phosphatase activity of PP1.

specific locations in the cell (see below). Upon binding to the regulatory subunits, cAMP triggers the release of the catalytic subunits, which phosphorylate numerous targets in the cytoplasm and nucleus. The actions of cAPK are terminated by the hydrolysis of cAMP to 5'-AMP by any one of a family of phosphodiesterases (Bentley and Beavo 1992). Many of these phosphodiesterases are controlled by other second messengers and/or phosphorylation-dephosphorylation, providing a mechanism for integrating several signal inputs through the manipulation of the occurrence, magnitude, and duration of a cAMP spike.

Other intracellular second messengers include cGMP, nitric oxide, and various lipid hydrolysis products such as inositol 1,4,5 triphosphate, arachidonic acid, or ceramide (Divecha and Irvine 1995, Hannun 1994, Koesling et al 1991). Another second messenger, Ca^{2+}, is released into the cytoplasm from intracellular or extracellular compartments through the action of regulated channels, rather than being synthesized (Clapham 1995). It acts in two ways, either directly through an allosteric Ca^{2+}-binding site on a protein, as is the case for phosphorylase kinase or PP2B, or by binding to an intracellular Ca^{2+} receptor protein such as CaM, whose Ca^{2+}-activated form proceeds to bind to and activate enzymes such as MLCK or CaM kinase II. One might also include in this class the receptor-modulated second messenger proteins, such as the soluble G-protein Ras.

4.2.3. COVALENT MODIFICATION OR PROTEIN KINASE (PHOSPHATASE) CASCADES.   Lastly, the activity of a protein kinase or protein phosphatase can be controlled through a covalent modification event such as phosphorylation or partial proteolysis. In the case of PP1, the regulatory phosphorylation events target its regulatory subunit, I-2, and its inhibitor protein, I-1. Phosphorylation of I-1 by protein kinases such as cAPK is required to unmask its inhibitory potential. This plays an important role in the control of glycogen synthase and glycogen phosphorylase activity (Figure 9-3). When epinephrine stimulates glycogen breakdown to provide energy, it does so through the activation of cAPK by cAMP. The requisite shifts in the phosphorylation states of glycogen synthase and glycogen phosphorylase are accomplished through the concomitant activation of the protein kinases that phosphorylate these enzymes, cAPK and phosphorylase kinase, and inactivation of the principal countervailing phosphatase, PP1. The latter is accomplished, in part, through the phosphorylation of I-1 by cAPK.

The MAP kinase cascade is responsible for modulation of the mitogenic effects of peptide hormones such as EGF and PDGF through stimulation of gene transcription events. A significant portion of this signal transduction cascade consists of a series protein phosphorylation events in which a protein kinase phosphorylates and activates a "downstream" protein kinase (Figure 9-4). The Ras-activated protein kinase Raf phosphorylates and activates MEK (alias MAP kinase kinase), which in turn phosphorylates and

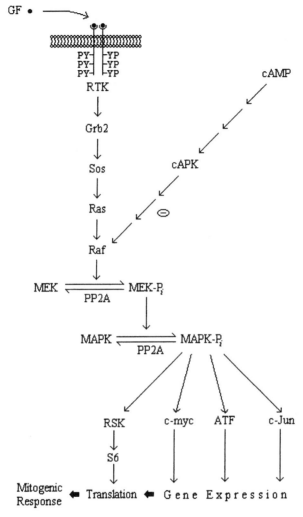

Figure 9-4. A simplified representation of the MAP kinase cascade responsible for mediating the stimulation of the mitogenesis in response to peptide growth hormones. Abbreviations used include GF, growth factor; RTK, receptor tyrosine kinase; MAPK, MAP kinase; MEK, MAPK/ERK kinase (a.k.a. MAP kinase kinase); and RSK, ribosomal protein S6 kinase. It should be noted that several protein phosphatases in addition to PP2A have been implicated in the dephosphorylation of MAP kinase.

activates MAP kinase, which phosphorylates and activates transcription factors such as c-myc, c-Jun, and ATF-2 that form some of the eventual targets of the cascade. MAP kinase also triggers the phosphorylation and activation of the 90-kDA ribosomal protein S6 kinase, RSK, which goes on to phosphorylate other transcription factors and ribosomal protein S6. The latter primes the cell's protein translation machinery to meet the impending wave of gene expression events (Blenis 1993).

Why use a series or cascade of kinase kinases? Why not simply activate the transcription factor kinases directly and cut out Raf, MEK, and the kinase activity of the receptor itself? When the first protein kinase cascade appeared, i.e. the cAPK/phosphorylase kinase cascade for mobilization of glycogen, it was postulated that the individual kinase kinase steps acted as catalytic amplifiers of the signal. Today we know that this multistep mechanism functions primarily as a means for signal branching and integration. These multiple protein kinase steps provide sites for signal feedback/integration, since they can serve as the target for the protein kinases and protein phosphatases that carry the "votes" of other cellular receptors. In this way the important and consequential decision of initiating mitosis or differentiation can be made with all relevant variables taken fully into account. For example, adenylate-cyclase-linked receptors can block the activation of Raf kinase through its phosphorylation by cAPK (Marx 1993), while at least one isoform of PKC appears to activate Raf kinase independently of growth factors (Seger and Krebs 1995). In addition, virtually all of the components of the MAP kinase pathway exist in multiple isozymic forms, adding another layer of complexity and capability. The MAP kinase pathway may in fact consist of parallel, perhaps interlocking, signaling cascades responsive to different variables in subtly different ways, some specific for cell division and others for cell differentiation, for example (Seger and Krebs 1995).

4.3. Recognizing Substrates, the Consensus Sequence Model

Many years ago it was observed that protein kinases such as phosphorylase kinase and cAPK could phosphorylate peptides derived from substrate proteins, and do so on the amino acid corresponding to that modified in the native protein. As the sequences of more and more phosphoproteins were determined, it was observed that the pattern of amino acids immediately surrounding the sites phosphorylated by a particular protein kinase exhibited similar patterns, particularly in the arrangement of acidic or basic amino acids. Experiments with synthetic peptides, in which these amino acids could be systematically substituted, confirmed that for many protein-serine/threonine kinases the presence and spacing of particular amino acids around the phosphoacceptor amino acid residue served as a primary recognition determinant, a sort of molecular zip code or consensus sequence

(Kennelly and Krebs 1991, Pinna and Ruzzene 1996). Under the consensus sequence model, the phosphorylation of disparate proteins by a single protein kinase could be readily explained by the presence of short (typically 4–7 residues), strategically positioned consensus sequence cassettes.

The small size of consensus sequences makes it possible to introduce multiple phosphorylation sites into a single protein, and thus create logic circuits for integrating the signals transmitted by different protein kinases. An "OR" gate can be created when two or more protein kinases phosphorylate the same amino acid residue. This often occurs, because many consensus sequences can be overlapped. PKC, for example, prefers a sequence containing basic residues on both sides of the phosphoacceptor serine or threonine residue, with a hydrophobic amino acid immediately N-terminal to the phosphoacceptor. cAPK recognizes the sequence Arg-Arg-Xaa-Ser/Thr. Thus, both protein kinases could recognize the serine within the sequence Arg-Arg-Ala-Ser-Phe-Lys, and the site would be phosphorylated when either cAPK or PKC was activated.

When the state of phosphorylation of one site affects that of another, "AND" or "NAND" gates can be created. This phenomenon is called hierarchal phosphorylation (Roach 1991). AND gates are formed when a particular protein kinase recognizes a phosphoamino acid as part of a consensus sequence. An example of this is the phosphorylation of glycogen synthase by glycogen synthase kinase-3 (GSK-3). GSK-3 targets the first four in a set of five clustered seryl residues that, when phosphorylated, inhibit catalytic activity. (Phosphorylation of the fifth serine does not affect activity). The sequence of this cluster is Ser-Xaa$_3$-Ser-Xaa$_3$-Ser-Xaa$_3$-Ser-Xaa$_3$-Ser. The consensus sequence for GSK-3 is Ser-Xaa$_3$-Ser(P), where Ser(P) stands for phosphoserine. If the last Ser in this cluster becomes phosphorylated, GSK-3 can go on to sequentially phosphorylate the remaining four serines, since each subsequent phosphorylation creates a new GSK-3 recognition site. The regulatory consequence of this behavior is that even when GSK-3 is activated, it can only phosphorylate and inhibit glycogen synthase activity if the protein kinase that targets the fifth serine, e.g. casein kinase II, also is active (and able to dominate any opposing protein phosphatase activities).

The introduction of a phosphate group can block phosphorylation by another protein kinase as well, either by introducing a strong negative determinant into the consensus sequence for another protein kinase or by causing conformational changes that somehow sequester the site away from its cognate protein kinase. An example of the former is the hormone-sensitive lipase, which contains two nearly adjacent phosphorylation sites, Ser[563] and Ser[565] (Yeaman 1990). In the dephosphorylated protein the first site can be phosphorylated by cAPK and the second by AMP-PK. Only phosphorylation of the first site has any direct functional consequences, in this case stimulation of lipase activity. However, phosphorylation of the two sites by these protein kinases is mutually exclusive. If Ser[565] is phosphorylated, cAPK will no longer recognize Ser[563], and activation of the enzyme is blocked. Thus, a simple but powerful logic gate of the form "cAMP but not AMP" is created (once again ignoring protein-phosphatase-borne signals for the sake of simplicity) within the hormone-sensitive lipase.

While successful in explaining the behavior of many serine/threonine kinases, the consensus sequence model apparently does not apply to either protein-tyrosine kinases or protein phosphatases. For these enzymes other mechanisms appear to hold center stage.

4.4. Protein Kinases as Their Own Substrates—Autophosphorylation

For many protein kinases, and some protein phosphatases as well, one of their principle targets is an (phospho)amino acid within their own structure, a phenomenon called autophosphorylation. In these instances it is often unnecessary for the targeted amino acid to be presented in the context of an optimum consensus sequence, since the physical tethering of a potential phosphoacceptor residue in close proximity to the active site creates a nearly infinite local substrate concentration. Such autophosphorylation events can have important functional consequences. For example, in CaM kinase II, autophosphorylation of a threonine located near the binding site for the activator molecule CaM renders the enzyme active in the absence of CaM (Hanson and Schulman 1992). This allows CaM kinase II to carry out catalysis long after the initial signal for its activation, release of the intracellular second messenger Ca^{2+} into the cytosol, has been attenuated. This, in turn, allows for various aspects of the cellular response to the Ca^{2+} signal to take place over different time frames.

In many receptor-tyrosine kinases, such as the EGF and PDGF receptors, the receptor represents the primary target of its own protein-tyrosine kinase activity. In this case, ligand binding induces receptor dimerization and triggers the transphosphorylation of numerous tyrosines on the receptor's cytoplasmic domains (Heldin 1995). The autophosphorylated tyrosine residues form the recognition sites for other proteins, particularly those containing SH2 domains (Pawson 1994), recruiting them onto the receptor scaffold (Figure 9-5). The result is the assembly of a signaling complex that interacts with other proteins to influence cellular events (Fry et al 1993). In the case of the EGF or PDGF receptor, one of the autophosphorylation sites binds the adapter protein Grb2, which in turn recruits the guanine nucleotide exchange protein SOS, which stands for "son of sevenless" (Seger and Krebs 1995). SOS activates the small GTP binding protein Ras, which activates the protein kinase Raf and a subsequent cascade of protein kinase phosphorylation/activation events (Figure 9-4). No exogenous protein is phosphorylated by the receptor tyrosine kinase during this process.

The high effective concentration of the tyrosyl residues that undergo autophosphorylation is important to this process, since each of the compo-

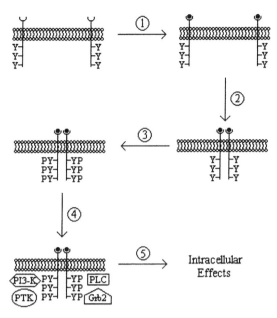

Figure 9-5. Assembly of cytoplasmic signal transduction complexes in response to growth factors. This takes place by a multistep mechanism. First, growth hormone binds to the extracellular domain of individual receptor monomers. Two, hormone-induced receptor dimerization takes place, followed by intermolecular autophosphorylation of several tyrosyl residues by the receptor's endogenous protein-tyrosine kinase. These phosphotyrosyl residues serve as the core of protein-protein recognition sites for SH2- and SH3-domain proteins such as phosphatidyl inositol 3-kinase (PI3-K), cytoplasmic protein tyrosine kinases (PTK), phospholipase C (PLC), and adaptor proteins such as Grb2 that recruit other molecules such as Sos to the complex. The binding and activation of these proteins, in turn, triggers a cascade of intracellular events (see Figure 9-4).

nents recruited to the complex recognizes a different type of SH2 domain (Pawson 1994). It enables the targeted tyrosines to be placed in the variety of sequence contexts needed for component protein binding without worrying about the need to simultaneously incorporate optimal protein kinase recognition features, as would be the case with an exogenous protein.

4.5. Subcellular Targeting, a Means for Controlling Both Where and When

One mechanism for controlling both what is (de)phosphorylated and when is the use of targeting subunits and anchoring proteins to selectively recruit protein kinases and phosphatases to particular microenvironments on demand. Targeting allows the use of catalytic units that are broadly specific, and hence versatile and flexible enough to interact with the multiple proteins required for signal branching, while providing a means of focusing this activity at the time and place where it is needed.

A prime example of the use of targeting is the protein-serine/threonine phosphatase PP1. In the test tube, PP1 displays relatively little ability to discriminate between one phosphoprotein and another. In the cell, however, very little PP1 is found floating freely in the cytosol where its promiscuous nature might lead to the inappropriate dephosphorylation of a phosphoprotein. Rather, the enzyme is found associated with glycogen particles, the endoplasmic reticulum, myofibrils, nuclei, etc., held via a series of targeting subunits (Hubbard and Cohen 1993). The binding of PP1 to these subunits is subject to dynamic control, generally by phosphorylation, allowing the pool of PP1 to be recruited to where it is needed on demand.

A large set of proteins, called AKAPs for "A Kinase Anchoring Proteins" (the term "A Kinase" refers to PKA, another pseudonym for the cAMP-dependent protein kinase), has been discovered that will anchor cAPK, PKC, and/or PP2B to particular sites in the cell (Dell'Acqua and Scott 1997). Unlike the targeting subunits for PP1, these proteins thus far appear to be limited to binding these enzymes in their inactive state, perhaps holding them ready to be released into a specific microenvironment at the proper moment. Many AKAPs are multivalent—able to bind multiple protein kinases and protein phosphatases—which creates the possibility for quite complex signaling activity. Other potential recognition/targeting domains found on protein kinases and protein phosphatases include hydrophobic lipid tails and diacylglycerol binding sites for targeting to cell membranes; protein-protein interaction domains such as SH2, SH3, pleckstrin homology domains; band 4.1 protein-like domains that may target them to membrane-cytoskeleton interfaces; nuclear localization signals; etc. (Barford et al 1995).

5. WHEN PROTEIN (DE)PHOSPHORYLATION GOES AWRY

Given that protein phosphorylation participates in virtually every aspect of the life of a human cell, from proliferation and differentiation to metabolism and cell-cell communication, it is not surprising that dysfunctions in protein phosphorylation-dephosphorylation processes can have serious consequences. Many poisons and pathogens target this portion of the cell's command and control network. Okadaic acid, the microbial toxin that inhibits PP1 and PP2A (MacKintosh and MacKintosh 1994), is the cause of diuretic shellfish poisoning, a sometimes fatal disease. In low doses it acts as a tumor promoter.

Other pathogens introduce their own protein kinases and protein phosphatases into the cell to disrupt normal signal transduction. A secreted protein-tyrosine phosphatase constitutes an essential virulence determinant for *Yersinia pestis*, the cause of the "black death" in medieval Europe (Guan and Dixon 1990). Many viral oncogenes are protein-tyrosine kinases that have been "borrowed" from the eukaryotic host and modified to escape normal cellular regulatory mechanisms and induce the inappropriate phosphorylation of proteins on tyrosine (Bishop 1987). A particularly clever strategy is employed by the SV40 virus. Instead of elaborating its own protein phosphatase, it infiltrates and co-opts one of the cell's own protein-serine/threonine phosphatases for its own use, PP2A. One of the immediate early gene products encoded by the virus, the small T antigen, can replace the regulatory B subunits of PP2A to create a new set of protein phosphatases customized for supporting viral infection (Sontag et al 1993). One of the visible manifestations of this is the stimulation of cell proliferation, a consequence of the stimulation of the MAP kinase pathway through the disturbance of the normal balance of the protein kinase and protein phosphatase activities that keep MEK and MAP kinase in a quiescent, dephosphorylated state (Figure 9-4).

Current interest in signal transduction phenomena has been fueled to a large degree by the desire to understand and intervene in the molecular mechanisms underlying diseases such as cancer, Alzheimer's, cystic fibrosis, diabetes, etc., mechanisms in which protein phosphorylation-dephosphorylation plays a prominent role. Much is known, but much remains to be discovered. For the last 40 years scientists have been uncovering the individual pieces of the cell's microprocessor and examining how these parts function individually. The challenge of the next decade is to uncover the architecture of the assembled network and use this knowledge to eliminate the "bugs" introduced by the toxins and pathogens that threaten human health, to subtly reprogram plants and microbes to increase their productivity in agricultural and biotechnological applications, etc.

REFERENCES

Barford D, Jia Z, and Tonks NK. 1995. Protein tyrosine phosphatases take off. Nature Structural Biol 2:1043–1053.

Barton GJ, Cohen PTW, and Barford D. 1994. Conservation analysis and structure prediction of the protein serine/threonine phosphatases. Sequence similarity with diadenosine tetraphosphatase from *Escherichia coli* suggests homology to the protein phosphatases. Eur J Biochem 220:225–237.

Bentley JK and Beavo JA. 1992. Regulation and function of cyclic nucleotides. Curr Opin Cell Biol 4:233–240.

Bishop JM. 1987. The molecular genetics of cancer. Science 235:305–311.

Blenis J. 1993. Signal transduction via the MAP kinases: Proceed at your own RSK. Proc Natl Acad Sci USA 90:5889–5892.

Bray D. 1995. Protein molecules as computational elements in living cells. Nature 376:307–312.

Chang C, Kwok SF, Bleecker AB, and Meyerowitz EM. 1993. *Arabidopsis* ethylene-response gene ETR-1: similarity of product to two-component regulators. Science 262:539–544.

Charbonneau H and Tonks NK. 1992. 1002 protein tyrosine phosphatases? Annu Rev Cell Biol 8:463–493.

Chu Y, Lee EYC, and Schlender KK. 1996. Activation of protein phosphatase 1. Formation of a metalloenzyme. J Biol Chem 271:2574–2577.

Clapham DE. 1995. Calcium signaling. Cell 80:259–268.

Cohen P. 1991. Classification of protein-serine/threonine phosphatases: Identification and quantitation in cell extracts. Meth Enzymol 201:389–398.

Cohen PTW and Cohen P. 1989. Discovery of a protein phosphatase activity encoded in the genome of bacteriophage λ. Probable identity with open reading frame 221. Biochem J 260:931–934.

Cozzone AJ. 1991. Analyzing protein phosphorylation in prokaryotes. Meth Enzymol 200:214–227.

Davie JR, Wynn RM, Meng M, Huang YS, Aslund G, Chuang DT, and Lau KS. 1995. Expression and characterization of branched-chain α-ketoacid dehydrogenase kinase from the rat. Is it a histidine-protein kinase? J Biol Chem 270:19861–19867.

Dell' Acqua ML, and Scott JD. 1997. Protein Kinase A anchoring. J. Biol. Chem., 272:12881–12884.

Divecha N and Irvine RF. 1995. Phospholipid signaling. Cell 80:269–278.

Duncan L, Alper S, Arigoni F, Losick R, and Stragier P. 1995. Activation of cell-specific transcription by a serine phosphatase at the site of asymmetric division. Science 270:641–644.

Fry MJ, Panayotou G, Booker GW, and Waterfield MD. 1993. New insights into protein-tyrosine kinase receptor signaling complexes. Protein Science 2:1785–1797.

Goldberg J, Huang HB, Kwon YG, Greengard P, Nairn AC, and Kuriyan J. 1995. Three-dimensional structure of the catalytic subunit of protein serine/threonine phosphatase-1. Nature 376:745–753.

Gottlieb TM and Jackson SP. 1994. Protein kinases and DNA repair. Trends Biochem Sci 19:500–504.

Graves DJ and Luo S. 1994. Use of photoacoustic fourier-transform infrared spectroscopy to study phosphates in proteins. Biochem Biophys Res Commun 205:618–624.

Graves DJ, Martin BL, and Wang JH. 1994. Co- and post-translational modification of proteins: Chemical principles and biological effects. New York: Oxford University Press. 348.

Guan K and Dixon JE. 1990. Protein tyrosine phosphatase activity of an essential virulence determinant in *Yersinia*. Science 249:553–556.

Guan K and Dixon JE. 1991. Evidence for protein-tyrosine-phosphatase catalysis proceeding via a cysteine phosphate intermediate. J Biol Chem 266:17026–17030.

Hanks SK and Hunter T. 1995. The eukaryotic protein kinase superfamily: Kinase (catalytic) domain structure and classification. FASEB J 9:576–596.

Hannun YA. 1994. The sphingomyelin cycle and the second messenger function of ceramide. J Biol Chem 269:3125–3128.

Hanson PI and Schulman H. 1992. Neuronal Ca^{2+}/calmodulin-dependent protein kinases. Annu Rev Biochem 61:559–601.

Hardie DG. 1992. Regulation of fatty acid and cholesterol metabolism by the AMP-activated protein kinase. Biochim Biophys Acta 1123:231–238.

Harris RA, Popov KM, Zhao Y, Kedishvili NY, Shimomura Y, and Crabb DW. 1995. A new family of protein kinases—The mitochondrial protein kinases. Advan Enzyme Regul 35:147–162.

Heldin CH. 1995. Dimerization of cell surface receptors in signal transduction. Cell 80:213–223.

Hubbard MJ and Cohen P. 1993. On target with a new mechanism for the regulation of protein phosphorylation. Trends Biochem Sci 18:172–177.

Hunter T. 1987. One thousand and one protein kinases. Cell 50:823–829.

Hunter T. 1995. Protein kinases and phosphatases: The yin and yang of protein phosphorylation and signaling. Cell 80:225–236.

Hurley JH, Dean AM, Thorsness PE, Koshland DE Jr, and Stroud RM. 1990. Regulation of isocitrate dehydrogenase by phosphorylation involves no long-range conformational change in the free enzyme. J Biol Chem 265:3599–3602.

Ingebritsen TS and Cohen P. 1983. The protein phosphatases involved in cellular regulation 1. Classification and substrate specificities. Eur J Biochem 132:255–261.

Jaquet K, Korte K, Schnackerz K, Vyska K, and Heilmeyer LMG Jr. 1993. Characterization of the cardiac troponin I phosphorylation domain by ^{31}P nuclear magnetic resonance spectroscopy. Biochemistry 32:13873–13878.

Jiang Z, Barford D, Flint AJ, and Tonks NK. 1995. Structural basis for phosphotyrosine peptide recognition by protein tyrosine phosphatase 1B. Science 268:1754–1758.

Johnson LN and Barford D. 1993. The effects of phosphorylation on the structure and function of proteins. Annu Rev Biomol Struct 22:199–232.

Kennelly PJ and Krebs EG. 1991. Consensus sequences as substrate specificity determinants for protein kinases and protein phosphatases. J Biol Chem 266:15555–15558.

Kennelly PJ and Potts M. 1996. Fancy meeting you here! A fresh look at 'prokaryotic' protein phosphorylation. J Bacteriol 178:4759–4764.

Koesling D, Bohme E, and Schultz G. 1991. Guanylyl cyclases, a growing family of signal-transducing enzymes. FASEB J 5:2785–2791.

Koonin EV. 1993. Bacterial and bacteriophage protein phosphatases. Mol Microbiol 8:785–786.

Klumpp DJ, Plank DW, Bowdin LJ, Stueland CS, Chung T, and LaPorte DC. 1988. Nucleotide sequence of *aceK*, the gene encoding isocitrate dehydrogenase/phosphatase. J Bacteriol 170:2763–2769.

Knighton DR, Zheng J, Ten Eyck LF, Ashford VA, Xuong NH, Taylor SS, and Sowadski JM. 1991. Crystal structure of the catalytic subunit of cyclic adenosine monophosphate-dependent protein kinase. Science 253:407–414.

Leng J, Cameron AJ, Buckel S, and Kennelly PJ. 1995. Isolation and cloning of a protein-serine/threonine phosphatase from an archaeon. J Bacteriol 177:6510–6517.

Levenson RM and Blackshear PJ. 1989. Insulin-stimulated protein tyrosine phosphorylation in intact cells evaluated by giant two-dimensional gel electrophoresis. J Biol Chem 264:19984–19993.

Li Y and Strohl WR. 1996. Cloning, purification, and properties of a phosphotyrosine protein phosphatase from *Streptomyces coelicolor* A3(2). J Bacteriol 178:136–142.

MacKintosh C and MacKintosh RW. 1994. Inhibitors of protein kinases and phosphatases. Trends Biochem Sci 19:444–448.

Maeda T, Wurgler-Murphy SM, and Saito H. 1994. A two-component system that regulates an osmosensing MAP kinase cascade in yeast. Nature 369:242–245.

Marks F. 1996. Protein phosphorylation. Chapter 1, The brain of the cell. New York: VCH Publishers. 1–35.

Marx J. 1993. Two major signal pathways linked. Science 262:988–999.

Missiakas D and Raina S. 1997. Signal transduction pathways in response to protein misfolding in the extracytoplasmic compartments of *E. coli:* Role of the two new phosphoprotein phosphatases PrpA and PrpB. EMBO J 16:1670–1685.

Omkumar RV, Darnay BG, and Rodwell VW. 1994. Modulation of Syrian hamster 3-hydroxy-3- methylglutaryl-CoA reductase activity by phosphorylation. Role of ser_{871}. J Biol Chem 269:6810–6814.

Omkumar RV and Rodwell VW. 1994. Phosphorylation of ser^{871} impairs the function of his^{865} of Syrian hamster 3-hydroxy-3-methylglutaryl-CoA reductase. J Biol Chem 269:16862–16866.

Pawson T. 1994. SH2 and SH3 domains in signal transduction. Adv Cancer Res 64:87–110.

Pinna LA and Ruzzene M. 1996. How do protein kinases recognize their substrates? Biochim Biophys Acta 1314:191–225.

Rajagopal P, Waygood EB, and Klevit RE. 1994. Structural consequences of histidine phosphorylation: NMR characterization of the phosphohistidine form of histidine-containing protein from *Bacillus subtilis* and *Escherichia coli.* Biochemistry 33:15271–15282.

Roach PJ. 1991. Multisite and hierarchal protein phosphorylation. J Biol Chem 266:14139–14142.

Samuel CE. 1993. The eIF-2alpha protein kinases, regulators of translation in eukaryotes from yeasts to humans. J Biol Chem 268:7603–7606.

Sanchez-Ruiz JM and Martinez-Carrion M. 1988. A fourier-transform infrared spectroscopic study of the phosphoserine residue in hen egg phosvitin and ovalbumin. Biochemistry 27:3338–3342.

Seger R and Krebs EG. 1995. The MAPK signaling cascade. FASEB J 9:727–735.

Shenolikar S and Nairn AC. 1991. Protein phosphatases: Recent progress. Adv Second Messenger Phosphoprotein Res 23:1–121.

Sontag E, Federov S, Kamibayashi C, Robbins D, Cobb M, and Mumby M. 1993. The interaction of SV40 small tumor antigen with protein phosphatase 2A stimulates the MAP kinase pathway and induces cell proliferation. Cell 75:887–897.

Sontag E, Nunbhakdi-Craig V, Bloom GS, and Mumby MC. 1995. A novel pool of protein phosphatase 2A is associated with microtubules and is regulated during the cell cycle. J Cell Biol 128:1131–1144.

Stock JB, Surette MG, McCleary WR, and Stock AM. 1992. Signal transduction in bacterial chemotaxis. J Biol Chem 267:19753–19756.

Swanson RV, Alex LA, and Simon MI. 1994. Histidine and aspartate phosphorylation: Two-component systems and the limits of homology. Trends Biochem Sci 19:485–490.

Tamura S, Lynch KR, Larner J, Fox J, Yasui A, Kikuchi K, Suzuki Y, and Tsuiki S. 1989. Molecular cloning of rate type 2C (IA) protein phosphatase mRNA. Proc Natl Acad Sci USA 86:1796–1800.

Tang WJ and Gilam AG. 1992. Adenylyl cyclases. Cell 70:869–872.

Taylor SS, Knighton DR, Zheng J, Sowadski JM, Gibbs CS, and Zoller MJ. 1993. A template for the protein kinase family. Trends Biochem Sci 18:84–89.

Thorsness PE and Koshland DE Jr. 1987. Inactivation of isocitrate dehydrogenase by phosphorylation is mediated by the negative charge of the phosphate. J Biol Chem 262:10422–10425.

Vogel HJ. 1989. Phosphorus-31 nuclear magnetic resonance of phosphoproteins. Meth Enzymol 177:263–282.

Westheimer FH. 1987. Why nature chose phosphates. Science 235:1173–1178.

Wo YY, Zhou MM, Stevis P, Davis JP, Zhang ZY, and Van Etten RL. 1992. Cloning, expression, and catalytic mechanism of the low molecular weight phosphotyrosyl protein phosphatase from bovine heart. Biochemistry 31:1712–1721.

Wong C, Faiola BF, Wu W, and Kennelly PJ. 1993. Phosphohistidine and phospholysine phosphatase activities in the rat. Potential protein-histidine and protein-lysine phosphatases? Biochem J 296:293–296.

Yeaman SJ. 1990. Hormone sensitive lipase, a multipurpose enzyme in lipid metabolism. Biochim Biophys Acta 1052:128–132.

Zhao Y, Boguslawski G, Zitomer RS, and DePaoli-Roach AA. 1997. *Saccharomyces cerevisiae* homologs of mamalian B and B' subunits of protein phosphatase 2A direct theenzyme to distinct cellular functions. J Biol Chem 272:8256–8262.

Part VI

APPLICATIONS OF SIGNAL TRANSDUCTION IN DISEASE AND DRUG ABUSE

10

Defects in Signal Transduction
Proteins Leading to Disease

N. GAUTAM

Human diseases can result from alterations in proteins involved in signaling pathways. Such proteins include receptors, G-proteins, and effectors. Proteins can be altered by extraneous agents or through mutations in the genes that encode them. Since the components of signaling pathways are large families of proteins, there are a vast number of genes in the human genome that can be the targets for debilitating mutations. There are two broad kinds of mutational alterations that can lead to disease: (1) heritable and (2) somatic. Heritable mutations are present in the germ line and are therefore inherited from one generation to another. Somatic mutations are not heritable, since they occur during the development of an organism and are restricted to a particular set of cells. Examples of such diseases are discussed below.

The nature of a mutation and its final effect are also variable. We now understand in many cases at the molecular level why mutations have different effects—recessive or dominant. These findings show that more than sixty years back, Muller was both prescient and broadly correct in his inferences about the genetic basis of the phenotypic effects of dominant mutations (Muller 1932).

Mutations can affect expression of the gene or the turnover of a signaling protein. Such mutations result in abnormally high or low levels of a protein, which can profoundly affect the physiological state of a cell. Another set of mutations is substitutions of amino acids by other residues possessing similar properties (e.g., a positively charged residue supplanted by another). In these cases, the altered protein may retain normal activity. Many mutations, however, lead to an inactive signaling protein, for various

Introduction to Cellular Signal Transduction
A. Sitaramayya, Editor
©1999 Birkhäuser Boston

reasons. The mutant protein may be folded improperly and incapable of effective interaction with a ligand or another protein in the signaling pathway. Since the human genome is diploid, mutant genes that encode these inactive proteins will be recessive to the normal wild-type allele. In contrast, several types of mutants are dominant. For instance, overexpression of a protein can lead to a signaling pathway functioning overactively. Similarly, underexpression of a protein can lead to a decrease of signaling below the threshold required for the normal phenotype (haplo-insufficiency). An altered protein can also possess a novel function such as constitutive activity or a higher than normal affinity for a protein partner.

Illustrative examples of signaling proteins affected functionally are discussed below. Many diseases are now known to be associated with mutations in signaling proteins. There is no attempt here to provide a comprehensive list of these diseases. This is available elsewhere (e.g., Spiegel 1996). The focus here is on a few altered proteins. These proteins help us understand at the molecular level how subtle alterations in signaling protein function can lead to diseases. Insights into the basis of faulty signal transduction also help us design therapies, as discussed at the end of this chapter.

AN INFECTIOUS AGENT ALTERS THE FUNCTION OF A SIGNALING PROTEIN

Cholera-toxin-mediated ADP ribosylation of the αs subunit

The bacterium *Vibrio cholerae* causes the devastating disease cholera. Cholera is still endemic to many parts of the world. Affected individuals suffer from acute diarrhea. Loss of fluid results from the extrusion of chloride ions and water from cells as a result of continually high cAMP levels. In intestinal cells, adenylate cyclase is normally regulated by a hormone, and the enzyme is inactive in its absence. In the affected cells, adenylate cyclase is constitutively active. We now know that this effect of cholera toxin is mediated by ADP ribosylation of the α subunit of the G-protein G_s (Moss and Vaughan 1988). Cholera toxin is a heteromultimeric ADP-ribosyltransferase. It is capable of transferring the ADP ribose group of NAD^+ specifically to the side chain of arginine. ADP ribosylation of an Arg residue at position 201 of αs is thought to disrupt the GTPase activity of αs (Figure 10-1). The ADP-ribosylated α subunit therefore remains bound to GTP and persistently activates adenylate cyclase in this state (Figure 10-2).

What is the molecular basis for the ADP-ribosylated Arg201 in αs preventing GTP hydrolysis? Recently, the three-dimensional structures of two G-protein α subunit types, αi1 and αt, were obtained in their ligand-bound forms—$\alpha \cdot GDP$, $\alpha \cdot GTP$, and $\alpha \cdot GDP/AlF^{4-}$ (Hamm and Gilchrist 1996). The last is equivalent to the transition state, the transitory state that the α subunit undergoes when it is hydrolyzing GTP to GDP. Comparison of

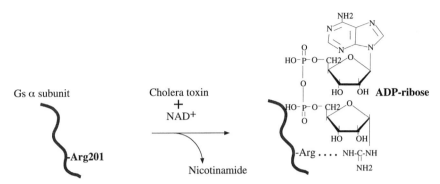

Figure 10-1. ADP ribosylation of the αs subunit is catalyzed by cholera toxin with NAD$^+$ as the substrate. It results in the addition of a large group to the Arg201 side chain that prevents the side chain from effectively interacting with the transition state of the nucleotide.

the active site of αil · GTP-γS with αil · GDP+AlF^{4-} shows that two residues in αil, Arg178 and Gln204, are oriented in strikingly different ways in these two ligand-bound forms. In the GTP-γS-bound form, the side chains of these two residues do not interact with GTP. However, when bound to GDP+AlF^{4-} they clearly interact with AlF^{4-} and GDP by reorienting their side chains (Figure 10-3) (Coleman et al 1994). These two residues are thus the most important in the active site for stabilization of the transition state. Since the G-protein α subunits are folded similarly, it is highly likely that the conserved Arg201 and Gln227 of αs play an identical role (Figure 10-3). Given this role for Arg201, it is reasonable to infer that the presence of the large ADP ribose group prevents the side chain of Arg from interacting appropriately with GTP on its way to being cleaved to make GDP. Thus, altering the structure of a side chain that is critical for a G-protein α subunit's function has a deadly physiological effect.

Considering the critical role of Arg201 and Gln227 in GTP hydrolysis, it is a reasonable surmise that mutations at these positions will also lead to diseases. A disease described below supports this premise.

ALTERATIONS IN SIGNALING PROTEINS THROUGH MUTATIONS IN GENES

Somatic Mutation

Mutations in αs are associated with pituitary tumors (Landis et al 1989).
In response to GH releasing hormone (GHRH), cells in the pituitary gland secrete growth hormone (GH). GHRH activates G$_s$ and, as a result, adeny-

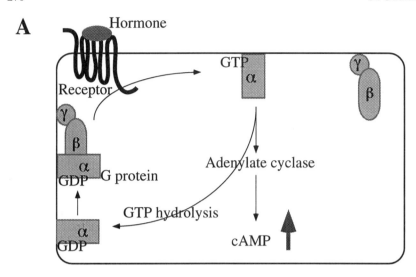

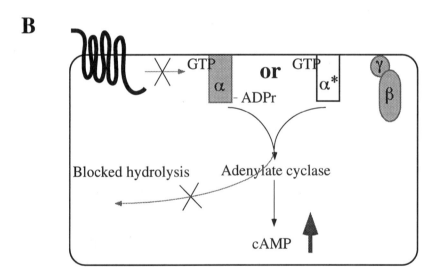

Figure 10-2. Modifications of Arg201 of αs result in constitutive signaling. GTP-bound α subunit stimulates some subtypes of adenylate cyclase, resulting in an increase in cAMP concentration in the cell. αs ADP ribosylated by cholera toxin or mutant αs with Arg201 substituted will activate adenylate cyclase in the absence of receptor activation. These α subunits are incapable of hydrolyzing GTP to GDP and remain constitutively active.

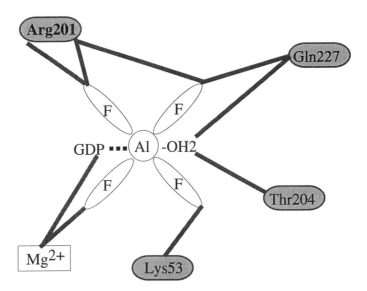

Figure 10-3. Diagram of the hypothetical αs active site bound to GDP and AlF^{4-}, thought to be the equivalent of the transition state during GTP hydrolysis (based on Coleman et al 1995). The crystal structure of the αi1 subunit with GDP and AlF^{4-} shows that the side chain of Arg178 (corresponding to Arg201 of αs) residue forms hydrogen bonds with two of the fluoride atoms. It also interacts with the α and β oxygens of GDP (not shown). The side chain of Gln204 (corresponding to Gln227 of αs) forms hydrogen bonds with one of the fluoride atoms and the water molecule. Neither of these residues binds GTP. They are reoriented during the formation of the transition state. On this basis, it is inferred that these residues stabilize the transition state and promote cleavage of GTP to GDP (Coleman et al 1995). Two other residues are involved in bonding with the substrate at the active site, and are shown. It is assumed here that αs would have a similar structure when bound to the same substrates. The conserved residues in αs corresponding to those in αi1 are shown.

late cyclase. The resulting increase in cAMP mediates the release of GH. When some human tumors originating in the pituitary gland were assayed for adenylate cyclase activity, the cyclase in these cells did not respond to GHRH with increased activity as in normal cells. Moreover, nonhydrolyzable analogs of GTP or AlF^{4-} (which activates GDP-bound α subunit) also failed to potentiate adenylate cyclase activity in these cells. Since the basal level of adenylate cyclase activity was also high, the simplest interpretaton for the unusual properties of adenylate cyclase in these cells was that the enzyme was constitutively active. One possible reason for this constitutive activity was a G_s α subunit that was incapable of breaking GTP to GDP and reaching the inactive state. To examine whether the G_s α subunit was mutated, RNA was prepared from different tumors and first-strand cDNA was synthesized. Since the nucleotide sequence of the αs subunit cDNA was known, primers specific to this sequence were synthesized, and PCR was performed on single-strand cDNA specific to pituitary tumor cells. The nucleotide sequence of the products was determined and compared with the sequence of the wild-type gene. The region of the αs cDNA chosen for examination included the codon for the residue Arg201, since cholera toxin ADP ribosylation of this residue leads to constitutive activity. The region including Gln227 was also amplified because it was suspected, on the basis of its homology to the small GTP-binding oncogenic protein Ras, that Gln227 was involved in GTP binding and hydrolysis.

One set of tumor cells that had constitutive adenylate cyclase activity possessed a mutant cDNA that was altered such that it encoded Cys or His instead of Arg at position 201. Cells from another tumor possessed a mutation of Gln227 to Arg. Both kinds of tumors also possessed cDNA that encoded the native residue (Arg or Gln) indicating (1) that the mutant copy of the αs gene was heterozygous and (2) that function of the mutant protein was dominant over that of the normal protein. When genomic DNA was examined from blood cells, only the wild-type sequences were seen in this region of the αs gene. This showed that the αs mutations were somatic. To examine the biochemical basis for the defects, the cDNA specific to αs was mutated by site-directed mutagenesis to introduce alterations mimicking those seen in tumor cell αs cDNA. The wild-type and mutant proteins were then expressed independently in cells that were deficient for the endogenous αs by subcloning the cDNAs into a retroviral expression vector. These cells possessed the β-adrenergic receptor that is coupled to adenylate cyclase through αs. Membranes from these cells were assayed for adenylate cyclase activity in the presence of an agonist or an antagonist for the β-adrenergic receptor. In membranes from cells that expressed the wild-type protein, strong stimulation of adenylate cyclase activity was noticed in the presence of the agonist. In the presence of the antagonist there was no activity. In marked contrast, membranes containing the mutant αs Arg201His and Gln227Arg showed strong stimulation of adenylate cyclase both in the presence of the agonist *and* the antagonist (Figure 10-2). Moreover, the basal adenylate cyclase activity—activity in the absence of the agonist—was considerably higher than for the wild-type. When the activity elicited by the Arg201His mutant was compared with the properties of wild-type αs treated with cholera toxin, they were strikingly similar. These findings confirmed the role of Arg201 and Gln227 in GTP hydrolysis. The results also pointed to the possibility of mutant G-protein α subunits being oncogenes.

A rigorous test for the ability of constitutively active αs to act as an oncogene is the induction of tumors in a whole animal through targeted expression of the mutant αs. This was tested by expressing the mutant αs (R201H) in transgenic mice under the control of a promoter that restricted the expression of the transgene in thyroid glands of mice (Michiels et al 1994). Thyroid glands from transgenic mice had tumors (adenomas) that were absent in nontransgenic mice. When the adenomas were examined for adenylate cyclase activity, they were found to possess significantly higher basal level of activity. These experiments provide strong evidence that mutations in the G-protein α subunit that lock it in the constitutively active form can cause tumors in certain endocrine tissue.

The intracellular context is clearly important for the mutant αs to display its ability to affect cell growth. As described above, αs that is ADP-ribosylated at Arg201 has a distinctly different effect on human cells than the same subunit type mutated at this position (e.g., Arg201). This is true despite the fact that the altered forms possess similar biochemical properties. The signaling environment in which these two forms function must therefore have a great impact on their final physiological effect. In thyroid cells, the hormone that activates G_s is different from that in intestinal cells. Similarly, in these two different cell types, the effectors regulated by cAMP are distinct. Thus, the final effect of a defective signaling protein on human physiology will depend not on its intrinsic properties but on other signaling activity in that cell.

Heritable Mutations

1. Mutations in the Ca^{2+}-sensing receptor are dominant and cause different diseases in heterozygotes and homozygotes (Pollak et al 1993). The Ca^{2+}-sensing receptor has the structure characteristic of G-protein-coupled receptors. It regulates a G-protein-mediated phospholipase C pathway that results in Ca^{2+} release from internal stores (Figure 10-4). This receptor is found in tissues such as parathyroid, thyroid, and kidney that play a role in the regulation of Ca^{2+} levels. The receptor is thought to respond to increases in Ca^{2+} levels in serum by activating a G-protein, which leads sequentially to the activation of phospholipase C, release of intracellular Ca^{2+}, and finally, inhibition of parathyroid hormone secretion, which, among other things, regulates resorption of Ca^{2+} by kidney.

A series of experiments began with the cloning of the cDNA for this receptor and culminated in the identification of defects in the gene for the receptor in families affected by diseases resulting in misregulation of serum Ca^{2+} levels. These findings make the link between mutations in a signaling protein and inherited diseases very clear.

The cDNA for the Ca^{2+} receptor was isolated by expression cloning—using the ability of *Xenopus* oocytes to express transmembrane proteins when injected with RNA specific to these proteins. Southern blots of

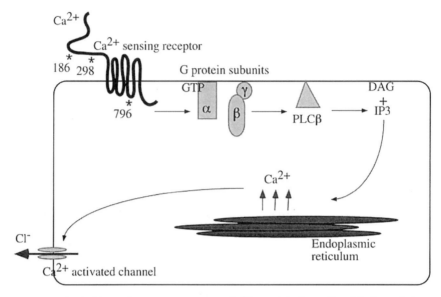

Figure 10-4. Ca^{2+}-sensing receptor senses Ca^{2+} outside the cell and induces intracellular Ca^{2+} release. Asterisks denote the position of mutations that inactive the receptor (see text).

restriction-enzyme-digested DNA from the human genome and a hamster/human hybrid cell containing chromosome 3 alone were probed with this cDNA. The probe hybridized to bands of the same size in both cases, indicating that the gene for the receptor was on chromosome 3. Previous linkage analysis had shown that a disease in which Ca^{2+} regulation was affected, familial hypocalciuric hypercalcemia (FHH), was linked to a gene on chromosome 3. A more extreme form of this disease—neonatal severe hyperparathyroidism (NSH)—was known to be related, because of its coincidence in families that exhibit FHH. Individuals with FHH have moderate elevations of serum Ca^{2+} and low excretion of urinary Ca^{2+}. NSH individuals have serum Ca^{2+} and PTH levels considerably elevated. FHH is relatively benign compared to NSH, which is lethal unless the parathyroid glands are removed.

To screen individuals suffering from these diseases, the gene for the receptor was isolated, using the cDNA as a probe. PCR was used on genomic DNA from the diseased individuals to amplify the exons. RNA probes specific to these exons from normal individuals were used in RNAase protection assays to identify defects in the genes of persons afflicted with FHH or NSH. Defective genes from three different families were examined by nucleotide sequence analysis. Affected individuals from each of these families showed independent alterations in a single amino acid residue of the

Ca^{2+} receptor. Two of these changes (Arg186→Glu, Glu298→Lys) were in the predicted N-terminal extracellular domain of the receptor (Figure 10-4). Since this domain is potentially involved in interacting with Ca^{2+} outside the cell, these mutations could affect the ability of the receptor to sense extracellular Ca^{2+} levels. The other mutation (Arg796→Trp) was in the predicted third cytoplasmic loop of the receptor. This loop has been implicated in the interaction of G-proteins with several receptors. It is therefore possible that the mutant receptor does not effectively activate the G-protein.

Several pieces of evidence pointed to the defects in the Ca^{2+}-sensing receptor as the cause of FHH and NSH: (1) When the defective gene from FHH individuals was examined by restriction enzyme digestion or by determining the nucleotide sequence of portions of the exon, the results indicated the presence of both wild-type and mutant copies of the gene. Therefore, they were heterozygous for the mutant. An NSH individual, however, possessed only the mutant gene, indicating homozygosity for the defect. This correlated well with the severity of the disease phenotype and with the evidence that NSH was prevalent in a family with consanguinous parentage. (2) Normal individuals within these families did not possess the mutant copies of the Ca^{2+}-sensing receptor gene. (3) When the receptor was mutated to mimic the defect in FHH individuals and assayed for activity, it was inactive. The Arg796→Trp mutation was introduced into the Ca^{2+}-sensing receptor cDNA by site-directed mutagenesis, and the RNA specific to this cDNA was expressed in *Xenopus* oocytes. Activity of the receptor was tested by measuring a Cl^- current after exposing the receptor to increasing concentrations of Ca^{2+}. In the case of the wild-type protein, the Cl^- current was activated by the release of intracellular Ca^{2+} (Figure 10-4). The mutant receptors were, however, inactive in this assay.

Why are the mutant receptors dominant if they are inactive? One possibility is that in heterozygous individuals who carry one copy of the gene, the threshold for sensing and responding to extracellular Ca^{2+} has increased because there are fewer active receptors (haplo-insufficiency) (Pollak et al 1993). Homozygous individuals would show the lethal symptoms of NSH because the Ca^{2+} levels that they respond to would be even higher than in the heterozygotes, since all the receptors would be inactive.

2. Constitutively active rhodopsin mutants lead to (1) night blindness and (2) retinal degeneration. Rhodopsin is the G-protein-coupled receptor in rod photoreceptor cells of the mammalian retina. When rhodopsin senses light it activates the G-protein G_t. The activated α subunit of G_t stimulates a cGMP phosphodiesterase. The resulting breakdown in cGMP leads to the closure of cation-conducting channels on the plasma membrane of the rod outer segments (Figure 10-5) (Baylor 1996). The change in membrane permeability to cations leads to hyperpolarization and a significant decrease in

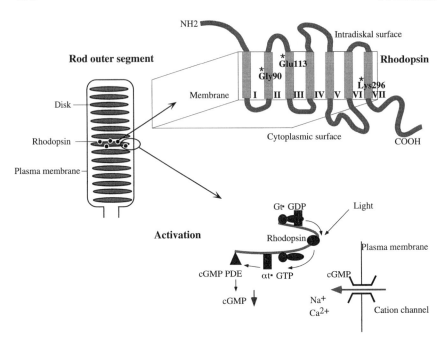

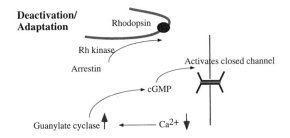

Figure 10-5. Signaling in the visual system. The outer segments of the rod photore-ceptors contain membranous disks. Rhodopsin is inserted in these membranes with the light-sensing face of the molecule inside the disk and the G_t-activating face outside. Asterisks on transmembrane helices denote the position of the mutations that result in diseases (see text and Figure 10-6B). **Activation.** Light stimulation of rhodopsin leads to the activation of the G-protein-G_t and cGMP phosphodiesterase (PDE) and to cGMP breakdown. The nucleotide-gated cation-conducting channel closes at low cGMP concentrations, inducing hyperpolarization and subsequent decrease in neurotransmitter release at the rod photoreceptor synapse. **Deactiva-tion/Adaptation.** Channel closure also lowers Ca^{2+}. Stimulation of guanylate syn-thase activity at low Ca^{2+} levels is one of the steps towards recovery and adaptation to the prevailing level of light intensity. Rhodopsin is switched off by phosphoryla-tion of rhodopsin by rhodopsin kinase, followed by arrestin binding.

neurotransmitter release. Thus, when a rod photoreceptor senses photons, an electrical response is triggered in the neurons downstream of the rod photoreceptors. In the human eye, rods mediate vision in dim light. In bright light, cone photoreceptors, which contain color pigments, mediate color vision. Since we need to be able to sense a signal above a background level of light, the rod photoreceptors contain mechanisms that allow the cell to adapt to a prevailing signal. When the cation-conducting channels close in response to rhodopsin activation, the concentration of calcium in the cell goes down (Figure 10-5). This decrease in calcium concentration in the rods plays an important role in adaptation by increasing guanylate cyclase activity, decreasing light stimulated cGMP phosphodiesterase activity, hastening rhodopsin deactivation, and increasing the affinity of the channel for its ligand, cGMP. Deactivation of rhodopsin is achieved by a specific set of proteins. Rhodopsin kinase phosphorylation and subsequent arrestin binding shut rhodopsin off after it has activated several G_t molecules.

Mutations in rhodopsin have now been identified that cause congenital night blindness or retinal degeneration (Rao and Oprian 1996). In the case of night blindness, it was initially observed that a dominant form was inherited. When affected individuals in a family were screened by amplifying the coding region for opsin from genomic DNA and determining the nucleotide sequence, a mutation that substituted Asp for Gly at position 90 cosegregated with night blindness (Sieving et al 1995). To examine the effect of this mutation on rhodopsin-activated signaling, different approaches—molecular and psychophysical—were used. In one experiment, an identical mutation was introduced into opsin cDNA, the protein was expressed in a cell line, and its coupling to the G-protein, G_t, was examined in the presence or absence of the chromophore, retinal (Rao et al 1994). In the absence of retinal, wild-type opsin did not stimulate binding of GTP-γS (a nonhydrolyzable analog of GTP) by G_t. In contrast, the mutant opsin stimulated GTP-γS binding by G_t. The mutation thus makes opsin constitutively active in the absence of retinal. In the presence of 11-*cis*-retinal, this activity is shut off.

Why does this mutant cause night blindness? Because of the intrinsic kinetics of retinal binding to opsin, a small proportion of rhodopsin is always free of retinal. Rod photoreceptors that possess the mutant opsin (Gly90Asp) will therefore always have a basal level of signaling activity. This will partially desensitize the rod photoreceptors and prevent them from sensing low levels of light. Individuals carrying the Gly90Asp mutant gene are therefore unable to see in dim light (Sieving et al 1995). The ability of the constitutively active mutant to partially desensitize the photoreceptors even in the presence of normal rhodopsin explains why the mutant is dominant in heterozygous individuals.

The nature of the Gly90Asp mutation also provides insight into the chemical basis of its effect on opsin (Rao and Oprian 1996). Inactive rhodopsin contains 11-*cis*-retinal and is stabilized by a salt bridge between the residues Glu113 and Lys296 (Figure 10-6). Light-activated rhodopsin contains all-*trans*-retinal and does not possess the salt bridge. Mutation of either of the residues that form the salt bridge, Glu113 or Lys296, to an uncharged residue disrupts the salt bridge and leads to constitutive activity. There is evidence that the Gly residue at position 90 is close to the Lys296 residue in the three-dimensional structure of rhodopsin (Figure 10-6). The Gly90Asp substitution is thus thought to form a salt bridge with Lys296 and as a result disrupt its bonding with Glu113. Since the salt bridge between Lys296 and Glu113 stabilizes rhodopsin in its inactive state, the surrogate bond between Gly90-Lys296 leads to constitutive activity.

If the Lys296 residue is important for keeping opsin in the inactive state, what happens when it is mutated? Families with individuals suffering from a severe form of retinal degeneration are now known to possess mutations of this residue. As expected, recombinant opsins containing similar mutations are constitutively active in an *in vitro* assay. It was thought that continuous constitutive activity by itself leads to cell death. However, transgenic mice that express the same mutant do not show signs of constitutive activity according to electroretinograms (Li et al 1995). Moreover, individuals with the Gly90Asp mutation, although bereft of normal vision in dim light, do not show signs of retinal degeneration (at least for many decades) (Sieving et al 1995). Why then do rod photoreceptors die when expressing the Lys296Asp mutant? A clue emanates from the transgenic mice expressing this mutant. The mutant opsin in these mice was found to be phosphorylated and tightly bound to arrestin. Arrestin's normal function is to deactivate rhodopsin by preventing dephosphorylation by phosphatase (Figure 10-5). Complexes of mutant opsin with inactivating proteins that clog the rod cells may then trigger off the pathway towards cell death.

Why does the Gly90Asp mutation have a much less severe effect on the retina in comparison to Lys296Glu—night blindness *versus* retinal degeneration? Rao and Oprian (1996) argue convincingly in favor of the following reasons. A striking difference between mutations at Gly90 and Lys296 is that opsin with the first mutation can bind 11-*cis*-retinal and be inactivated, while the second cannot bind this chromophore.

As mentioned before, 11-*cis*-retinal-bound opsin is inactive. Since a small proportion of rhodopsin is releasing retinal at any given time, as a result of the intrinsic kinetics of retinal binding to opsin, the Gly90 mutant is constitutively active only during those periods when it is not bound to 11-*cis*-retinal. It is therefore not deactivated permanently. In marked contrast, the Lys296 mutant behaves differently because its constitutive activity cannot be switched off by 11-*cis*-retinal. This presumably encourages proteins involved in the deactivation pathway to cling to the mutant opsin continuously. The resulting large protein aggregates initiate cell death.

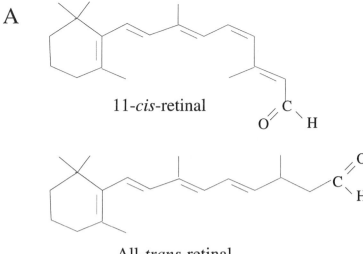

11-*cis*-retinal

All-*trans*-retinal

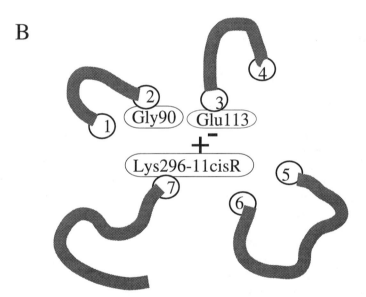

Figure 10-6. The two forms of retinal and interactions between residues that hypothetically stabilize inactive rhodopsin. 11-*cis*-retinal is covalently bound to inactive rhodopsin, forming a protonated Schiff base with the side chain of Lys296. This positive charge forms a salt bridge with the negative charge of Glu113. When light activates rhodopsin, 11-*cis*-retinal is transformed to all-*trans*-retinal, the Schiff base proton is lost, and the salt bridge dissolves. On the basis of the crude crystal structure for rhodopsin, it has been hypothesized that the Gly90 residue will be sufficiently close to Lys296. When Gly90 is substituted with a residue containing a negative side chain, a competing bond can form, disrupting the bond with Glu113. This can lead to constitutive activity. (Based on Rao et al 1994).

THERAPY

Understanding the molecular basis of signaling protein function has helped explain the roots of defective signaling and its damaging impact on human physiology. Ironically, some defective proteins also provide insights into the function of their normal counterparts. Examples are constitutively active G-proteins and receptors, as described above.

The molecular biology of signaling protein function can also help in designing appropriate therapies to control the effects of defective signaling. For instance, identifying the posttranslational processing events that lead to the attachment of a prenyl moiety to Ras has helped identify a drug that blunts the growth-promoting effects of oncogenic *ras*.

Finally, therapeutic approaches by themselves may also allow us to discern fundamental properties of signaling systems that were previously unknown. An example is the identification of inverse agonists, which helped determine that receptors exist in at least two distinct states.

Inhibitors of Posttranslational Processing

Ras is a 21-kDa GTP-binding protein that is a central component of growth factor signaling pathways that control cell proliferation (Lowy and Willumsen 1993). When growth factors activate a tyrosine kinase receptor, a cascade of interactions between various proteins is set off, finally leading to transcriptional changes in the cell. A pivotal switch for this pathway is the binding of GTP in place of GDP by the Ras protein. The GTP-bound Ras protein is active, while the GDP-bound Ras is inactive. Ras switches off when its intrinsic GTPase activity breaks the GTP to GDP. Several mutants of *ras*, many of them with the native GTPase incapacitated, will transform cells and induce uncontrolled proliferation. About 30% of all cancers are now known to be associated with *ras* mutations. Mutant *ras* is also present in cells from the majority of pancreatic and colon cancers.

Ras is posttranslationally modified by a sequence of enzymatic reactions (Figure 10-7). First, a C-terminal Cys is farnesylated, then the last three residues downstream of this Cys are proteolytically cleaved, and finally the C-terminus is carboxy methylated (Schafer and Rine 1992). Farnesyl is a 15-carbon isoprenoid, a lipid. An enzyme farnesyl transferase attaches the farnesyl group to the C-terminus of the Ras protein. The specific Cys that is modified is part of a Cys-a-a-x- (a: aliphatic residue; x: any residue) motif that is conserved in all proteins that are farnesylated. When the Cys at the C-terminus is mutated in *ras*, the mutant protein is not farnesylated. When an oncogenically active mutant *ras* is not farnesylated, it does not induce cell proliferation. Farnesylation is thus essential for the ability of oncogenic *ras* to transform cells. This has led to focused efforts to obtain molecules that will inhibit the farnesyl transferase as potential anticancer drugs.

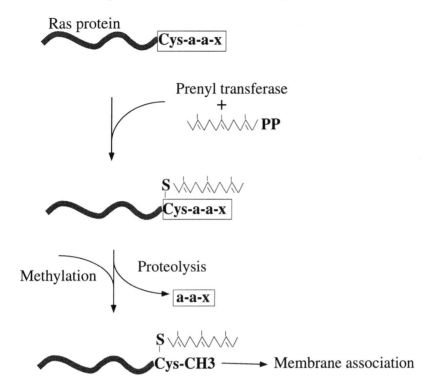

Figure 10-7. Posttranslational modification of Ras proteins. Prenyl transferase transfers the prenyl group from prenyl diphosphate to the Cys in the Cys-a-a-x box at the C-terminus of Ras. Cys forms a thioether linkage with the 15-C lipid. This step is followed by proteolytic removal of the last three residues of the protein and carboxymethylation. The modified protein then associates with the plasma membrane where most of the processed form is usually found.

Such drugs have now been, developed, with designs based on the structures of the substrates for the farnesylating enzyme—farnesyl diphosphate and tetrapeptides with the sequence Cys-a-a-x (Kohl et al 1995). One drug based on the structure of the tetrapeptide inhibits the farnesylating enzyme effectively *in vitro* and *in vivo*. It is also highly selective, inhibiting the farnesylating enzyme, but not related enzymes. When transgenic mice express an oncogenic form of *ras*, they develop tumors. When these mice are injected with the inhibitor of the farnesylating enzyme, the tumors regress. Surprisingly, the inhibitor does not have any pleiotropic effects, although several other important proteins in the cell are farnesylated. Identifying the molecular requirements for *ras* oncogenicity has thus yielded a potentially powerful drug to quell Ras constitutive activity and its outcome.

Inverse agonists

When the β2-adrenergic receptor, which is a G-protein-coupled receptor, was overexpressed specifically in the heart of transgenic mice, a significant proportion of the receptors were spontaneously active in the absence of the agonist (Bond et al 1995). When a previously characterized antagonist was examined for its ability to bind to these overexpressed receptors in heart tissue, it bound preferentially to the inactive receptors and shifted the equilibrium of the entire population of β-adrenergic receptors towards inactive receptors. In contrast, an agonist shifted the equilibrium towards active receptors. The antagonist thus behaved as an agonist would, if its function were inverted. Antagonists that behave in this manner are now called inverse agonists (Bond et al 1995). On the basis of these findings, it is now thought that G-protein-coupled receptors are found in two states—an active conformation (R*) that couples to the G-protein in the absence of the agonist and an inactive conformation (R) that requires agonist-induced change to activate the G-protein (Figure 10-8). In this scenario, substances classically defined as agonists have a high affinity for R*. They bind to R* and shift the equilibrium in the population of receptors toward the R* form. Inverse agonists however, have a high affinity for R and induce an increase in the population of R. Neutral antagonists have equal affinity for both forms of the receptor and have no effect on the population size of R and R*. They can therefore compete with both agonists and inverse agonists for receptor binding.

This model for receptor states predicts that mutations that increase the overall expression of a receptor will lead to constitutive signaling and, perhaps, to disease states (Figure 10-8). The first part of this prediction is borne out by the constitutive signaling noticed in transgenic mice overexpressing the β-adrenergic receptor. In rhodopsin, as described earlier, mutations that affect the structure of a receptor can also lead to constitutive activity. Inverse agonists can be of great value in shutting off such constitutive signaling. They will bind preferentially to the R form and shift the equilibrium towards the inactive form of the receptor. This will effectively reduce the number of active receptors and, as a result, the level of signaling.

The potentially powerful therapeutic role that such inverse agonists can play is illustrated in the rhodopsin system, where 11-*cis*-retinal binds to opsin and keeps it inactive, essentially behaving as an inverse agonist. On the basis of this property of 11-*cis*-retinal, it was predicted that compounds that have a similar structure may bind to rhodopsin mutants that are constitutively active and inhibit their activity. Although the Lys296Glu mutant that is found in individuals suffering from retinal degeneration (autosomal dominant retinitis pigmentosa) is incapable of binding 11-*cis*-retinal, it does bind a retinylamine compound, and its constitutive activity is irreversibly inactivated by this compound (Rao and Oprian 1996). This compound does

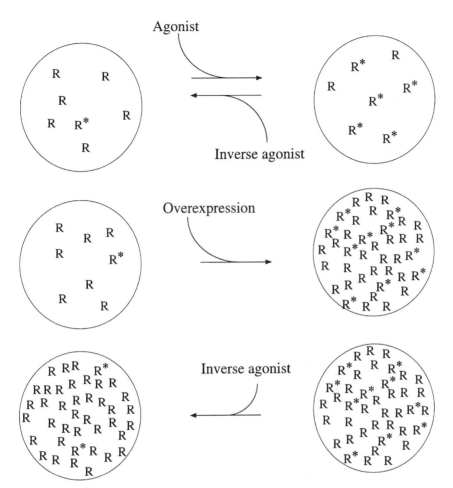

Figure 10-8. The two states of G-protein-coupled receptors. Agonists increase the proportion of active receptors, while inverse agonists increase the proportion of inactive receptors. Overexpression leads to the appearance of a larger number of active receptors. This can result in a high basal level of constitutive activity that has potential to cause disease. Inverse agonists can decrease the concentration of active receptors and act therapeutically.

not bind to normal opsin. It has the potential therefore to selectively inhibit the constitutive activity of a mutant opsin in a heterozygous individual without affecting signaling by the normal protein.

Many drugs previously classified as antagonists may now prove to be inverse agonists. These drugs can be of considerable value in treating diseases caused by constitutive activity.

FUTURE

Complex networks of signaling pathways control everything from our cardiovascular function to mental health. It is very likely that many more flaws
in signaling will be identified as the bases of different diseases. In contrast
to mutant proteins that chronically affect signaling, defective signaling may
also be transient, as in the case of mood disorders such as depression and
anxiety. Unraveling the molecular mechanisms that dictate the specific interaction of signaling proteins in a cell can provide novel approaches towards therapy. For instance, the design of drugs can be based on the
structure of domains involved in functional interaction between receptor-G-
protein or G-protein-effector. Such drugs may be able to overcome defective signaling by inhibiting or enhancing signaling activity.

REFERENCES

Baylor D. 1996 How photons start vision. Proc Natl Acad Sci U S A 93:560–565.
Bond RA et al. 1995 Physiological effects of inverse agonists in transgenic mice with
    myocardial over-expression of the β2-adrenoreceptor. Nature 374:272–275.
Coleman DE et al. 1994 Structures of active conformations of Giα1 and the mecha-
    nism of GTP hydrolysis. Science 265:1405–1412.
Hamm HE and Gilchrist A. 1996 Heterotrimeric G-proteins. Current Opinion in
    Cell Biology 8:189–196.
Kohl NE et al. 1995 Inhibition of farnesyltransferase induces regression of mammary
    and salivary carcinomas in *ras* transgenic mice. Nature Medicine 1:792–797.
Landis CA et al. 1989 GTPase inhibiting mutations activate the alpha chain of Gs
    and stimulate adenylyl cyclase in human pituitary tumours. Nature 340:692–696.
Li T et al. 1995 Constitutive activation of phototransduction by K296E opsin is not
    a cause of photoreceptor degeneration. Proc Natl Acad Sci U S A 92:3551–3555.
Lowy DR and Willumsen BM. 1993 Function and regulation of ras. Annu Rev
    Biochem 62:851–891.
Michiels FM et al. 1994 Oncogenic potential of guanine nucleotide stimulatory
    factor alpha subunit in thyroid glands of transgenic mice. Proc Natl Acad Sci U
    S A 91:10488–10492.
Moss J and Vaughan M. 1988 ADP-ribosylation of guanyl nucleotide-binding regu-
    latory proteins by bacterial toxins. Advances in Enzymology & Related Areas
    of Molecular Biology 61:303–379.
Muller HJ. 1932 Further studies on the nature and causes of mutations. Proceedings
    of the Sixth International Congress of Genetics. 1:213–52. From: Studies in
    Genetics, the Selected Papers of H.J. Muller. 1962. Bloomington, IN: Indiana
    University Press. 105–125.
Pollak MR et al. 1993 Mutations in the human Ca^{2+} sensing receptor gene cause
    familial hypocalciuric hypercalcemia and neonatal severe hyperparathyroidism.
    Cell 75:1297–1303.

Rao VR and Oprian DD. 1996 Activating mutations of rhodopsin and other G-protein coupled receptors. Annu Rev Biophys Biomol Struct 25:287–314.

Rao VR, Cohen GB, and Oprian D D. 1994 Rhodopsin mutation G90D and a molecular mechanism for congenital night blindness. Nature 367:639–642.

Schafer WR and Rine J. 1992 Protein prenylation: genes, enzymes, targets and functions. Annu Rev Genet 26:209–237.

Sieving PA et al. 1995 Dark-light: model for night blindness from the human rhodospin Gly90→Asp mutation. Proc Natl Acad Sci U S A 92:880–884.

Spiegel AM. 1996 Defects in G-protein coupled signal transduction in human disease. Annu Rev Physiol 58:143–170.

11

Intracellular Messengers in Drug Addiction

MAARTEN E. A. REITH

1. DRUGS OF ABUSE: WHAT DO THEY HAVE IN COMMON?

The major classes of compounds with drug abuse liability are the pyschostimulants, such as cocaine, amphetamine, and methamphetamine; the opiates, heroin and morphine; the sedatives, including barbiturates and benzodiazepines; the hallucinogens, lysergic acid diethylamine (LSD[1]) and phencyclidine (PCP[1]); marijuana; inhalants from various sources; ethanol; and nicotine, which has recently been labeled an addictive drug by the U.S. Food and Drug Administration. Clearly, various sites and mechanisms of action are involved in the way these different classes of compounds affect the brain (for reviews see Di Chiara and North 1992, Katzung 1995, Hardman et al 1996). For example, cocaine and amphetamine both act to release dopamine (DA[1]) from electrically active DA neurons, by separate mechanisms. Cocaine acts by blocking uptake of released DA into nerve cells (for review see Reith et al 1997), and amphetamine increases extracellular DA through a genuine releasing effect (Reith et al 1991). In contrast, opiates have a major site of action on γ-aminobutyric acid (GABA[1]) neurons, causing disinhibition of the normal, tonic influence of GABA over DA cell activity (for review see Di Chiara and North 1992). One of the important actions of barbiturates and benzodiazepines is the binding to GABA$_A$ receptors, which facilitates the opening of ligand-gated chloride channels (for reviews see Katzung 1995, Hardman et al 1996). Whereas LSD action involves serotonin (5-HT[1]) receptors, PCP can act on σ-opiate receptors as well as N-methyl-D-aspartic acid (NMDA[1]) receptors and can increase DA concentration by interfering with neuronal uptake of DA. Marijuana acts on specific cannabinoid receptors, which have been cloned recently. Inhalants

Introduction to Cellular Signal Transduction
A. Sitaramayya, Editor
©1999 Birkhäuser Boston

are too varied in their sources, usually industrial solvents, to allow us to pinpoint a common neuronal mechanism. Ethanol acts on $GABA_A$, glutamate, and $5\text{-}HT_3$ receptors, and, finally, nicotine stimulates nicotinic cholinergic receptors (for reviews see Katzung 1995, Hardman et al 1996).

Despite their wide variety of actions, most of these classes of compounds share an acutely rewarding, hedonic action. Indeed, behavioral and neuropharmacologic research has identified a system of reward pathways in the brain. One pathway is a major core of mesolimbic DA nerve fibers that is acted upon by drugs of abuse from many different classes. Mesolimbic DA increases have been implicated in the hedonic action of all of the above classes of drugs (for reviews see Wise and Bozarth 1987, Huston-Lyons and Kornetsky 1992), including marijuana (Gardner and Lowinson 1991), except for the action of sedatives and LSD. In the case of the sedatives, it is unknown whether the DA system is implicated in the rewarding effects. LSD is probably more a hallucinogenic than euphoric compound. The mesolimbic DA system has its neuronal cell bodies in the ventral tegmental area (VTA[1]) in the midbrain with the nerve endings located in the nucleus accumbens (NAc[1]) in the forebrain and prefrontal cortex (see Figure 11-1). The mesolimbic DA system is intimately connected with "pleasure pathways," which animals like to self-stimulate electrically through stategically placed electrodes when given the chance (for review see Milner 1989).

The rewarding action of drugs of abuse through the DA system is concordant with the proponent theory that says that reinstatement of drug use occurs because of the hedonic effect of the drug. The opposite view is the opponent theory. This theory predicts that a drug-dependent subject who has a new level of homeostasis in brain reward pathways continues taking the drug to alleviate unpleasant withdrawal symptoms (for references see Wise 1996). In a third view, also known as the incentive-sensitization theory of addiction (Robinson and Berridge 1993), repeated drug exposure causes an increase in the incentive to use the drug ("wanting"), even if the rewarding effect of the drug ("liking") is decreasing over time. In these concepts, the rewarding or incentive activity of drugs is mediated by mesolimbic DA, whereas the induction of withdrawal symptoms may involve other systems. Craving for a drug could be thought of as the wish for renewed positive reinforcement, the desire to diminish the withdrawal symptoms, or the result of enhanced incentive. Most likely, all these mechanisms play a role to some extent, depending on the drug and the context in which it is being used. A complicating factor is that repeated drug use can lead to tolerance to some symptoms, i.e. a reduction in the drug action, and to sensitization to other symptoms, i.e. enhanced responsiveness. For instance, chronic morphine administration induces tolerance to its pain-reducing effect (for reviews see Katzung 1995, Hardman et al 1996), and repeated cocaine exposure results in tolerance to its euphoric effect (Robinson and

Berridge 1993). On the other hand, long-term cocaine self-administration can sensitize humans to experience paranoia (Satel et al 1991).

The research on intracellular messenger involvement in drug addiction is only beginning to address the plastic changes that occur in brain circuits, and it cannot, at this point in time, fit those changes into the framework of the complex behavioral patterns. From among the different brain circuits involved in the action of drugs of abuse, this chapter will focus on the mesolimbic DA system because of its central importance. From among the different drugs of abuse, this chapter will focus on the psychostimulants cocaine and amphetamine, and on the opiate morphine because these compounds have been well studied with regard to the role of intracellular messengers in their mechanism of action.

2. DRUGS OF ABUSE: SECOND MESSENGERS

In response to neurotransmitters acting as first messengers, enzymes that produce second messengers can be activated or inhibited or ion channels can be activated, allowing the passage of Ca^{2+} (see Table 11-1). Among the many second messengers such as cAMP, cGMP, Ca^{2+}, inositol triphosphate, nitric oxide (NO^1), and arachidonic acid, most attention has been devoted to the cAMP systems as a target of drugs of abuse. This is not surprising if one considers the following: (1) DA circuits are important in drug reinforcement, (2) DA acts on two main classes of receptors, the D_1-type DA receptors (encompassing the D_1 and D_5 subtypes) and the D_2-type DA receptors (encompassing the D_2, D_3, and D_4 subtypes), and (3) the D_1 class is linked with G_s, a G-protein that stimulates the formation of cAMP by adenylate cyclase, and the D_2 class is linked with G_i, which reduces cAMP formation.

Recent studies have shown that repeated exposure of rats to cocaine or morphine causes an up-regulation of the cAMP system in the NAc, as evidenced by reduced levels of G_i and increased levels of adenylate cyclase and cAMP-dependent protein kinase (PKA^1) measured at the time the next drug dose is due (Nestler et al 1993) (see heavy arrows in Figure 11-1). When drugs that mimic these changes are locally applied into the NAc, the reinforcing potency of self-administered drugs appears to be reduced. This finding is consistent with the cAMP system being involved in drug reward. For example, intra-NAc application of pertussis toxin, which inactivates G_i, cholera toxin, which activates G_s, or the PKA activator Sp-cAMPS can reduce the reinforcing properties of cocaine or heroin as measured by self-administration (Self et al 1994, Self and Nestler 1995). Conversely, application of the PKA inhibitor Rp-cAMPS increases the reinforcement in drug self-administration. Activation of the cAMP system in the NAc also affects the stimulation of locomotor activity by cocaine and amphetamine (Self and Nestler 1995). Taken together, the data suggest that cAMP pathways in the NAc are involved in the action of drugs of abuse. It has been suggested that

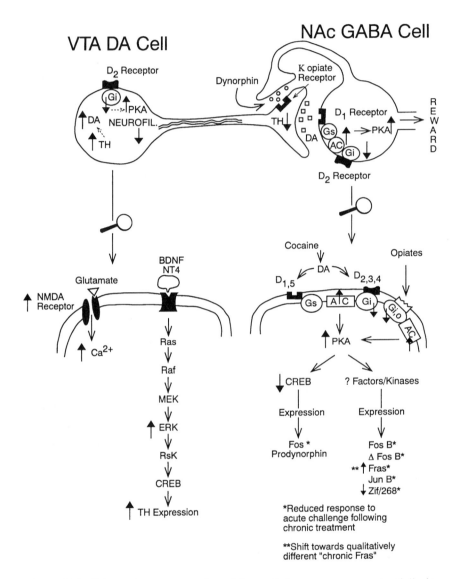

Figure 11-1. Schematic representation of intracellular messengers potentially involved in the action of drugs of abuse. The pathways shown are examples and are not meant to be all-inclusive. Light arrows indicate sequence of events upon acute activation or inhibition; heavy arrows denote changes (up or down) upon chronic drug treatment. The VTA DA cell is depicted to project to a GABA cell in the NAc that also expresses dynorphin. This peptide acts on κ-opiate receptors in DA terminal plasma membranes in the NAc to inhibit DA release; this constitutes a feedback mechanism that dampens further DA release through DA acting on D_1 receptors with the activation of adenylate cyclase (AC), causing enhanced PKA phosphorylating

the up-regulation of the cAMP system in the NAc upon chronic exposure to cocaine or heroin makes the organism less susceptible to the effect of renewed drug intake, representing a tolerance phenomenon (Nestler 1994, Self and Nestler 1995, Nestler et al 1996).

Repeated exposure to drugs of abuse can also lead to plastic changes in the VTA, the DA cell body region from which the axons project to the NAc. For example, chronic morphine administration reduces neurofilament protein and axonal transport (Nestler et al 1993, Nestler et al 1996), possibly underlying the decrease in NAc tyrosine hydroxylase, the rate-limiting enzyme in DA synthesis (see heavy arrows in Figure 11-1). The latter also occurs with chronic exposure to stimulants (Beitner-Johnson et al 1992) and is expected to lead to reduced DA release in the NAc, which is exacerbated by excess dynorphin peptides acting on κ-opiate receptors on DA terminals to reduce DA release (Hyman 1996) (see Figure 11-1 and also Section 4

CREB, in turn enhancing expression of prodynorphin, the precursor of dynorphin (see arrow with magnifying glass pointing to the zoomed-in area on the right). PKA also facilitates the expression of various transcription factors, through CREB or other transcription factors acting on their promoters, and through intranuclear PKC altering the phosphorylation state of transcription factors as they interact with DNA. Proteins of the Jun and Fos family can form hetero- or homodimers in many different combinations with unique DNA-binding and regulatory properties; in some cases the combinations have lost the ability to bind to DNA, representing inhibitory variants of transcription factors (Grzanna and Brown, 1993). In the VTA, the interaction between Ca^{2+} and the neurotrophin-activated pathways is depicted with tyrosine hydroxylase (TH) as the endpoint (see arrow with magnifying glass pointing to zoomed-in area on the left). Heavy arrows (up or down) indicate the changes (increases or decreases) observed upon chronic administration of psychostimulants and opiates at the time the next drug dose is due. In the VTA, chronic drug exposure increases NMDA receptors, which through elevated Ca^{2+} influx leads to higher ERK activity and TH expression. The increased TH in the VTA in the drug-dependent state leads to enhanced DA release (see dotted arrow) and subsensitive D_2 receptors (G_i down and PKA up, see dotted arrow). In addition, neurofilament proteins are down, probably causing decreased TH in the NAc, which leads to reduced DA release and, in turn, supersensitive D_1 receptors on NAc GABA cells, reduced $G_{i,o}$, and enhanced AC as well as PKA activity. The latter causes desensitization of CREB, explaining the reduced response in Fos production in the NAc upon repeated drug exposure. Desensitization of drug-induced production of Zif/268 and Fras also occurs, with the latter phenomenon being associated with the production of different "chronic Fras." For references and details see text, the reviews of Rogue and Malviya (1994), Hyman and Nestler (1996), Nestler et al (1996), and Reith et al (1997), and publication of Berger and colleagues (Tolliver et al 1996).

below). Thus, the drug-addicted state combines a reduced DA release in the NAc with a supersensitivity of adenylate cyclase coupled through G_s to the D_1 receptor. The latter could be an attempt to compensate for the chronically decreased activation by DA. If this compensation is incomplete, it might explain the dysphoric state accompanying drug withdrawal. However, more work is needed to understand the processes underlying drug craving and tolerance or sensitization.

There is also evidence for an up-regulated cAMP system in the VTA as a result of repeated treatment with psychostimulants such as cocaine and amphetamine (for references see Tolliver et al 1996). Thus, G_i is reduced in the VTA (Nestler et al 1990), and intra-VTA application of cholera toxin, an activator of adenylate cyclase through G_s, sensitizes animals to the locomotor activating effect of amphetamine, which can be blocked by the PKA inhibitor H8 (Tolliver et al 1996). Although this would suggest sensitization rather than tolerance to pychostimulant action, as suggested above for the NAc phenomena, there is not necessarily a discrepancy. Thus, locomotor activity does not always correspond to measures obtained in self-administration studies, probably reflecting nonidentical underlying brain circuitries, and the changes observed in the VTA in the study of Tolliver et al (1996) were not accompanied by changes in NAc DA release as measured by microdialysis. Another change in the VTA upon chronic morphine administration is the increase in tyrosine hydroxylase activity (see heavy arrows in Figure 11-1), which, by elevating DA and its subsequent release, leads to D_2 receptor subsensitivity consonant with the observed reduction in G_i (Nestler et al 1993, Nestler et al 1996) and increase in PKA activity (Tolliver et al 1996).

There is also evidence for the involvement of NO in the action of drugs of abuse. NO produced by activation of NMDA receptors, located on non-DA neurons (Bredt and Snyder 1989), can affect DA transmission, presumably by NO acting as a retrograde neurotransmitter by virtue of its gaseous nature. In previous *in vitro* studies NO has been reported to stimulate DA release from neighboring DA terminals (Hanbauer et al 1992, Zhu and Luo 1992), but in most *in vivo* studies applying microdialysis NO has been found to have an inhibitory effect on DA efflux (Bugnon et al 1994, Silva et al 1995). In agreement with the latter, NMDA-induced DA release in the striatum is potentiated by NO synthase inhibition (Shibata et al 1996). An important finding in the context of drugs of abuse is that the sensitization of the locomotor-stimulating effect of repeated cocaine injections in rats is attenuated by pretreatment with MK-801, an NMDA receptor antagonist, or with N^G-nitro-L-arginine, an NO synthase inhibitor (Pudiak and Bozarth, 1993). It is possible that the NMDA-NO pathway is one of the links involved in the development of the behavioral sensitization, so that blockade of this pathway, either by NMDA receptor blockade or by NO-synthase inhibition, interferes with the responsiveness to subsequent cocaine administration.

Our own work on arachidonic acid gives an example of a different mode of interaction of second mesengers with drug action. In addition to being directly involved in the pathways mediating drug effects, these messengers can act on important sites of drug action. Thus, the neuronal DA transporter is a major target for cocaine, and is also required for amphetamine action; our recent work shows it to be affected by arachidonic acid in a complex manner (Zhang and Reith 1996). Low concentrations of arachidonic acid can enhance, and high concentrations can decrease, its function. This could have an impact on the action of cocaine or amphetamine under conditions in which arachidonic acid levels are changing.

3. DRUGS OF ABUSE: THIRD MESSENGERS

Third messengers are generally transcription factors that are phosphorylated by kinases, which are in turn under the influence of second messengers (see Table 11-1). For instance, the second messenger cAMP is required for the capability of PKA to phosphorylate cAMP response element binding protein (CREB[1]), which binds to the CRE. The CRE binding motif is an 8-bp palindrome element 5'-TGACGTCA-3', which confers full cAMP responsiveness (for review see Hyman and Nestler 1996). cAMP can also lead, through PKA, to phosphorylation of the third messenger serum response factor (SRF[1]) (for review see Rogue and Malviya 1994). The second messenger Ca^{2+} (produced intracellularly by opening of Ca^{2+} channels coupled by G-proteins to receptors or by influx through the cation channel of the NMDA receptor) stimulates the activity of the SRE, and Ca^{2+}/calmodulin-dependent kinase II (CaCM KII[1]) phosphorylates CREB and SRF. The second messenger diacylglycerol (DAG[1]) (formed by activation of receptors linked to the phospholipid signaling system) activates protein kinase C (PKC[1]), which affects the phosphorylation state and activity of several transcription factors. Activated PKC can also stimulate the Ras (small G-protein) pathway (see review of Rogue and Malviya 1994) (see Section 5 below). Because all of these pathways are affected by DA, they are potential targets for drugs of abuse that activate dopaminergic circuits in the brain. In this context, members of the steroid receptor family (e.g., for progesterone or vitamin D, or the orphan receptor COUP-TF) can also be regarded as third messengers, since these receptors can be phosphorylated by PKA, which is stimulated by DA-induced increases in cAMP (see Rogue and Malviya 1994). These cytoplasmic receptors function as ligand-regulated transcription factors upon translocation to the nucleus.

Research on the involvement of third messengers in the mechanism of action of drugs of abuse has focused mostly on CREB. Chronic morphine administration has been shown to result in a decrease in CREB immunoreactivity in rat NAc (but not in striatum or frontal cortex) (Widnell et al 1996). This stands in contrast to the up-regulation of PKA activity observed

Table 11-1. Examples of pathways directly regulated by second messengers with involvement in the action of drugs of abuse

Link	Messengers (First = Neurotransmitter) and their Links				
	Second	Link	Third	Link	Fourth
Cation channel of NMDA receptor, or receptor linked by G-protein to Ca²⁺ channel	Ca^{2+}	CaCM KII	SRF	Binding to SRE	Fos
	Ca^{2+}	CaCM KII	CREB	Binding to CRE	Fos, Prodynorphin
D_1, G_s or D_2, G_i	cAMP	PKA	SRF	Binding to SRF	Fos
	cAMP	PKA	CREB	Binding to CRE	Fos, Prodynorphin
	cAMP	PKA	IRBP	Binding to IRE	Jun B
	cAMP	PKA	Steroid Receptor	Binding to RE	
Receptor linked with PLC	DAG	PKC*	Various Transcription Factors	Binding to RE	Fos and others
	DAG	PKC	IRBP	Binding to IRE	Jun B
	DAG	PKC	Raf	MEK	
Receptors as above	cAMP,DAG	PKA, PKC*	Various Transcription Factors	Binding to RE	Fos B, ΔFos B, Fras, Jun, Zif/268
Receptor → Ca^{2+} → PKC → PLA_2	AA	DA transporter			
Receptor (NMDA) → Ca^{2+}, calmodulin → NO synthase	NO	Guanylyl cyclase	cGMP	PKG	

* Phosphorylation of various transcription factors to be elucidated. PKC is also present endogenously in the nucleus, where it affects the phosphorylation state and activity of various transcription factors (for review see Rogue and Malviya 1994).

The pathways shown are examples of what is currently known (see text) but are not meant to be all-inclusive. AA, arachidonic acid; IRBP, inverted repeat element binding protein; IRE, inverted repeat element; for other abbreviations see footnote 1.

following chronic morphine treatment (see Section 2 above), suggesting that the latter up-regulation occurs through a CREB-independent mechanism. It has been proposed that the up-regulated PKA activity mediates the CREB down-regulation in the NAc as a compensatory response (Widnell et al 1996). If this also occurs with repeated cocaine administration, it would explain the desensitization of cocaine-induced Fos production over time (Fos being the messenger subsequent to CREB, see Table 11-1 and Section 4 below). Possibly, the reduced CREB activity upon chronic morphine administration underlies the observed decrease in $G_{i\alpha}$, perhaps through a direct effect of CREB on $G_{i\alpha}$ expression, or through a mechanism involving other intermediates (Widnell et al 1996). In a drug-dependent organism with reduced G_i in the NAc, renewed exposure to the drug of abuse would be expected to reduce the reinforcing potency of the drug, as observed following intra-NAc application of pertussis toxin which inactivates G_i (see Section 2 above and Self et al 1994). Of course, D_1 (coupled to G_s) and D_2 (coupled to G_i) receptors may be localized in different subsets of DA neurons, and how this affects drug reward will depend on the interlinkages and endpoints involved. For example, challenging an already up-regulated PKA system (through reduced CREB and G_i and increased adenylate cyclase) in the chronically treated animal may or may not increase downstream events above levels seen by challenging a naive animal with the drug of abuse. Furthermore, D_1- and D_2-mediated pathways may subserve different behavioral endpoints. Thus, in a recent study, Self et al (1996) show that D_2-like, but not D_1-like, agonists trigger a relapse of cocaine-seeking behavior in rats with prior cocaine experience. In addition, D_2-like agonists enhance cocaine-seeking behavior induced by cocaine itself, whereas D_1-like agonists prevent this behavior (Self et al 1996). A final consideration is that, in addition to lowering the second messenger cAMP, activation of D_2-like receptors can also increase the second messengers Ca^{2+} and DAG through activation of phospholipase C (PLC), which then stimulates phosphatidylinositol 4,5-bisphosphate (PIP_2) hydrolysis (see Rogue and Malviya 1994). This PIP_2 cascade can then activate PKC, which may exert a feedback control on D_2 receptor function by uncoupling it from the G_i protein (Rogue et al 1990). Clearly, the different signaling pathways associated with D_1 and D_2 receptors will play important roles in the action of DA agonists in drug-seeking behaviors.

4. DRUGS OF ABUSE: FOURTH MESSENGERS

The third messenger CREB induces the expression of immediate early genes (IEG's[1]) that encode transcription factors such as Fos (the protein product of the *c-fos* gene) as a fourth messenger (see Table 11-1). For example, it has been shown that CREB antisense treatment in the NAc attenuates the production of Fos induced by acute cocaine administration

(Widnell et al 1996). The *c-fos* promoter carries two CREs, one of which is sensitive to both Ca^{2+} and cAMP (CaRE/CRE[1]) (see review of Rogue and Malviya 1994). Acute exposure to amphetamine or cocaine induces phosphorylation of CREB and expression of *c-fos* and related IEGs coding for fos-related antigens (Fras[1]) and Jun B (the protein product of *jun B*), which can be blocked by D_1 receptor antagonists (for review see Hyman 1996). The production of IEG transcription factors by psychostimulants such as amphetamine or cocaine regulates the expression of target genes, which probably plays a role in the acute behavioral effects of these drugs. For example, both amphetamine and cocaine can induce expression of preprotachykinin mRNA and prodynorphin mRNA and protein in striatonigral neurons (see Hyman 1996). Prodynorphin expression appears to require stimulation of D_1 receptors by DA, since D_1 receptor knockout mice show decreased prodynorphin gene expression (Xu et al 1994). As mentioned above (see Section 2), dynorphin peptides can act as an autofeedback system tempering DA overactivity by stimulating κ-opiate receptors on DA nerve terminals in the striatum or NAc to decrease DA release. An intriguing observation is that selective stimulation of either D_1 or D_2 receptors does not induce *c-fos* expression. Combined administration of low doses of D_1 and D_2 agonists, or an a-selective $D_{1,2}$ agonist by itself, is needed for the effect, suggesting a synergistic D_1 and D_2 receptor interaction (see also Rogue and Malviya 1994). The *c-fos* induction is attenuated by D_1 antagonism, but not by a D_2 antagonist, which by itself actually induces *c-fos* by reducing the inhibitory G_i linkage with adenylate cyclase. A complex D_1-D_2 interaction has also been reported for the induction of the IEG *zif/268* (giving the protein product Zif/268) by amphetamine and methamphetamine (Wang et al 1995, Wang and McGinty 1995). Acute cocaine administration increases striatal *zif/268* mRNA (Daunais and McGinty 1995) and induces *c-fos* and *jun B*, but not *c-jun* (or *jun D*, which is constitutively expressed) (Moratalla et al 1993). Most likely, cocaine increases *c-fos* expression through cAMP (D_1 receptor) or Ca^{2+} (potentially through D_1 receptors linked to the inositol signaling system) by acting on the CaRE/CRE element. The *jun B* expression is stimulated through an inverted repeat element in the *jun B* promoter, which responds to PKA and PKC, whereas *c-jun* does not respond, perhaps because it is actively repressed by cAMP or *jun B* (Moratalla et al 1993).

In contrast to these acute effects, animals withdrawn from chronic cocaine administration show a reduced forebrain *zif/268* expression (Bhat et al 1992), and are less sensitive to the activity of cocaine to enhance striatal Fos (Rosen et al 1994) and accumbal Fos, Fos B, Jun, Jun B, and Zif/268 (Hope et al 1992). In addition, upon chronic cocaine treatment the pattern of Fras shifts from "acute Fras" to "chronic Fras," with a prolonged half-life in the brain (Hyman and Nestler 1996). "Chronic Fras," part of the AP-1 complex that binds to the AP-1 site on the DNA, may build up over time as a result of the repeated increases during cocaine treatment (Hope et al

1994) and could mediate long-lasting changes in brain reward circuitries (Nye et al 1995). Recent studies with transgenic mice, in which the *fos B* gene has been disrupted by homologous recombination, show that the chronic Fras are products of the *fos B* gene, specifically isoforms of ΔFos B, a truncated splice variant of Fos B (Hiroi et al 1997).

5. DRUGS OF ABUSE: OTHER CHEMICAL MESSENGERS

The brain also makes use of signaling pathways that are not directly regulated by G-proteins and/or second messengers (see Table 11-2). For example, extracellular signal regulated kinases (ERKs[1]) and mitogen-activated protein (MAP[1]) kinases are two protein serine/threonine kinases that are not under the direct influence of second messengers (see Nestler 1994). In addition, the brain has protein tyrosine kinases associated with receptors for growth factors (trk proteins) or other intracellular tyrosine kinases (src kinases). There are indirect links between these pathways and the second-messenger-dependent chains of events. In addition, the combination of two kinases from the Janus kinase (JAK[1]) family, or, in the case of some growth factors, of one JAK-type kinase together with one receptor tyrosine kinase, can phosphorylate transcription factors such as *sis*-inducible factor (SIF[1]), which binds to the SIE[1] (for review see Rogue and Malviya 1994).

Among the pathways not directly regulated by second messengers, the growth-factor-regulated chains of events have been studied in most detail in the context of the mechanism of action of drugs of abuse. Two growth-factor-regulated pathways are prominent. First, binding of neurotrophins, such as brain-derived neurotrophic factor (BDNF[1]), neurotrophin-4 (NT4[1]), or NT3 to their receptors (Trk[1]) activates Ras, a small G-protein that enhances the activity of Raf (a cytoplasmic serine/threonine protein kinase), which phosphorylates MAP kinase kinase (MEK[1]). This in turn phosphorylates MAP kinase (also called ERK), which activates a variety of substrate proteins (see Hyman and Nestler 1996). The latter include reactive phosphatase-treated S6 Kinase (Rsk) 2 which phosphorylates CREB and SRF. Second, binding of ciliary neurotrophic factor (CNTF[1]), another growth factor belonging to the cytokine class of compounds, to its receptor activates the cytoplasmic protein kinase JAK, which phosphorylates the signal transducers and activators of transcription (STAT[1]) family of proteins regulating expression of various genes.

Application of BDNF or NT4 into the VTA prevents morphine from up-regulating the cAMP pathway in the NAc, supporting the concept that the plastic changes in the DA terminal region upon chronic drug treatment are secondary to alterations occurring in the DA cell body area (Berhow et al 1995, 1996), as argued above (Section 2) for NAc tyrosine hydroxylase. Chronic treatment with morphine or cocaine increases the phosphorylation state and activity of ERK in the VTA without changing the total amount of

Table 11-2. Examples of pathways not directly regulated by second messengers with involvement in the action of drugs of abuse

Extracellular	Receptor	Link	Kinase Chain	Target/Transcription Factor
BDNF, NT4, NT3	Trk	Ras	Raf, MEK, ERK, Rsk	CREB, SRF
BDNF, NT4, NT3	Trk	Ras	Raf, MEK, ERK	Various Substrate Proteins
EGF, PDGF	Trk			SIF
CNTF	CNTF α subunit	two β units of CNTF receptor	JAK	STAT

The pathways shown are examples of what is currently known (see text) but are not meant to be all-inclusive. The lines connecting Trk and JAK and pointing to SIF indicate possible combinations of receptor tyrosine kinase and JAK-type kinase, or of two JAK-type kinases, which can phosphorylate SIF. Not shown here is the activity of CREB in regulating expression of Fos and dynorphin. For abbreviations see footnote 1.

kinase. The underlying mechanism for this could be that up-regulation of glutamate receptors upon drug treatment causes enhanced Ca^{2+} influx to activate the ERK pathway (Nestler et al 1996). The latter activation could be thought to enhance the activity of the protein kinase Rsk, which phosphorylates CREB, which binds to the CRE in the tyrosine hydroxylase promoter, resulting in tyrosine hydroxylase activation in the VTA. Another growth factor pathway affected by drugs of abuse is that containing the JAK and STAT proteins. Chronic cocaine administration increases the amount of JAK2 in the VTA without enhancing STAT activity (Nestler et al 1996). When the VTA of animals that have been treated chronically with cocaine is infused with CNTF, STAT binding is greatly enhanced when compared with the effect observed in naive rats. It is not known how this up-regulation of the JAK-STAT pathway contributes to the chronic effects of cocaine. The promoter for tyrosine hydroxylase, for instance, may not contain a STAT binding site, and it has been suggested that additional transcription factors could be involved, or that JAK2 in glial cells could affect cocaine action through as yet unknown glial-neuronal cell interactions (Nestler et al 1996). The combined data indicate important effects of drugs of abuse on intracellular pathways outside the classical second messenger cascades. Neurotrophic factors such as BDNF, NT4, and NT3 can counteract the plastic changes induced by chronic cocaine or morphine treatment in the VTA and NAc (Berhow et al 1996) and may potentially open up novel therapeutic approaches for drug addiction.

6. CONCLUDING REMARKS

The messenger pathways discussed above are schematically presented in Figure 11-1. There are of course many processes that are involved in the effects of drugs of abuse but are not part of the depicted events. In addition, the emphasis is on effects that are shared between the psychostimulants and opiates (heavy arrows up or down indicating increases or decreases following chronic drug treatment). Clearly, these are simplifications that overlook differences among various drugs of abuse. In addition, the pathways shown are only as complete as our current knowledge of intracellular neuronal signaling mechanisms. The changes shown strengthen the concept that chronic exposure to drugs of abuse leads to plastic neuronal changes that can result in long-lasting alterations of neuronal function. These changes may affect the response to subsequent drug exposure. There is substantial evidence that *c-fos* and other IEGs play an important role in the transition between short-term signaling and long-term changes in the brain. Regulation of gene expression is an integral component in this process, and it has become abundantly clear that different IEGs interact with each other as well as with other second and third messengers to change transcription in concert. As we learn more about these complex interactions that alter gene expression, the underlying mechanisms involved in plasticity upon exposure to drugs of abuse will become more evident.

ACKNOWLEDGMENTS:

I would like to thank the National Institute on Drug Abuse (DA 08379 and DA 03025) for supporting the experiments in which we study the involvement of first and second messengers in the action of cocaine, and J.C. Milbrandt, X.-D. Wang, Y. Salch, C. Xu, and L.L. Coffey for discussions regarding this manuscript.

ABBREVIATIONS

BDNF, brain-derived neurotrophic factor
CaCM KII, Ca^{2+}/calmodulin-dependent kinase II
CNTF, ciliary neurotrophic factor
CREB, cAMP response element binding protein
CRE, cAMP response element
DA, dopamine
DAG, diacylglycerol
EGF, epidermal growth factor
ERK, extracellular signal regulated kinase
Fra, fos-related antigen
GABA, γ-aminobutyric acid
5-HT, serotonin
IEG, immediate early gene
IRBP, inverted repeat element binding protein
IRE, inverted repeat element
JAK, Janus kinase
LSD, lysergic acid diethylamine
MAP, mitogen-activated protein
MEK, MAP kinase kinase
NMDA, N-methyl D-aspartic acid
NAc, nucleus accumbens
NT3, neurotrophin-3
NT4, neurotrophin-4
NO, nitric oxide
PCP, phencyclidine
PDGF, platelet-derived growth factor
PIP_2, phosphatidylinositol 4,5-bisphosphate
PKA, cAMP-dependent protein kinase
PKC, protein kinase C
PLC, phospholipase C

RSK, reactive phosphatase-treated S6 Kinase
SIE, *sis*-inducible factor response element
SIF, *sis*-inducible factor
SRE, serum response factor response element
SRF, serum response factor
STAT, signal transducers and activators of transcription
VTA, ventral tegmental area

REFERENCES

Beitner-Johnson D, Guitart X, and Nestler EJ. 1992. Common intracellular actions of chronic morphine and cocaine in dopaminergic brain reward regions. Ann NY Acad Sci 654:70–87.

Berhow MT, Russell DS, Terwilliger RZ, Beitner-Johnson D, Self DW, Lindsay RM, and Nestler EJ. 1995. Influence of neurotrophic factors on morphine- and cocaine-induced biochemical changes in the mesolimbic dopamine system. Neuroscience 68:969–979.

Berhow MT, Hiroi N, and Nestler EJ. 1996. Regulation of ERK (extracellular signal regulated kinase), part of the neurotrophin signal transduction cascade, in the rat mesolimbic dopamine system by chronic exposure to morphine or cocaine. J Neurosci 16:4707–4715.

Bhat RV, Cole AJ, and Baraban JM. 1992. Chronic cocaine treatment suppresses basal expression of zif268 in rat forebrain: in situ hybridization studies. J Pharmacol Exp Ther 263:343–349.

Bredt DS and Snyder SH. 1989. Nitric oxide mediates glutamate-linked enhancement of cGMP levels in the cerebellum. Proc Natl Acad Sci USA 86:9030–9033.

Bugnon O, Schaad NC, and Schorderet M. 1994. Nitric oxide modulates endogenous dopamine release in bovine retina. Neuroreport 5:401–404.

Daunais JB and McGinty JF. 1995. Cocaine binges differentially alter striatal preprodynorphin and zif/268 mRNAs. Mol Brain Res 29:201–210.

de Groot RP, Auwert J, Karperien M, Staels B, and Kruijer W. 1991. Activation of junB by PKL and PKA signal transduction through a novel cis-acting element. Nucl Acids Res 19:775–781.

Di Chiara G and North RA. 1992. Neurobiology of opiate abuse. Trends Pharmacol Sci 13:185–193.

Gardner EL and Lowinson JH. 1991. Marijuana's interaction with brain reward systems: update 1991. Pharmacol Biochem Behav 40:571–580.

Grzanna R and Brown RM (eds.). 1993. Activation of immediate early genes by drugs of abuse. NIDA Research Monograph 125. Rockville, MD: U.S. Department of Health and Human Services, 54–71.

Hanbauer I, Wink D, Osawa Y, Edelman GM, and Gally JA. 1992. Role of nitric oxide in NMDA-evoked release of [^{3}H]-dopamine from striatal slices. Neuroreport 3:409–412.

Hardman JG, Limbird LE, Molinoff PB, Ruddon RW, Goodman and Gilman A.

1996. Goodman & Gilman's The pharmacological basis of therapeutics. New York: McGraw-Hill.

Hiroi N, Brown JR, Haile CN, Hong Y, Greenberg ME, and Nestler EJ. 1997. FosB mutant mice: loss of chronic cocaine induction of Fos-related proteins and heightened sensitivity to cocaine's psychomotor and rewarding effects. Proc Natl Acad Sci 94:10397–10402.

Hope B, Kosofsky B, Hyman SE, and Nestler EJ. 1992. Regulation of immediate early gene expression and AP-1 binding in the rat nucleus accumbens by chronic cocaine. Proc Natl Acad Sci USA 89:5764–5768.

Hope BT, Nye HE, Kelz MB, Self DW, Iadarola MJ, Nakabeppu Y, Duman RS, and Nestler EJ. 1994. Induction of a long-lasting AP-1 complex composed of altered Fos-like proteins in brain by chronic cocaine and other chronic treatments. Neuron 13:1235–1244.

Huston-Lyons D and Kornetsky C. 1992. Brain-stimulation reward as a model of drug-induced euphoria: comparison of cocaine and opiates. In: Lakoski JM, Galloway MP, and White FJ eds. Cocaine: pharmacology, physiology, and clinical strategies. Boca Raton: CRC Press. 47–71.

Hyman SE. 1996. Addiction to cocaine and amphetamine. Neuron 16:901–904.

Hyman SE and Nestler EJ. 1996. Initiation and adaptation: a paradigm for understanding psychotropic drug action. Am J Psychiatry 153:151–162.

Katzung BG. 1995. Basic and clinical pharmacology. Norwalk, CT: Appleton and Lange, 478–491.

Milner PM. 1989. The discovery of self-stimulation and other stories. Neurosci Biobehav Rev 13:61–67.

Moratalla R, Vickers EA, Robertson HA, Cochran BH, and Graybiel AM. 1993. Coordinate expression of c-fos and jun B is induced in the rat striatum by cocaine. J Neurosci 13:423–433.

Nestler EJ, Terwilliger RZ, Walker JR, Sevarino KA, and Duman RS. 1990. Chronic cocaine treatment decreases levels of the G-protein subunits Gi alpha and Go alpha in discrete regions of rat brain. J Neurochem 55:1079–1082.

Nestler EJ, Hope BT, and Widnell KL. 1993. Drug addiction: a model for the molecular basis of neural plasticity. Neuron 11:995–1006.

Nestler EJ. 1994. Molecular neurobiology of drug addiction. Neuropsychopharmacology 11:77–87.

Nestler EJ, Berhow MT, and Brodkin ES. 1996. Molecular mechanisms of drug addiction: adaptations in signal transduction pathways. Molecular Psychiatry 1:190–199.

Nye HE, Hope BT, Kelz MB, Iadarola M, and Nestler EJ. 1995. Pharmacological studies of the regulation of chronic FOS-related antigen induction by cocaine in the striatum and nucleus accumbens. J Pharmacol Exp Ther 275:1671–1680.

Pudiak CM and Bozarth MA. 1993. L-NAME and MK-801 attenuate sensitization to the locomotor-stimulating effect of cocaine. Life Sci 53:1517–1524.

Reith MEA, Jacobson AE, Rice KC, Benuck M, and Zimanyi I. 1991. Effect of metaphit on dopaminergic neurotransmission in rat striatal slices: involvement of the dopamine transporter and voltage-dependent sodium channel. J Pharmacol Exp Ther 259:1188–1196.

Reith MEA, Xu C, and Chen N-H. 1997. Pharmacology and regulation of the neuronal dopamine transporter. Eur J Pharmacol 324:1–10.

Robinson TE and Berridge KC. 1993. The neural basis of drug craving: an incentive-sensitization theory of addiction. Brain Res Rev 18:247–291.

Rogue P, Zwiller J, Malviya AN, and Vincendon G. 1990. Phosphorylation by protein kinase C modulates agonist binding to striatal dopamine D2 receptors. Biochem Int 22:575–582.

Rogue P and Malviya AN. 1994. Regulation of signalling pathways to the nucleus by dopaminergic receptors. Cell Signal 6:725–733.

Rosen JB, Chuang E, and Iadarola MJ. 1994. Differential induction of Fos protein and a Fos-related antigen following acute and repeated cocaine administration. Brain Res Mol Brain Res 25:168–172.

Satel SL, Southwick SM, and Gawin FH. 1991. Clinical features of cocaine-induced paranoia. Am J Psychiatry 148:495–498.

Self DW, Terwilliger RZ, Nestler EJ, and Stein L. 1994. Inactivation of Gi and G(o) proteins in nucleus accumbens reduces both cocaine and heroin reinforcement. J Neurosci 14:6239–6247.

Self DW, Barnhart WJ, Lehman DA, and Nestler EJ. 1996. Opposite modulation of cocaine-seeking behavior by D1- and D2-like dopamine receptor agonists [see comments]. Science 271:1586–1589.

Self DW and Nestler EJ. 1995. Molecular mechanisms of drug reinforcement and addiction. Annu Rev Neurosci 18:463–495.

Shibata M, Araki N, Ohta K, Hamada J, Shimazu K, and Fukuuchi Y. 1996. Nitric oxide regulates NMDA-induced dopamine release in rat striatum. Neuroreport 7:605–608.

Silva MT, Rose S, Hindmarsh JG, Aislaitner G, Gorrod JW, Moore PK, Jenner P, and Marsden CD. 1995. Increased striatal dopamine efflux in vivo following inhibition of cerebral nitric oxide synthase by the novel monosodium salt of 7-nitro indazole. Br J Pharmacol 114:257–258.

Tolliver BK, Ho LB, Reid MS, and Berger SP. 1996. Evidence for involvement of ventral tegmental area cyclic AMP systems in behavioral sensitization to psychostimulants. J Pharmacol Exp Ther 278:411–420.

Wang JQ, Smith AJ, and McGinty JF. 1995. A single injection of amphetamine or methamphetamine induces dynamic alterations in c-fos, zif/268 and preprodynorphin messenger RNA expression in rat forebrain. Neuroscience 68:83–95.

Wang JQ and McGinty JF. 1995. Differential effects of D1 and D2 dopamine receptor antagonists on acute amphetamine- or methamphetamine-induced up-regulation of zif/268 mRNA expression in rat forebrain. J Neurochem 65:2706–2715.

Widnell KL, Self DW, Lane SB, Russell DS, Vaidya VA, Miserendino MJ, Rubin CS, Duman RS, and Nestler EJ. 1996. Regulation of CREB expression: in vivo evidence for a functional role in morphine action in the nucleus accumbens. J Pharmacol Exp Ther 276:306–315.

Wise RA. 1996. Neurobiology of addiction. Curr Opin Neurobiol 6:243–251.

Wise RA and Bozarth MA. 1987. A psychomotor stimulant theory of addiction. Psychol Rev 94:469–492.

Xu M, Moratalla R, Gold LH, Hiroi N, Koob GF, Graybiel AM, and Tonegawa S. 1994. Dopamine D1 receptor mutant mice are deficient in striatal expression of dynorphin and in dopamine-mediated behavioral responses. Cell 79:729–742.

Zhang L and Reith MEA. 1996. Regulation of the functional activity of the human dopamine transported by the arachidonic acid pathway. Eur J Pharmacol 315:345–354.

Zhu XZ and Luo LG. 1992. Effect of nitroprusside (nitric oxide) on endogenous dopamine release from rat striatal slices. J Neurochem 59:932–935.

Keyword Index